205
Advances in Polymer Science

Advances in Polymer Science
Recently Published and Forthcoming Volumes

Polysaccharides II

Volume Editor: Dieter Klemm

With contributions by

T. Heinze · N. Heßler · B. Heublein · S. Hornig · M. Hornung
D. Klemm · P. Kosma · F. Kramer · T. Liebert · S. Marsch · Y. Nishio
A. Potthast · T. Rosenau · H.-P. Schmauder · D. Schumann

 Springer

The series *Advances in Polymer Science* presents critical reviews of the present and future trends in polymer and biopolymer science including chemistry, physical chemistry, physics and material science. It is adressed to all scientists at universities and in industry who wish to keep abreast of advances in the topics covered.

As a rule, contributions are specially commissioned. The editors and publishers will, however, always be pleased to receive suggestions and supplementary information. Papers are accepted for *Advances in Polymer Science* in English.

In references *Advances in Polymer Science* is abbreviated *Adv Polym Sci* and is cited as a journal.

Springer WWW home page: springer.com
Visit the APS content at springerlink.com

Library of Congress Control Number: 2006930119

ISSN 0065-3195
ISBN-10 3-540-37102-8 Springer Berlin Heidelberg New York
ISBN-13 978-3-540-37102-1 Springer Berlin Heidelberg New York
DOI 10.1007/11776895

Springer is a part of Springer Science+Business Media

springer.com

Cover design: WMXDesign GmbH, Heidelberg
Typesetting and Production: LE-TEX Jelonek, Schmidt & Vöckler GbR, Leipzig

Printed on acid-free paper   02/3100 YL – 5 4 3 2 1 0

Volume Editor

Prof. Dieter Klemm
Institut für Organische Chemie
und Makromolekulare Chemie
Friedrich-Schiller-Universität Jena
Humboldtstraße 10
07743 Jena, Germany
dieter.klemm@uni-jena.de

Advances in Polymer Science
Also Available Electronically

For all customers who have a standing order to Advances in Polymer Science, we offer the electronic version via SpringerLink free of charge. Please contact your librarian who can receive a password or free access to the full articles by registering at:

springerlink.com

If you do not have a subscription, you can still view the tables of contents of the volumes and the abstract of each article by going to the SpringerLink Homepage, clicking on "Browse by Online Libraries", then "Chemical Sciences", and finally choose Advances in Polymer Science.

You will find information about the

– Editorial Board
– Aims and Scope
– Instructions for Authors
– Sample Contribution

at springer.com using the search function.

Preface

As the most important skeletal component in plants and also a substance produced by certain bacteria, polysaccharide cellulose is an almost inexhaustible polymeric raw material with a fascinating structure and interesting properties. The fusion of both carbohydrate and polymer chemistry in a macromolecule biosynthetically composed of repeating glucose units generates surprising specificity and impressively diverse supramolecular/morphological structures, reactivities, and functions.

Since the discovery of cellulose by Anselme Payen in 1838, the start of large-scale cellulose ester, ether, and fiber production (which started more than 100 years ago), and the extensive scientific investigations of cellulose by Hermann Staudinger in 1920, cellulose research and product development have been redefined by the increasingly more significant impact of such work.

During the past 5–10 years a considerable increase in knowledge of the structure, chemistry, and processing of cellulose, as well as development of innovative cellulose products, has been observed. New frontiers involve sophisticated methods of structural analysis, environmentally safe cellulose-fiber technologies, as well as progressive work with bacterial nanocellulose, (bio)materials, and a broad spectrum of cellulose composites.

Associated with this trend, related polysaccharides like hemicelluloses, dextran, chitin, and starch have also attracted growing interest.

Illustrative examples of this exciting evolution are presented in five selected contributions from international experts in their fields in this special volume titled *Polysaccharides II* within the series *Advances in Polymer Science*. This serves as a follow-up to *Polysaccharides I* (vol. 186). The comprehensive reviews stem from the authors' own research experience and demonstrate clearly the latest scientific results and their practical applications as well as the interdisciplinary character of the subject and its importance to polymer science.

The first chapter, written by Antje Potthast, Thomas Rosenau, and Paul Kosma, summarizes the state of the art of the reliable and sophisticated analysis of oxidized functionalities in cellulose. The carbonyl and carboxyl groups in question are generated as alien structure units by pulping, bleaching, irradiation, and aging processes. The combination of group-selective fluorescence labeling with GPC and its application for monitoring oxidative processes of cellulose production and handling are pointed out.

The second chapter by Dieter Klemm, Dieter Schumann, Hans-Peter Schmauder, and coworkers focuses on the recent knowledge of cellulosics characterized by a property-determining supramolecular nanofiber structure. Topics in this interdisciplinary contribution are the types of nanocelluloses and their use in technical membranes and composites as well as in the development of medical devices, in veterinary medicine, and in cosmetics.

The material functionalization of cellulose and related polysaccharides via different types of microcompositions is the subject of the following chapter written by Yoshiyuki Nishio.

In the fourth chapter, Thomas Rosenau, Antje Potthast, and Paul Kosma describe current investigations on reactive intermediates and reaction mechanisms in cellulose chemistry. The chemical modification and regeneration of cellulose requires its activation and/or dissolution in media like sodium lye, N-methylmorpholine-N-oxide (NMMO), N, N-dimethylacetamide (DMAc), or in carbanilation mixtures containing dimethylsulfoxide (DMSO). In these media – very important for research and large-scale processes – the trapping and characterization of the intermediates has been carried out.

The final chapter by Thomas Heinze, Tim Liebert, Brigitte Heublein, and Stephanie Hornung represents the important interaction of cellulose research and development with other related polysaccharides. The authors describe the sources, structure, properties, and applications of dextran and functionalized dextranes including their use as bioactive materials.

We hope that this volume will not only become an important reference for those in the cellulose and polysaccharide field, but also serve as source of information and motivation for the readers from the entire polymer community.

Jena, August 2006   *Dieter Klemm*

Contents

Contents of Volume 186

Polysaccharides I

Volume Editor: Thomas Heinze
ISBN: 3-540-26112-5

Adv Polym Sci (2006) 205: 1–48
DOI 10.1007/12_099
© Springer-Verlag Berlin Heidelberg 2006
Published online: 30 August 2006

Analysis of Oxidized Functionalities in Cellulose

Antje Potthast · Thomas Rosenau · Paul Kosma (✉)

Department of Chemistry, Christian-Doppler-Laboratory of Pulp Reactivity,
University of Natural Resources and Applied Life Sciences,
Member of the European Polysaccharide Network of Excellence (EPNOE),
Muthgasse 18, 1190 Vienna, Austria
paul.kosma@boku.ac.at

Abstract This review gives an overview on oxidized functionalities in celluloses, i.e., carbonyl and carboxyl groups, with regard to their chemical structure, the different ways of introduction, and their analytics. Starting from different processes introducing oxidative modifications into celluloses a survey on the chemical nature of these functionalities is given and analytical approaches towards their determination are discussed. Special emphasis is placed on recent developments which combine group-selective fluorescence labeling with multi-detector GPC analysis to provide carbonyl and carboxyl group profiles according to the CCOA and FDAM method, respectively. Examples of monitoring the oxidation state of celluloses and its changes during processing stages are given, for example bleaching, aging, dissolution or irradiation procedures.

Keywords CCOA method · Cellulose · Carbonyl groups · Carboxyl groups · FDAM method · Fluorescence labeling · GPC

Abbreviations

ADAM	9-Anthryldiazomethane
AGU	Anhydroglucose unit
CCOA	Carbazole-9-carboxylic acid [2-(2-aminooxy-ethoxy)ethoxy]amide
DAM-MC	4-Diazomethyl-7-methoxycoumarine
Dansyl	5-(Dimethylamino)-1-naphthalenesulfonic acid
DMAc	N,N-Dimethylacetamide
DP	Degree of polymerization
DS	Degree of substitution
DS_{CO}	Degree of substitution of carbonyl groups
DS_{COOH}	Degree of substitution of carboxyl groups
ECF	Elemental chlorine free
FAD	Flavin adenine dinucleotide
FDAM	$9H$-Fluoren-2-yl-diazomethane
GPC	Gel permeation chromatography
HexA	Hexenuronic acid
Lyocell	Fiber production process based on direct dissolution of cellulose in NMMO
MALLS	Multi-angle laser light scattering
M_n	Number average molecular weight
M_w	Weight average molecular weight
MW	Molecular weight
MWD	Molecular weight distribution
NMM	N-Methylmorpholine
NMMO	N-Methylmorpholine N-oxide
P	Peroxide bleaching stage
PDAM	1-Pyrenyl-diazomethane
PHK	Prehydrolysis kraft pulp
PDI	Polydispersity index
REG	Reducing end group(s)
TEMPO	2,2,6,6-Tetramethylpiperidine-1-oxyl

TCF   Totally chlorine free
TMP   Thermo-mechanical pulp
TOF-SIMS Time-of-flight secondary ion mass spectrometry
TTC   2,3,5-Triphenyltetrazolium chloride
Z   Ozone bleaching stage

1
Oxidized Groups in Cellulose

1.1
A Survey of Oxidative Modifications

Cellulose as synthesized by nature can be considered a quite perfect molecule: anhydroglucose units (AGU) are connected by β-1,4-glycosidic linkages resulting in a homopolymer with three hydroxyl groups per AGU and a terminal aldehyde masked as hemiacetal at the reducing end. While such cellulose synthesized in vitro or in vivo represents the ideal polymer molecule in terms of chemical purity, processing steps, such as isolation and purification, as well as natural conditions, such as exposure to environmental stress and aging, are factors for the introduction of additional oxidized functionalities.

Due to the 3 hydroxyl groups available for oxidation within one anhydroglucose unit and due to the polymeric character of the cellulose a great variety of structural modifications and combinations is possible. As with other types of chemical changes at the cellulose molecule also in this case the oxidation can affect different structural levels differently. Depending on the oxidative stress imposed on the cellulose, the individual hydroxyls within the AGU and within the polymer chain are involved to varying extent and may respond to further treatment and reactions in a specific way. Despite their low concentration in the μmol/g range, oxidative functionalities are one of the prime factors to determine macroscopic properties and chemical behavior of cellulosic materials (Fig. 1).

The main causes for the formation of carbonyl and carboxyl groups in cellulose are isolation and purification procedures besides natural aging. This applies in particular to cellulosic pulps from wood, which has undergone a number of processing steps to be freed from lignin, hemicelluloses, and extractives.

Although we are far away from being able to fully analyze an oxidized cellulose with all these different possibilities of oxidative modifications with regard to their type and exact location, cellulose chemists have taken major steps towards developing methodology to address those problems. Major obstacles on the way to a comprehensive analysis of oxidized functionalities are certainly the low concentration of these structures in cellulosic materials, the characterization of their position within the molecular weight distribu-

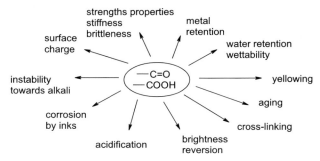

Fig. 1 Possible effects of carbonyl and carboxyl groups on macroscopic properties of cellulose

tion (MWD), and the problems generally faced in analytics of celluloses, such as limited solubility, the polymeric character, morphological structure and limited accessibility.

This review summarizes the generation, nature and determination of oxidized structures in celluloses as well as recent developments in their analytics.

1.1.1
Pulping

The first process to bring about oxidative changes onto cellulose is pulping. In acidic sulfite pulping, generally a larger number of carbonyl groups is introduced as compared to prehydrolysis kraft pulping. In celluloses from kraft pulping procedures, most carbonyls actually originate from the hemicellulose fraction. The alkaline conditions of kraft pulping favor a more narrow distribution of the molecular weight as well as the dissolution of low molecular weight material, which has a high relative carbonyl content. During sulfite pulping, hydrolysis of cellulose and hence the generation of more reducing end groups (REG) is promoted. Aldonic acids as final oxidation products have been isolated from sulfite cooking liquors [1]. If paper pulps are considered the situation is mainly influenced by the content of hemicelluloses, which per se contain a larger number of carbonyls due to the smaller molecular weight and a higher number of carboxyl groups due to the side chain uronic acids. While the oxidation of reducing ends of dissolved monomers and oligomers from hemicellulose and cellulose during sulfite pulping is rather pronounced, the oxidation of the reducing end of the non-dissolved remaining cellulose was found to be rather small. However, pulping with bisulfite, having a higher process pH, favors the oxidation of the REG to the corresponding aldonic acids [2]. After alkaline pulping, metasaccharinic and gluconic acid residues are typical carboxyl groups in cellulose [3]. Metasaccharinic acid is the final product of the stopping reaction after alkali-induced peeling and rearrangement reactions. Under alkaline conditions, gluconic acid carboxyl groups are crucial with regard to cellulose stabilization, since they hamper degradation by the peeling process.

1.1.2
Bleaching

Bleaching of pulp represents the main origin of additional carbonyl groups in cellulose. While pulping proceeds in the absence of strong oxidants, oxidizing agents are deliberately used in bleaching to achieve the desired brightness effect.

Chlorine dioxide (D stage) is considered rather selective for lignin moieties. Cellulose integrity is not affected to a large extent [4]. Chlorine dioxide primarily reacts with the reducing end groups of carbohydrates converting them to carboxylic acids. However, depending on the pH other chlorine-containing species are formed, e.g., hypochlorous acid and chlorate. In that respect, also effects of those oxidants have to be considered. Hypochlorite formerly used directly as H stage is being phased out mainly due to environmental concerns. If HOCl is present in a pH range between 2 and 8, considerable amounts of carbonyls are introduced into cellulose, whereas under alkaline conditions chiefly carboxyl groups are formed. An H stage in all pH ranges is accompanied by a decrease in the molecular weight, which depends linearly on the oxidative damage done, i.e., on the carbonyls created, but not on the pH, as proposed by Lewin and Epstein [5]. The oxidation potential and selectivity in hypochlorite stages is strongly governed by the conditions chosen [6, 7], as the active species depend on the pH: in acidic solution the equilibrium between free halogen and hypohalous acid is shifted to the side of free halogen, whereas under alkaline conditions the hypohalous acid is the effective and also more selective species. In the pH range between 2 and 8, where hydroxyl groups are predominantly oxidized to keto and aldehyde groups, the drop in DP is most severe, whereas the oxidative power of hypohalite at a higher pH is merely sufficient to oxidize aldehyde groups to the corresponding acids, leaving the DP largely unaltered.

The oxygen-based bleaching chemicals of TCF bleaching sequences exhibit a lower selectivity towards lignin and residual chromophores so that oxidative damage involves also the carbohydrates. In particular, the heavy metal management during pulping becomes an important issue, since a number of very reactive radical species, such as hydroxyl and hydroperoxyl radicals, are formed in the presence of the bleaching agents and transition metal ions (Fenton and Haber–Weiss cycles). Other sources of radicals are disproportionation reactions under alkaline conditions, which applies especially to hydrogen peroxide, secondary radical formation in homolytic reactions and autoxidation. Gierer [8] and Gratzl [9] state the generation of different oxygen-derived radical species in all TCF sequences as the main cause of the limited selectivity. According to Chirat and Lachenal the poor selectivity of ozone is mainly due to the oxidant ozone itself [10].

Keto groups are introduced during an ozone (Z) stage [11, 12]. Besides the action of hydroxyl radicals, which can be generated in an ozone stage either by slow decomposition of ozone in water or be triggered in the presence of metal ions, also an ionic pathway by the common 1,3-insertion mechanism of ozone has been proposed [13, 14], leading to the eventual formation of lactones.

Such processes are always accompanied by a DP loss, either by electrophilic attack of ozone, by an ozone-catalyzed cleavage of the glycosidic bond or by attack of secondary radical species [15]. Residual lignin also plays a crucial role in ozone bleaching. Model studies showed that lignin with free phenolic hydroxyl groups accelerated carbohydrate oxidation, probably by activation of oxygen via phenoxyl radicals, whereas etherified phenolic model compounds had a protective effect [16, 17].

During a hydrogen peroxide (P) stage, basically the same applies as for an ozone stage, if radicals are present and their subsequent reactions are allowed to proceed. In their absence, however, a P stage introduces predominantly keto groups, which neither cause a pronounced DP loss nor a decreased brightness. Hence, the keto groups were thought to be generated at position C3. Even though chain scission can occur also in this case, a non-reducing chain and a diketo derivative, subsequently leading to a stable saccharinic acid derivative, are formed [18, 19]. A peroxide stage followed by alkaline extraction (E) significantly reduces the total amount of carbonyl groups, leading to a pronounced brightness stability. One reason for this is also the removal of 2,3-diketo structures by reaction with the hydroperoxide anion (HOO^-), which can add to keto and enone structures as well. The formed acids are then extracted within the E stage.

During an alkaline oxygen stage the formation of keto structures by oxidation of ketols is postulated, finally leading to acids. Sodium permanganate treatments yield a moderate increase in both, carbonyls and carboxyl groups at short reaction times [20].

In ECF and TCF bleaching the same reactions that lead to the formation of carbonyl groups can also account for the generation of carboxyl groups, especially if radical species are involved.

1.1.3
Irradiation

High energy radiation, for instance γ-irradiation or β-irradiation (electron beaming), causes a considerable increase in carbonyl groups, mainly through the action of radical species being generated. Both procedures are accompanied by a DP loss, which can be used to adjust the molecular weight of the cellulose prior to utilization in the viscose process [21]. The number of carbonyls introduced by high energy electron beams is directly correlated with the applied dosage [22]. e-Beaming is also considered as a means for cellulose activation, since the radiation can enter also highly ordered regions in cellulose, changing the morphology for subsequent reactions in a favorable way [21]. Also γ-irradiation shows a linear relationship between dose and carbonyl content [23]. The presence of lignin during γ-irradiation does not prevent cellulose degradation [24].

1.1.4
Aging Processes

The presence of oxygen, often also in the presence of light, causes a number of autoxidative reactions to proceed. Accelerated aging linearly increases with the partial pressure of oxygen [25]. Once the process has started, radical reactions come into play that finally result in the formation of hydroperoxyl structures [26, 27], which in turn can activate oxygen to generate very reactive radical species, such as hydroxyl or hydroperoxyl radicals. These radicals react with cellulose under H-atom abstraction, finally leading to formation of carbonyl or carboxyl groups, to chain cleavage, and thus also to a loss in DP and fiber strength. Additional factors triggering oxidation reactions during aging are transition metal ions, especially iron and copper from writing and painting media, chromophores that serve as activators, as well as elevated temperature, air pollutants and light. Light-induced processes have been extensively studied on high-yield pulps containing considerable amounts of lignin [28–30], but have recently also been investigated in the case of fully bleached pulp samples [31].

Natural aging, as experienced in historic documents, drawings and cellulosic fabrics is thought to be caused by two major parallel processes, hydrolytic cleavage [32–34] of the glycosidic bond by acid of various origins and oxidation [35–37], triggered by external factors such as metal ions, air, light, or pollutants. While the hydrolytic pathway is well investigated [35], fewer studies report on the oxidation reactions [38].

Thermo- and photo-oxidative degradation in the context of paper conservation science was lately summarized and investigated [39]. Methylene blue dyeing was used to investigate reactions occurring at wet-dry interfaces [40]. Investigation on different pulps has shown that there is indeed an influence of carbonyl groups during the aging of alkaline pulps [41]. Reactions occurring under alkaline conditions in aged papers as they are deliberately induced during deacidification treatments are increasingly addressed [42].

1.2
Carbonyl Groups in Cellulose

1.2.1
The Reducing End Group (REG)

The reducing end groups in cellulose are the only naturally occurring carbonyl functionalities in this material. Cellulose from *Acetobacter xylinum* [43] contains an amount of carbonyl groups which corresponds approximately to the number of reducing end groups, and can be considered as rather genuine material.

The reducing ends are the starting points for the well-known peeling reaction. The conditions for the peeling to proceed require alkaline media and a free 2-OH group. In case of a substituted hydroxyl at C2, as it occurs for instance in hemicelluloses, the peeling rate is drastically reduced.

Reducing end groups can be readily further oxidized to the corresponding aldonic acids.

The reducing ends are very likely to be present as hemiacetals in pyranose units, but only to a small extent as aldehydes and aldehyde hydrates [44].

1.2.2
2,3-Dialdehyde Structures

Oxidation with periodate under acidic conditions, the *Malaprade* reaction, is mainly used to introduce a large number of aldehydes into cellulose [45]. Whether a similar reaction proceeds also under conditions of natural or accelerated aging conditions has not been clarified, but corresponding processes have been postulated to occur [46]. The oxidized groups introduced are either used to further functionalize the cellulose, e.g., by reaction with

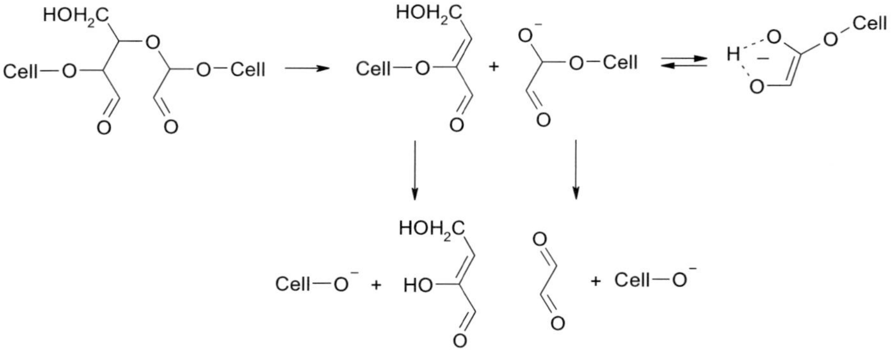

Scheme 1 β-Elimination starting from C2,C3-dialdehyde structures in cellulose [93]

substituted amino functionalities, oxidation to the corresponding acids, or reduction. All of those procedures significantly change the properties of the cellulose.

The oxidation proceeds predominantly at C2 and C3 under aldehyde formation, with concomitant ring opening cleavage between the C2–C3 bond. The periodate oxidation is accompanied by a decrease in crystallinity [47, 48]. Kim et al. [49] investigated the influence of this oxidation in the crystalline domains. Besides the decrease in crystallinity an uneven distribution of aldehyde groups was shown by gold labeling. From this result, it was proposed that the oxidation proceeds highly heterogeneously, forming isolated oxidized domains. The mechanism for β-elimination at 2,3-dialdehyde celluloses under alkaline conditions is shown in Scheme 1.

1.2.3
6-Aldehyde Structures

Celluloses oxidized to an aldehyde at C6 are found for instance as intermediates of the TEMPO oxidation [50]. Depending on the reaction conditions a large number of such groups may survive in the final products, the polyglucuronic acids, and also in partially TEMPO-oxidized pulps materials it is highly likely that a large number of carbonyl groups are present as C6-aldehyde.

Cellulose can be oxidized to different degrees of C6-aldehyde content by photolysis of the 6-azido-6-deoxy derivatives [51]. The β-elimination reaction of the 6-aldehyde (Scheme 2) may lead to terminal double bonds exhibiting a UV absorption ($\lambda_{max} = 250$ nm) [52, 53]. With model C6-aldehyde celluloses, β-elimination started at 30 °C at pH 9; at elevated temperatures (80 °C) β-elimination was detected already at very low pH levels (pH 3.5). Similar results were obtained with HOCl-oxidized pulps, with β-elimination starting at room temperature at pH 8.5 (unpublished results).

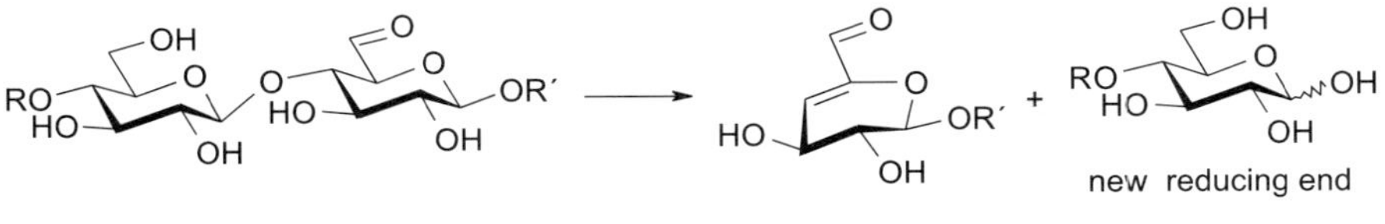

Scheme 2 β-Elimination starting from C6-aldehyde structures in cellulose

1.2.4
Keto Groups

Keto groups can be introduced at position C2 or C3, and they may be present also as 2,3-diketo structures. According to keto-enol tautomerism the position of such carbonyls may fluctuate (Scheme 3). Their reaction products formed under alkaline conditions differ significantly depending on the original position of the keto group.

While a keto group at C2 leads to the generation of a new reducing end group (Scheme 4) and this way at least to a starting point for further peeling reactions, a keto at position C3 is considered a more innocent carbonyl function [18]. Also in this case the cellulose chain will be cleaved, however, the resulting fragments contain a novel non-reducing end group and an acid formed by rearrangement (Scheme 5). Both structures can be considered rather stable in subsequent reactions.

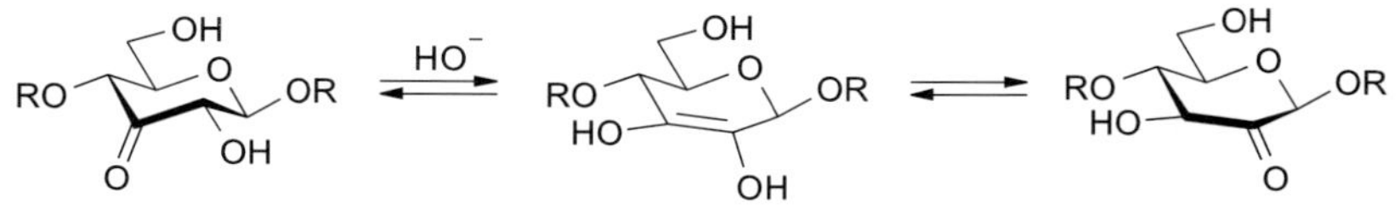

Scheme 3 Keto-enol tautomerism at carbonyl moieties in cellulose

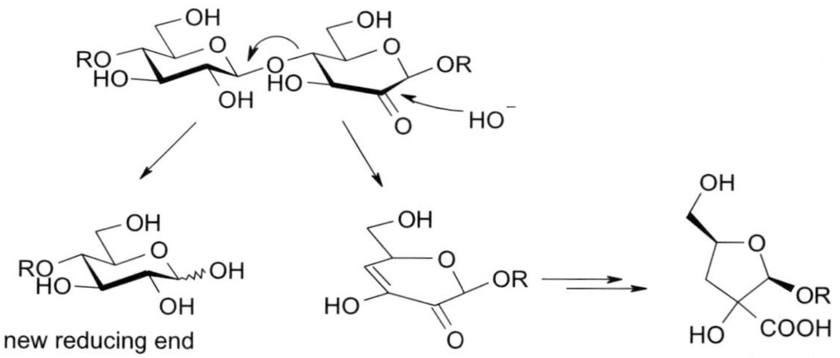

new reducing end

Scheme 4 Cleavage of the glycosidic bond by β-elimination starting from C2-keto structures along the cellulose chain

non-reducing

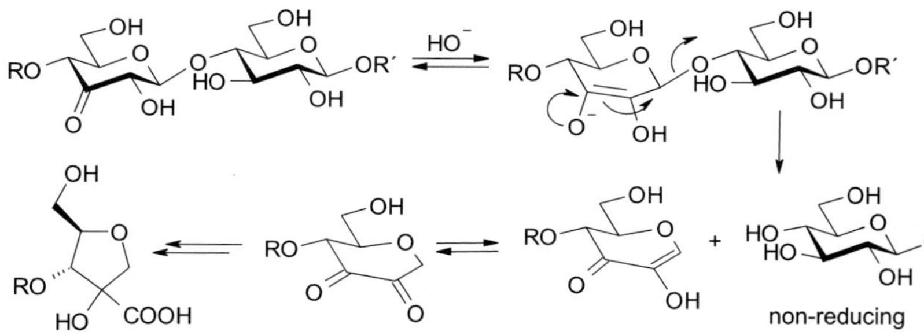

Scheme 5 Cleavage of the glycosidic bond by β-elimination starting from C3-keto structures along the cellulose chain

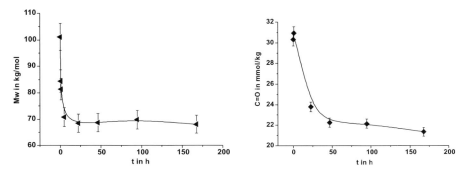

Fig. 2 Effect of alkaline conditions on oxidized cotton linters. *Left*: Changes in M_w (error bars 5%). *Right*: Changes in carbonyl groups (error bars 2%)

How β-elimination reactions and subsequent processes under alkaline conditions affect the carbonyl groups and the M_w of an oxidized cotton linters is shown in Fig. 2. The initial concentration of carbonyls is 30 µmol/g, introduced by hypochlorite treatment at pH 7. The sample was treated with a buffer solution of pH 11 at 40 °C. The M_w suffers a steep decrease within the first 5 hours, the drop in carbonyls proceeds at a significantly lower rate. While the changes in M_w level off after approx. 20 hours of treatment, the carbonyl groups are slightly decreasing further (unpublished results).

1.2.5
The Nature of Carbonyls in Cellulose

Carbonyl groups are not always present in their usual "double-bond" form, but also exist as the tautomeric enols or as hydrates upon addition of water, which was demonstrated by means of differently oxidized carbohydrate model compounds by NMR [102] (Scheme 6). A recent study revealed a much higher percentage of hydrated aldehyde groups compared to the pure acyclic form for all aldohexoses in water [44].

The hydration/(hemi)acetalization state of aldehyde and keto groups in celluloses is largely unknown. Non-conjugated carbonyl structures generally absorb in the near UV, around 270–280 nm, for aldehyde and keto groups. Even though high degrees of oxidation can be obtained, e.g., by periodate oxidation, the carbonyl groups can hardly be detected by spectroscopic methods such as FTIR spectroscopy [54], likely due to hydration and acetalization effects. Severe drying down to 7% relative humidity and high temperatures eventually increased the $C = O$ vibrations [55, 56], which indicated a large extent of hydration of the carbonyl groups in these substrates. Other explanations are strong cross-linking by the formation of hemicaetals with neighboring hydroxyl groups. Such intra-molecular and inter-molecular cross-linking has been postulated [57], but only be proven indirectly by changes in macro-

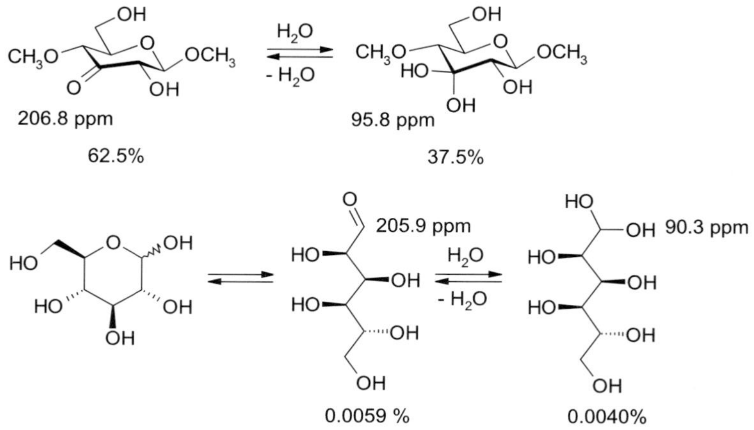

Scheme 6 Examples of the equilibrium between keto and hydrate forms in aqueous solution, including the respective ^{13}C NMR shifts of the carbonyl/hydrate atoms. Glucose data from [44]

scopic properties, e.g., solubilities. The interchange between hydrated carbonyls and those present in the sp^2-hybridized form can be investigated combining the CCOA method (see next chapter) and the resonance Raman technique. While the CCOA method detects all carbonyl groups independent of hydration or hemiacetalization/hemiketalization, resonance Raman records only sp^2-hybridized carbonyl groups. It was demonstrated, that carbonyl functions in cellulose are present to a large extent as hydrates and/or hemiacetals/hemiketals in addition to their free, double-bonded form [58].

1.3
Carboxyl Groups in Cellulose

Only second to carbonyl groups, carboxyls are a very important oxidized function in celluloses (Scheme 7). Whereas hemicelluloses inherently contain a high number of different acid groups, carboxyl groups in cellulose are artifi-

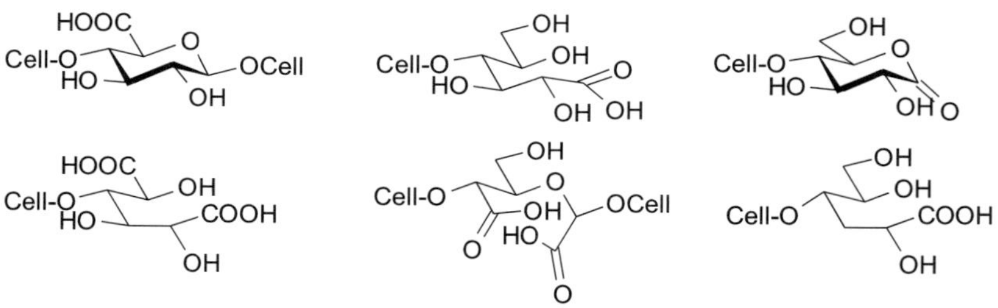

Scheme 7 Examples of different carboxyl structures in cellulose

cially introduced by pulping and bleaching processes. The primary positions for oxidative attack are the reducing end groups, which are oxidized to the corresponding gluconic acid end groups. Oxidation at the primary C6 hydroxyl group leads to glucuronic moieties. The pyran ring structure of an AGU and the integrity of the cellulose may be destroyed by cleavage at the ring-oxygen or by formation of acids at C2 and/or C3 (Scheme 7).

1.3.1
Gluconic Acids

Oxidation of the reducing end group leads to the corresponding gluconic acid. Since the REG is the starting point for the peeling process under alkaline conditions, an oxidation represents a stabilization of the polymer chain. Especially under kraft pulping conditions additives or modifications aimed at improving the process yield often involve the oxidation of the REG (e.g., polysulfide, AHQ) [59].

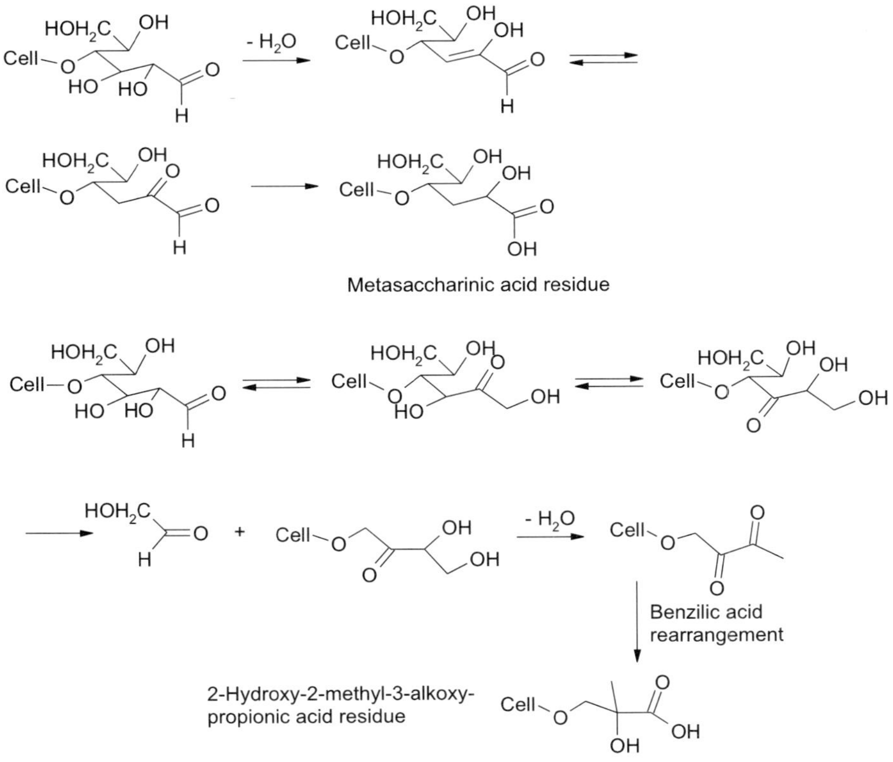

Scheme 8 Formation of acidic residues, stopping the peeling process of cellulose under alkaline conditions

As a result of the peeling/stopping process, metasaccharinic acid end groups formed by the Cannizzaro reaction, or 2-hydroxy-2-methyl-3-alkoxy-propanoic acid residues are generated (Scheme 8) [60].

1.3.2
Glucuronic Acids

Depending on the type and the origin of the pulp a major source for glucuronic acids are the hemicelluloses. The glucuronoxylans of hardwoods and the (arabino)glucuronoxylans of softwoods contain α-D-glucuronic acids and/or their 4-O-methyl derivatives attached to position C2 of the xylan backbone [61]. Chemical generation of uronic acids in cellulose can be accomplished by TEMPO oxidation [62], or, less selectively, by nitrogen oxide treatments [63, 64].

Also under conditions of ozone bleaching, further oxidation of 6-aldehyde cellulose to glucuronic acids proceeds. Decarboxylation of the resulting C6-carboxyl group is made responsible for CO_2 formation and DP loss during this process [4, 14] (Scheme 9).

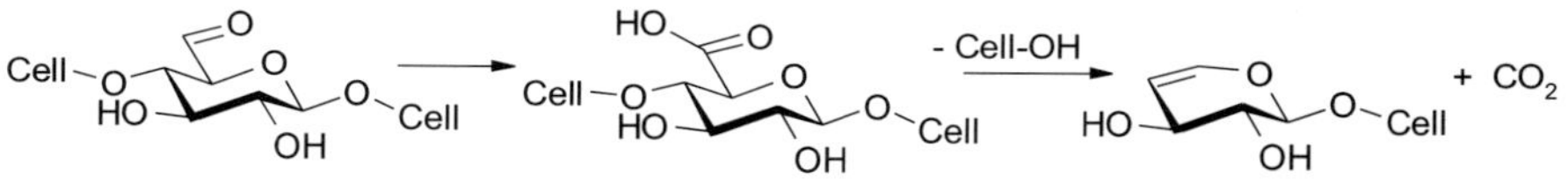

Scheme 9 Decarboxylation of acids under the conditions of ozone bleaching

1.3.3
Hexenuronic Acids

Hexenuronic acid (i.e., 4-deoxy-L-*threo*-hex-4-enopyranosyl-uronic acid) is formed under alkaline conditions by elimination of methanol from side chain residues in xylans [65] (Scheme 10). The reaction is promoted by both increasing alkali concentration and temperature [66, 67]. After kraft pulping, only

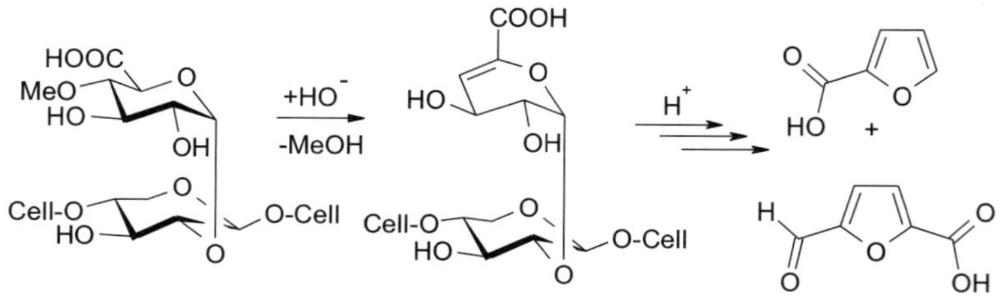

Scheme 10 Formation of HexA residues from xylan under alkaline conditions and degradation to furan products in acidic media

about 12% of the remaining carboxyl groups in accessible xylan are still of the 4-O-methylglucurono type [68]. Hexenuronic acids add to the total carboxyl group content and also to the kappa number. HexAs are seen as being partly responsible for diminished brightness stabilities of bleached pulps. Under acidic conditions hexenuronic acids are unstable (Scheme 10), so that a mildly acidic treatment can thus be used to selectively remove HexA from the pulp [69].

1.3.4
Lactone Entities

Within oxidized polysaccharides, carboxylic units may occur at the former reducing end as gluconic groups as well as along the chain in the form of uronic acid residues or at the C-2 and C-3 positions in periodate-cleaved and oxidized products. Under appropriate conditions induced by elevated temperatures, dehydration or low pH, the carboxylic groups are prone to intra-molecular ester formation, which in case of cellulose includes the 5-OH group of the proximal cellobiose moiety of the cellulose chain or the 2-OH group of a glucose unit preceding the glucuronic acid residue within the polysaccharide backbone. Whereas in the former case a six-membered lactone moiety in a distorted half-chair conformation will be generated, an inter-residue lactone in a sterically feasible seven-membered ring will be present in the latter case. This arrangement resembles the positioning of the hydrogen-bond pattern stabilizing the cellulose I structure by hydrogen bonding from OH-2' to O-6. In the literature [133], the occurrence of 6,3-lactone units has also been discussed. The formation of this intra-residue lactone, however, requires a transition from the $^4C_1$ conformation into a boat conformation which would meet with considerable steric restraints (Scheme 11).

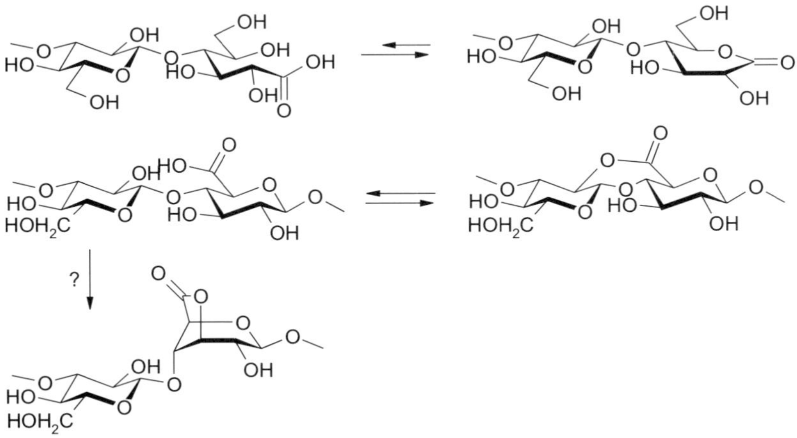

Scheme 11 Formation and hydrolysis of glucono-1,5-lactone and 2',6- and 3,6-lactone entities

1.3.4.1
D-Glucono-1,5-Lactone Formation and Stability

In contrast to D-glucono-1,4-lactone, the 1,5-lactone exhibits pronounced instability. Depending on the applied conditions, the lactone will be in equilibrium with the corresponding gluconic or cellobionic acid moiety, respectively,
which in solution is shifted to the open-chain form. Semiempirical calcu-
lations of the hydrolysis of 1,5-gluconolactone indicated preferential cleavage of the C1-oxygen linkage, which was supported by ^{13}C NMR data of
a hydrolyzate obtained from gluconolactone in $H_2^{18}O$ at pH 7.5 [70]. Thus,
isolation of cellobiono-1,5-lactone has formerly only been achieved by removal from the mixture by selective crystallization [71]. The formation of
gluconic acid lactones in cellulosic substrates has been implicated in various FAD-assisted enzymatic reactions involved in cellulose degradation. As
an example, in addition to the well-studied cellobiose dehydrogenases, a glu-
cooligosaccharide oxidase has been described, which also accepts cellodex-
trins as substrate [72, 73]. Mechanistic and crystal structure data showed
the formation of the glucono-1,5-lactone intermediate in the oxidation step,
which subsequently hydrolyzes spontaneously to give the open-chain form.
Due to the inherent lability of the 1,5-lactone, a cellobionolactame analogue
has therefore been used for cocrystallization experiments with a cellobiose
dehydrogenase, and these lactone mimetics are of general interest as potential glycosidase inhibitors [74]. In addition to the crystallographic data, direct
^{1}H NMR-spectroscopic evidence has been obtained for the enzymatic conversion of cellobiose into cellobionic acid [75].

Cellobionic acid was also found as the major product arising from ozonation of cellobiose and was isolated by HPLC and fully characterized by ^{1}H
and ^{13}C NMR spectroscopy [76]. Ozonation of methyl β-D-glucopyranoside
afforded a lactone of arabinonic acid arising via a Ruff-type degradation of
gluconic acid followed by subsequent oxidation [16]. Alkaline conditions,
such as those which have been applied to open lactone rings, are prone to
generate additional carboxylic groups. Kraft pulps, which have undergone

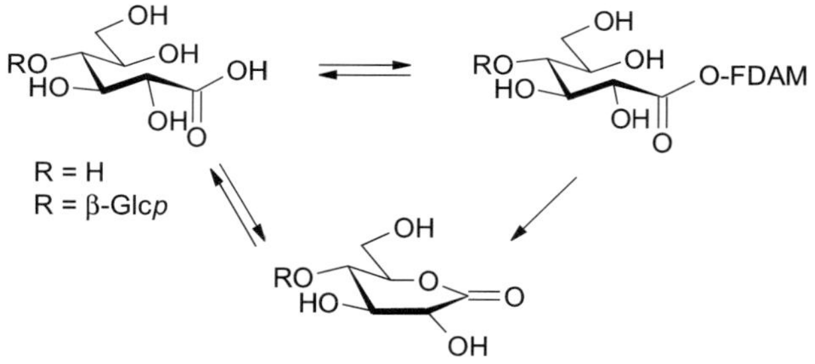

Scheme 12 Formation and hydrolysis of FDAM-labeled aldonic acids

alkaline treatments contain metasaccharinic end units in addition to glu-
cono lactones. These end groups have been reported to form stable lactones,
which are fairly resistant towards iodine-induced lactone opening [77]. Attempts to label gluconic acid residues with the fluorescence label FDAM
(see forthcoming chapter) reflect the ambiguous reactivity of 1,5-lactones
(Scheme 12). In model labeling experiments with xylonic, gluconic and cellobionic acid, formation of the labeled products could be monitored by TLC.
Isolation of the derivatives by column chromatography, however, provided
trace amounts only due to direct hydrolysis or transesterification reactions,
respectively [78].

1.3.4.2
Formation of Intra-Chain Lactones

The formation and hydrolytic stability of intra-chain lactones has been in-
ferred from oxidized cellulose samples (oxycelluloses). Little is known on
the structure and occurrence of lactones in bleached pulps. The amount of
glucuronic acid in bleached wood pulps is rather small. The potential in-
volvement of lactones and ester bridges in inter-chain cross-linking has been
controversially discussed in the literature [79]. Nevertheless carboxylic and
lactone groups have been linked to numerous processes and product properties such as hornification, yellowing, aging or tensile strength [79, 80].
Model studies for the labeling reaction and for lactone formation in solution
have been performed using synthetic methyl β-D-glucopyranosyl-(1 $\rightarrow$ 4)-β-
D-glucopyranosiduronic acid. In situ NMR experiments at pD 1–2 did not
indicate the presence of lactones [81]. The situation in a solid matrix, however, might be different where additional parameters such as accessibility,
Donnan effects, hydration and the steric environment are operative. Further-
more, in striking contrast to the labeling experiments with aldonic acids, the
FDAM-labeled uronic acid derivative could be isolated in preparative yield
(Scheme 13).

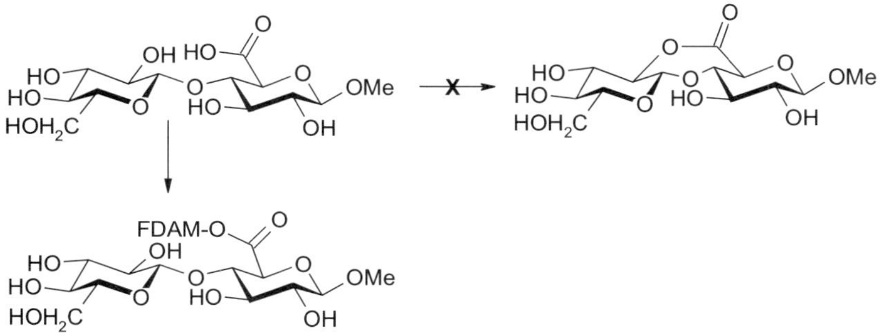

Scheme 13 Labeling of a disaccharide uronic acid derivative

2
Determination of Oxidized Functionalities

The quantitation of oxidized groups can in principle be performed at three levels of information. The variant mostly utilized so far is the determination of the total amount of a respective functional group as a sum parameter. Examples of different methods are given in the following section. To gain more detailed information in which part of the polymer these functional groups are situated a combination of group-selective labeling and size exclusion chromatography can be used. The third level is the determination of the type of functional group within the AGU. For cellulose no methods are currently available to quantify different types of oxidized groups within the AGU. Such an investigation would require for instance total hydrolysis and analysis of the fragments obtained, and determination despite the extremely low concentration of differently oxidized units remains a future challenge.

2.1
Carbonyl Functionalities (Aldehyde and Keto Groups)

2.1.1
Conventional Methods for Carbonyl Analysis in Celluloses

The quantitation of carbonyl groups in cellulose was so far limited to the measurement of the total carbonyl content by different methods, which are summarized in Table 1.

The so-called copper number is still the method of choice in the pulp and paper industry in process control, but sometimes also in the research lab. The reducing power of cellulose is measured by reaction with an alkaline Cu^{II}-salt under defined conditions, the formed Cu^{I} ions can be titrated after re-oxidation [82]. The underlying reaction mechanisms are still not entirely understood, neither are the types of oxidized structures recorded. However, even though the copper number is only a sum-parameter, the data of which cannot be directly linked to the quantity of a specific oxidized function, it remains a valuable parameter for control in a number of industrially relevant processes.

Cyrot [83] proposed the reaction with hydroxylamine to the corresponding oxime as a means of monitoring, since oximation was more sensitive than reaction with similarly reacting reagents such as hydrazine. The degree of nitrogen fixation can be measured either by a Kjeldahl procedure, or, as more recently applied, by elemental analysis. According to Rehder et al. [84], running the reaction in a zinc acetate buffer eliminated side-reactions with acidic groups and minimized those with lactones. According to Lewin [85] the formation of cyanohydrins by reaction of carbonyl groups with cyanide can also be taken as a measure of oxidized functionalities. Excess cyanide is removed

Table 1 Classical methods for determination of carbonyl groups in cellulose

Method	Reaction	Detection	Comment	Refs.
Copper number	(Unspecific) Reducing power	Titration	Only relative data, mechanism ill-defined	[82]
Hydrazinium-salt	Formation of charged groups on cellulose	Photometric	Only semi-quantitative	[88]
Hydroxyl-amine	Oximation	Elemental analysis, photometric, titration	Sensitivity depends highly on detection mode	[83, 89]
Cyanide	Formation of cyanohydrins	Titration	Toxic reagents, overestimation	[85]
NaBH$_4$	Reduction to the alcohol	Titration	Rather insensitive	[86, 87]
TTC-reaction	Formation of a red dye from reducing ends	Photometric	Only for REG, cellulose degradation, overestimation	[90, 91]

from the pulp and determined by titration with AgNO$_3$. This cyanide method often yields too large values due to adsorption phenomena, and is, moreover, less often used today because of the toxicity of the reagents.

The consumption of sodium borohydride (reduction method) upon reduction of carbonyl groups can also be used for their quantification. Remaining NaBH$_4$ is quantified by the amount of hydrogen formed after reaction with acid [86, 87].

Determination of dialdehydes in periodate cellulose can also be based on consumption of hydroxyl ions by the Cannizzaro reaction [92]. The β-elimination reaction of 2,3-dialdehyde celluloses in combination with DP-determination is proposed as a means to roughly determine the extent of oxidation [93].

An interesting procedure for quantification of C6-aldehyde groups of highly substituted C6-aldehydo derivatives by reduction with NaBD$_4$ and MS analysis of the hydrolyzed material is given by Clode and Horton [51].

2.1.2
Differentiation Between Keto and Aldehyde Groups

An important issue is the differentiation between aldehyde and keto groups in cellulosic materials. Different approaches are theoretically conceivable: a selective derivatization of either aldehyde and keto groups, a mathematical calculation of the reducing end groups and its subtraction from the over-all amount of keto structures determined, or a selective "masking" of either of the two oxidized species by oxidation or reduction [94, 95]. Classical approaches follow the latter mode. However, all of these approaches give only a rough estimate of the aldehyde/keto ratio. An interesting approach was followed by Sihtola et al. [96] who correlated different rates of oximation with different carbonyl types present, but without making assignments of the rates to underlying structures, a similar principle was later applied by Blaha et al. [97]. Different oximation rates have been observed for differently bleached samples and aged pulps. However, the extent of morphological influences in this procedure is quite difficult to assess.

An elegant approach would employ a fluorophore, of which the wavelength depended strongly on the type of carbonyl it is attached to. So far, such a differentiation of keto and aldehyde groups by labeling is not available. The mathematical calculation of reducing end groups from the number average molecular weight data is possible, and would directly result in a distribution of keto and aldehyde groups relative to the molecular weight. Further chemical manipulation of the cellulose is not necessary here. However, also the calculation approach has considerable drawbacks: the error of the measurement of M_n is rather large and depends highly on the data evaluation of the GPC measurements, and so does the calculated REG content. In addition, pulp samples and also aged papers contain oxidized reducing end groups in the form of acid residues, since the REG is a primary site of oxidative attack. However, if molecular weight data (M_n) are available, the number of REG can be roughly estimated.

The oxidation of reducing end groups of cellulose to the corresponding aldonic acids by chlorous acid was thought to proceed selectively enough to be used as a means of differentiation between keto and aldehyde groups. A second determination of either remaining keto carbonyls or newly formed carboxyl groups would finally yield the keto/aldehyde ratio [98]. However, the oxidation with chlorous acid proved to be not sufficiently selective (unpublished results). Another method was based on the work by Siggia and Maxcy [99] who estimated aldehydes by bisulfite addition and titration of the excess bisulfite with alkali. As this addition proceeds also with keto groups, albeit at a slower rate, the method must be applied with great care only [100].

The reducing power of aldehyde groups can be used to convert TTC (2,3,5 triphenyltetrazolium chloride) into triphenylformazane, a red dye, which can be quantified spectrophotometrically [90]. As the reaction is carried out in aqueous alkaline media, it induces β-elimination and hence results in overestimation of aldehyde groups. The same principle was applied by Strlic and Pihlar [90] in homogeneous solution (DMAc/LiCl) yielding results more reproducible than those according to the heterogeneous procedure in aqueous solution. However, also under these modified conditions, most industrial pulps suffer considerable DP-loss which results in an overestimation of REGs.

Only pulps from cotton linters and rag paper proved to be stable enough to be safely subjected to this method. A combination of the TTC protocol and the CCOA method to obtain pure keto group profiles is given by Nagel et al. [177].

2.1.3
The CCOA Method for Carbonyl Quantification

The classical procedures described above yield the carbonyl value as a sum parameter with varying accuracy. An additional drawback is the large sample amount required to obtain sufficient sensitivities. The comparably high reactivity of the carbonyl groups allows the application of carbonyl-selective reagents being UV-active or fluorescent compounds. This already lowers the sample demand significantly (see *Hydrolysis Method* in this chapter). The actual advantage, however, represents the opportunity of combining fluorescence labeling with a separation technique (see *Combination with GPC*). A suitable carbonyl-selective fluorescence label was synthesized: the reagent carbazole-9-carboxylic acid [2-(2-aminooxy-ethoxy)ethoxy]amide (carbazole-carbonyl-oxyamine, CCOA) (Scheme 14).

The CCOA label contains a fluorophore and a flexible spacer. The reactive anchor group is an oxyamine. Oxyamines showed a higher reactivity towards carbonyl groups as compared to hydroxyl amines [101]. Due to the fluorescence characteristics ($\lambda_{ex} = 340$ nm in DMAc) no interference with MALLS detection ($\lambda = 370$ nm or higher) is encountered. CCOA can be applied both in aqueous buffer systems and in the solvent DMAc/LiCl, as tested with model compounds [102], and gives generally neat and quantitative conversion of carbonyl functions into the corresponding O-substituted oximes (Scheme 15).

The label is completely stable in this solvent system at ambient temperature. Calibration, which is required for quantification of the carbonyl content,

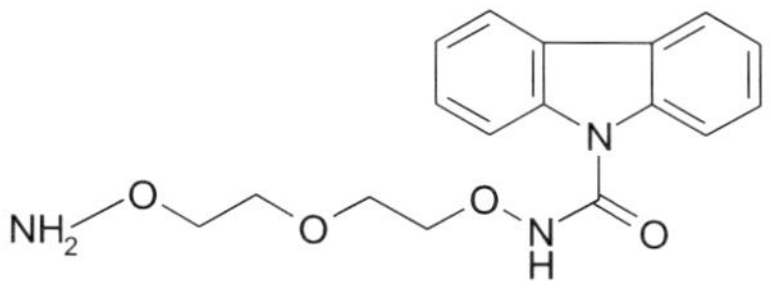

Scheme 14 Structure of the fluorescence label CCOA for carbonyl-detection in cellulose

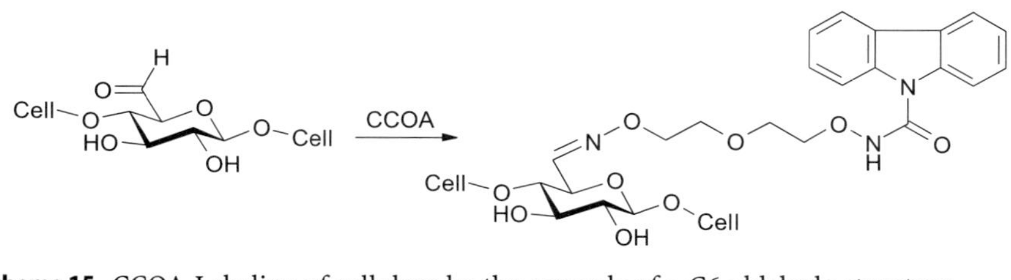

Scheme 15 CCOA-Labeling of cellulose by the example of a C6-aldehyde structure

can be performed with the label CCOA (Scheme 14) which has exactly the same fluorescence properties as the labeled compounds. The integrity of the cellulose was not affected upon labeling, i.e., no degradation occurred.

The advantages of the CCOA method are:

- Relatively low sample amounts required (5–25 mg)
- High selectivity and high sensitivity
  (LOD: 0.01 μmol/g; LOQ: 0.05 μmol/g)
- Yields all MWD data in addition to carbonyl profiles, combined with GPC
- MWD does not change after labeling (no aggregation phenomena)
- Separate numeric analysis of different MW regions is possible
- Practical on a routine basis

The following points can be considered as drawbacks of the method:

- The label is commercially unavailable.
- There is no differentiation between aldehyde and keto groups.
- There is no differentiation between lignin-carbonyls and cellulose-carbonyls.
- The combination with GPC is limited to pulps and papers soluble in DMAc/LiCl.

2.1.4
The CCOA-Hydrolysis Procedure

The hydrolysis procedure is based on heterogeneous carbonyl-selective fluorescence labeling with CCOA, which is subsequently released with triflic

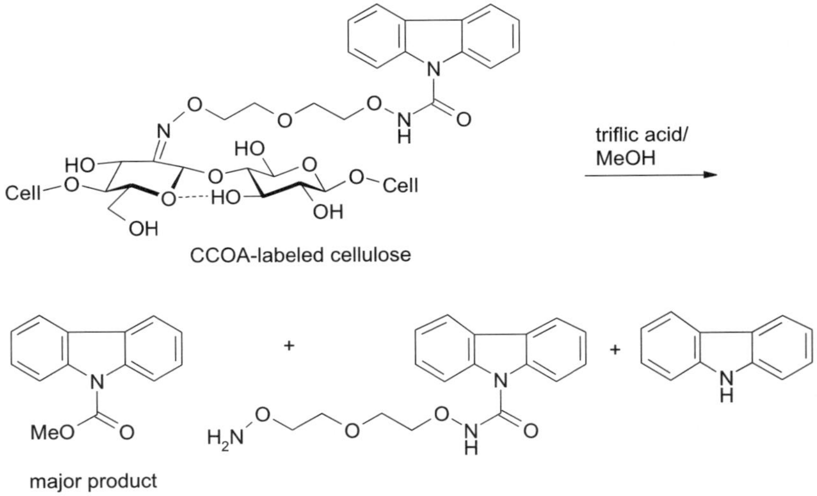

triflic acid/
MeOH

CCOA-labeled cellulose

major product

Scheme 16 Hydrolytic cleavage of CCOA-labeled cellulose by triflic acid in methanol under release of N-(methoxycarbonyl)carbazole, CCOA, and carbazole acid from the labeled pulp in a quantitative manner, and the concentration of CCOA and CCOA-derived products are determined by HPLC with fluorescence detection [103] (Scheme 16). The procedure requires material in the mg range only and can be applied to pulp and paper samples which do not readily dissolve in DMAc/LiCl. Calibration can be performed against DMAc/LiCl-soluble standard pulps. The data obtained by this approach correlate well with data from the traditional CCOA procedure. Limit of detection (3σ) and determination (10σ) are very low with $0.042\,\mu mol/g$ and $0.14\,\mu mol/g$, respectively, as compared to conventional methods. Scheme 16 shows the reaction and the products obtained after hydrolysis.

2.2
Carboxyl Groups

2.2.1
Conventional Methods for Carboxyl Analysis in Cellulose

Two procedures for the determination of carboxyl groups predominate today: titration of the acids under formation of salts with different cations and utilization of the anion exchange capacity of pulps, which is a consequence of the presence of carboxyl groups. The acidic groups are transferred into their salt forms, using either an inorganic cation as in the former case, or an organic cationic dye, such as methylene blue or crystal violet, in the latter. The cation is subsequently quantified either after recovery from the pulp or by depletion analysis of the remaining species in solution. Also decarboxylation and determination of the freed carbon dioxide has been applied to quantify cellulose carboxyls [104, 105]. Investigations of Wilson [106] showed a strong dependence of the ion exchange capacity of carboxyl groups in pulp on the ion strengths. Various modifications of the method have been published [107–109]. Samuelson and Törnell [110] studied the determination of carboxyls in the presence of carbonyl groups. The pH of the alkaline sodium chloride solution to obtain the sodium salt of the carboxyl groups did not exceed pH 8 in order to prevent β-eliminiation reactions of the carbonyl groups. Sihtola [111] used a potentiometric determination after desalting by electrodialysis, displacement by potassium ions, and subsequent titration to pH 7.0. The iodide-iodate method [112] has the advantage of detecting carboxyl groups, lactones and enediol groups. The conductometic titration allows also differentiating between strongly acidic groups (e.g., sulfonic acids) and carboxyls. Table 2 gives an overview on conventional methods to estimate carboxyl groups in cellulosics.

Table 2 Classical methods for determination of carboxyl groups in cellulose

Method	Reaction	Detection	Refs.
(Reversible) Methylene blue	Ion exchange	Photometric	[113–116]
Sodium bicarbonate/NaCl	Neutralization	Titration	[117, 118]
Zinc acetate	Ion exchange	Gravimetric, complexometric	[119]
Crystal violet base	Acid-base reaction in benzene	Photometric	[120]
HCl	Neutralization	Conductometric	[121, 122]

Problems frequently encountered with the above described methods are:

- Inaccuracy due to unspecific binding of the cations
- Large sample demand
- Limited sensitivity
- Lactones have to be "opened" prior to the measurement [84, 89] but exhibit different rates of saponification depending on their type.

For a recent review on a critical evaluation on classical methods see Fardim et al. [123]. In addition to the wet chemical methods, FTIR spectroscopy and ESCA [124, 125] techniques have been applied to quantify carboxyl groups in cellulose and pulp. In case of FTIR, satisfactory results in comparison to wet chemical analysis could only be obtained with carboxyl-rich samples [126–128].

2.2.2
Determination of Lactones

In most cases, uronic acids are liberated from acidic polysaccharides by hydrolysis leading to irreproducible concomitant formation of lactones. Several methods to circumvent this problem have been published describing conversion of the uronic acid into methyl esters followed by reduction with borohydride or borodeuteride reagents and subsequent hydrolysis and GC-MS detection [129]. Other techniques are based on the liberation and quantification of carbon dioxide. Direct determination of uronic acid residues in hydrolyzates has frequently been performed according to colorimetric assays, which are rather insensitive and have thus mostly been replaced by high-performance anion exchange chromatography (HP-AEC) methods [130–132].

Due to the inherent chemical propensity of lactones with respect to hydrolysis and reformation, the determination has mostly relied on the total quantification of carboxyl groups (assuming a complete hydrolysis of lactones under alkaline conditions and negligible regeneration under acidic conditions needed for the deionization of carboxyl groups) followed by measurement of free carboxylic groups using a variety of alkalimetric or complexometric protocols. The direct and unambiguous detection and differentiation of lactone moieties has not been accomplished thus far. An estimation of the lactone content has been gained by a variant of the iodometric titration of carboxyl groups [133, 134]. Hydrolysis of lactones under non-alkaline conditions is achieved by $KI - KIO_3$ which releases iodine – albeit at a slower rate than free carboxylic groups – which is scavenged in the presence of thiosulfate and quantified by an indirect iodometric titration [135]. The different types of lactone moieties, however, display varying degrees of reactivity towards hydrolysis.

2.2.3
Determination of Hexenuronic Acids

The following gives a brief compilation of procedures to determine hexenuronic acids in cellulosic pulp samples. The common methods are based on hydrolysis of HexA moieties from pulp, either enzymatically or chemically, with subsequent quantification of the hydrolysis products either directly or after chemical conversion into UV active compounds. A comparison of these three methods is given by Tenkanen et al. [136]. For comparison rather than exact determination of HexA, e.g., during bleaching stages, the diffuse reflection UV VIS method can be applied [137]. A photoacoustic FTIR procedure based on chemometric analysis has been described as well [138]. In Table 3, the available methods to analyze HexA moieties in cellulosic material are summarized.

Table 3 Methods for the determination of hexenuronic acids in cellulose

Method	Principle	Detection	Refs.
VTT method	Enzymatic hydrolysis	HPAEC-PAD, CE	[139, 140]
HUT method	Acid hydrolysis	Photometric	[69]
KTH method	Hg-salt, hydrolysis	Photometric, HPLC	[141, 142]
Resonance Raman	Direct detection of double bond	Resonance Raman	[143, 144]
Cadoxen method	Direct detection of double bond	UV	[145]

2.2.4
The FDAM Method for Carboxyl Group Determination

An approach different from the classical methods described in the previous chapter is the mild esterification of carboxylic acids as described in

protein chemistry [146, 147] or utilized for fatty acid analytics [148–150].
However, most of these methods are not applicable towards quantification of
sugar acids in polysaccharides due to insufficient yields. So far, the only appropriate reaction for sugar acids turned out to be that with diazomethane
reagents [151]. Aromatic diazomethane derivatives do not require an activation step, no auxiliaries are necessary, traces of water do not interfere
with the conversion, and the reaction proceeds at ambient temperatures to
afford stable esters. Hydroxyl groups do not react under these conditions.
If a fluorophore is linked to the diazomethane anchor group, the detection
limit can be lowered significantly. In addition, a combination with other
detection modes becomes possible. 9-Anthryldiazomethane (ADAM) [152–
154], 1-pyrenyl-diazomethane (PDAM) [155, 156], and 4-diazomethyl-7-
methoxycoumarine (DAM-MC) [157] have been described in the analytics of
fatty acids and similar analytes, but could not be used in analytics of cellulosic carboxyls due to the limited stability of the product esters and because
of the interference of the fluorescence label with the MALLS detection in GPC
analysis.

For carboxyls in cellulose the FDAM label (Scheme 17) was found to be
most suitable in combination with GPC-MALLS analysis [158].

The reaction was tested on carbohydrate model compounds and was
shown to provide quantitative yields [78, 81] (Scheme 18). Following from
the reaction mechanism, the carboxyl groups must be present in free form
(COOH) for the labeling to proceed. Carboxylates (salts), lactones or esters

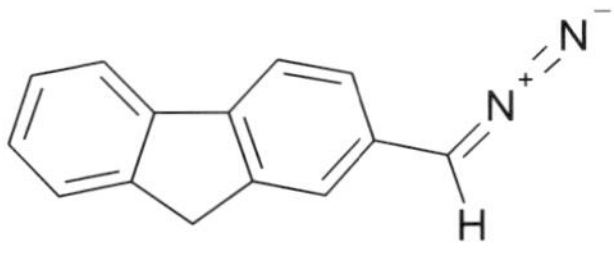

Scheme 17 Structure of the fluorescence label FDAM for carboxyl-detection in cellulose

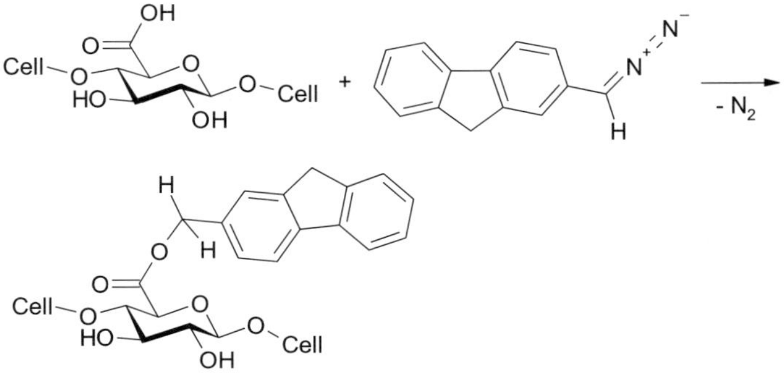

Scheme 18 FDAM-labeling of cellulose by the example of a C6-carboxylic (glucuronic)
acid structure

are not detected. At present, the statement of neat labeling with FDAM and reliable GPC detection must be restricted to uronic acids. Even though preparative experiments showed a conversion also for aldonic acids, the primarily formed fluoren-2-ylmethyl onates appeared to behave like activated esters under normal phase chromatography conditions, promoting the formation of lactones (see Scheme 12 above) with concomitant release of fluoren-2-yl-methanol from the FDAM label.

The reaction conditions for FDAM labeling were optimized for cellulose to fit into the usual activation protocol prior to dissolution in DMAc/LiCl (9%). FDAM can be prepared in the GPC eluant DMAc from 9H-fluoren-2-yl-carboxaldehyde via its hydrazone by oxidation with excess manganese dioxide [78].

The advantages of the FDAM method are:

- Relatively low sample amounts required (5–25 mg)
- Yields all MWD data in addition to carboxyl profiles in combination with GPC
- Separate numeric analysis of different MW regions is possible
- Practical on a routine basis
- FDAM label is easily and neatly synthesized prior to derivatization

The following points can be considered as drawbacks of the method:

- There is no differentiation between lignin-carboxyl and cellulose-carboxyls
- Combination with GPC is limited to pulps and papers soluble in DMAc/LiCl
- Degree of labeling of gluconic acid residues at present is unknown

2.3
Combination of Group-Selective Fluorescence Labeling with GPC

A substantial progress in analytics of oxidized groups in celluloses was achieved with the possibility to locate carbonyl and carboxyl groups along the cellulose macromolecule [159, 160]. This technique combined group-selective fluorescence labeling with GPC analysis. Addition of a fluorescence detector reports the amount of fluorescence in every slice of the MWD, and provides directly the concentration of the respective functional group after signal calibration. The standard GPC setup used in such approaches is given in Fig. 3.

From the integrated signals of the RI detector, which is concentration-sensitive, and the fluorescence detector, which is functional group-sensitive, the total amount of carbonyl or carboxyl groups can be calculated.

The MALLS detector allows the absolute determination of the molecular weight, provided that the dn/dc value of the polymer in solution is known. The minute concentration of functional groups (labeled carbonyls or carboxyl groups) does not influence the dn/dc value. From the MALLS and the RI signal the MWD is calculated. Combination of both, the fluorescence signal and the MWD allows the calculation of the functional group profile. Figure 4

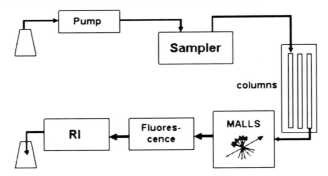

Fig. 3 GPC setup for functional group profiling. RI: refractive index detector; Fluorescence: fluorescence detector; MALLS: multi-angle laser light scattering detector

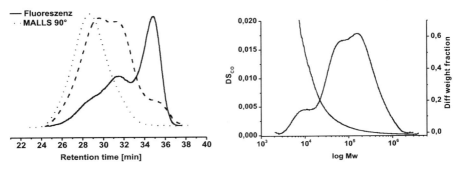

Fig. 4 *Left*: Fluorescence, MALLS (90°) and RI signals from a CCOA-labeled cellulose sample. *Right*: Differential molecular weight distribution and degree of substitution of carbonyl groups (DS_{CO}) as calculated from the detector outputs

shows an example of the signals obtained during GPC and the resulting MWD and carbonyl profile.

This approach requires a sensitive means of tagging the functional group since the absolute amount analyzed by GPC ranges in the μg range. The sensitivity needed is achieved with the above described fluorescence labels. However, in order to combine the fluorescence labeled material with GPC in combination with MALLS detection, the label must not emit light close to the laser wavelength of the MALLS device, which puts some restrictions to the chromophore of the marker. Classical carbonyl labels such as dansyl hydrazine [161], dansyl oxyamine [101], phenylhydrazine [162] do react with cellulose carbonyl groups as well [163] but are not suitable in combination with GPC-MALLS. The CCOA and FDAM labels described above have been designed to work in combination with GPC detection.

In cellulose chemistry, the term "degree of substitution" (DS) usually denotes the number of substituted OH groups per anhydroglucose unit; with this meaning it is most frequently used for cellulose ethers or ethers. The DS thus reflects the completeness of a chemical modification at the hydroxyl groups of the polysaccharide. However, the term DS was also used to describe the average content of CO or COOH groups per anhydroglucose unit, hence the terms DS_{CO} and DS_{COOH} have been used by analogy to illustrate the amount of functional groups in relation to the MWD.

The outputs of the respective detector signals are used as described above to calculate the DS profiles, i.e., the concentration of the respective functional group in relation to the molecular weight distribution. Figure 5 presents examples of DS_{CO} and DS_{COOH} profiles, obtained from a sulfite dissolving pulp. The DS_{CO} generally increases with decreasing molecular weight due to the increasing number of reducing end groups. The DS_{COOH} profile differs significantly from the DS_{CO} profile. In the example shown, the number and distribution of carboxyl groups is mainly governed by glucuronic acid moieties in the hemicelluloses. The signal passes through a maximum and decreases towards both small and large MW fragments.

ΔDS plots, which simply give the difference between two DS curves, facilitate the comparison of two samples with regard to their carbonyl or carboxyl contents relative to the molecular weight. ΔDS plots are very suitable graphic representations to report even slight differences in carbonyl or carboxyl contents in order to better visualize changes occurring during specific treatments starting from the same material. They allow, for instance, to analyze in a very straightforward way, how a chemical treatment increases or decreases functional groups in certain molecular weight ranges. Examples of ΔDS plots are given in the following application chapter.

Carbonyl or carboxyl group profiles can also be evaluated numerically. This way, the amount of functionalities in preselected regions of the MWD are compared, provided that the samples have similar molecular weights. Certain characteristics of the cellulose state of oxidation or the distribution of

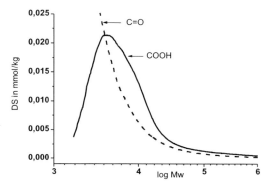

Fig. 5 Degree of substitution plots for the molecular weight dependent distribution of carbonyl groups (DS_{CO}) and carboxyl groups (DS_{COOH}) for an example pulp

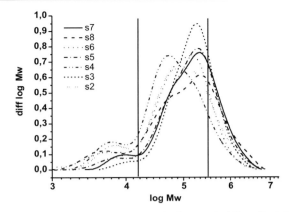

Fig. 6 MWD of different dissolving pulps (s2–s8), the *vertical lines* limit DP ranges of 100 and 2000, corresponding to the carbonyl contents in Fig. 7 below

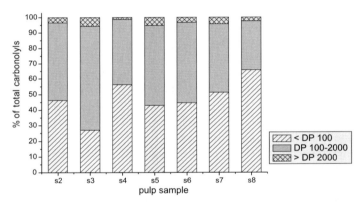

Fig. 7 Distribution of carbonyl groups over three MW regions (DP < 100, 100 < DP < 2000, and DP > 2000), corresponding to the MWD of the pulps in Fig. 6 above

hemicelluloses can easier be visualized. Figure 6 gives the MWD of different dissolving pulps, vertical lines indicating the DP 100 and DP 2000 limits. The corresponding percentages of carbonyls in these regions are displayed in Fig. 7. The data obtained may be further correlated with other pulp parameters.

3
Applications of the CCOA and FDAM Methods: Monitoring Oxidative Processes

Through combination of MWD analysis and functional group analysis it also became possible to examine how functional groups in certain regions of the molecular weight behave during specific treatments [164]. Often the over-all trend of functional groups outweighs effects occurring in specific areas (low MW or high MW parts). These effects can be examined in detail using functional group profiling. If the pulp is degraded during the treatment the profiles report both hydrolysis, by generation of new REG and decrease of M_w, and oxidation. If the MWD does not change significantly, only oxidative changes are considered. The CCOA and FDAM method are thus very powerful tools to follow cellulosic material through different processing steps with regard to oxidative changes occurring thereupon. In the following application chapter, several examples are given: different bleaching treatments, processing of Lyocell dopes, alkalization of pulps, irradiation treatments of cellulose and cellulose aging. In all of these cases, CCOA- and FDAM-monitoring were used to clarify which oxidative changes occur in which molecular weight region. The additional dependence on reaction conditions and reaction time provided mechanistic insights much more detailed than possible by any means available so far.

3.1
Bleaching Treatments

Bleaching can be considered as the major process introducing functional groups, especially carbonyls, into cellulosics. Yellowing and limited brightness stability caused problems in the early days of hypochlorite bleaching, and became an issue again with the introduction of ozone in TCF bleaching sequences. In the following, the effect of different bleaching types and agents on the carbonyl group profiles and the MWD of the bleached celluloses will be summarized.

In Fig. 8 the carbonyl DS and ΔDS plots for ozone-bleached beech sulfite pulps are shown. With increasing bleaching intensity, the ΔDS_{CO} was increased for medium and high-molecular weight regions, but dropped below the value of the starting material for the low MW range. The boundary region between DS increase and DS decrease was around M_w 10 000–20 000 g/mol. Thus, the ozone treatment comprises an interplay of carbonyl-generating and carbonyl-consuming processes according to the MW regions.

Reducing end groups represent the major part of the total amount of carbonyl groups in cellulose chains with a low DP. Upon ozone treatment, these reducing end groups are oxidized to lactones and carboxylic acids, so that an ozone bleaching stage lowered the carbonyl DS in the lower molecular weight regions. In cellulose chains with a high molecular weight, however, the contribution of reducing end groups to the total amount of carbonyls is much smaller, so that the oxidation of reducing end groups is overcompensated by the introduction of new carbonyl functions. The generation of carbonyls predominantly in higher-molecular weight material might explain the well-known observation that ozone-bleached pulps suffer a severe DP loss

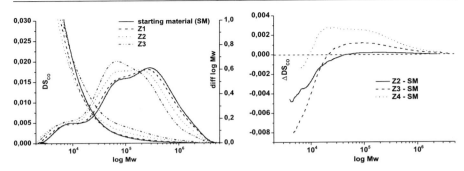

Fig. 8 Ozone bleaching (Z) of a beech sulfite pulp, increasing numbers reflect increasing intensity of the ozone treatment. *Left*: differential MWD und carbonyl DS. *Right*: ΔDS_{CO} plots showing the differences in carbonyl content between the bleached sample and the starting material (SM). Reprinted with permission from Biomacromolecules (2002) 3:969. Copyright (2002) American Chemical Society

in a subsequent P stage: when the carbonyls introduced upon ozone treatment are reduced by NaBH$_4$, the pulp is rendered largely insensitive towards the subsequent peroxide treatment [165].

With the pulp used, a very low ozone charge did not significantly increase the carbonyl content, since preferably the residual lignin present in the material was attacked by the oxidant [16]. The progressing degradation of the polysaccharide material upon further oxidation, as reflected by the decreasing M_w, is clearly visible in Fig. 8.

The effect of a peroxide bleaching stage is demonstrated in Fig. 9, which displays again differential MWD, DS$_{CO}$ and ΔDS_{CO} graphs. A standard beech sulfite pulp was subjected to an ozone treatment followed by a peroxide bleaching stage. Again, it was obvious that the initial ozone treatment increased the carbonyl content at molecular weights above 20 000 g/mol, and decreased it below this value. Also the simultaneous DP loss was evident. The subsequent peroxide stage has a quite beneficial effect with regard to the C = O content: in all molecular weight ranges the carbonyl content was decreased as compared to the ozone-bleached material. The DS profile drops even below the level of the unbleached starting material at molecular weights below 10^5, and ranged only slightly above the curve for the initial pulp above 10^5 g/mol.

As a last example for the applicability of the CCOA method in bleaching chemistry, the effect of an intensive hypochlorite treatment of pulp (H stage) at pH 7.0, aimed at producing highly oxidized pulps, is presented (Fig. 10). The DP loss was progressing with increasing bleaching intensity. The carbonyl DS generally ranged above the values for the genuine pulp, but increased only slightly with enhanced hypochlorite charge in higher-molecular weight regions. Carbonyl functions were mainly introduced into

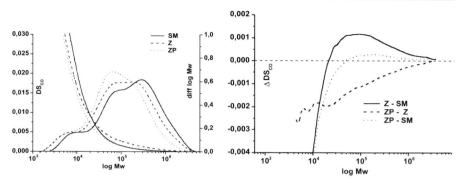

Fig. 9 Ozone bleaching (Z) of a beech sulfite pulp, followed by a peroxide stage (P). SM: starting material, Z: ozone treatment, ZP: ozone – peroxide treatment. *Left*: differential MWD and carbonyl DS. *Right*: ΔDS_{CO} plots showing the differences in carbonyl content between the bleached sample and the respective starting material. Reprinted with permission from Biomacromolecules (2002) 3:969. Copyright (2002) American Chemical Society

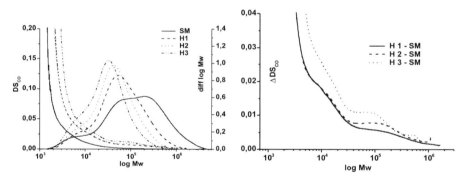

Fig. 10 Hypochlorite treatment (H) of a beech sulfite pulp, increasing numbers reflect increasing intensity. SM: starting material. Reprinted with permission from Biomacromolecules (2002) 3:969. Copyright (2002) American Chemical Society

shorter chains below a MW of about 30 000. This is a major difference to the above-discussed ozone and peroxide treatments of pulps.

In summary, ozone-bleaching increased the DS_{CO} for medium and high-molecular weight regions, but decreased the DS_{CO} for the low-molecular weight range. A peroxide treatment after ozonation decreased the carbonyl content throughout all MW regions. Hypochlorite treatments introduced carbonyl functions mainly into low-molecular weight regions, in contrast to ozone and peroxide treatments. In all three bleaching types a significant decrease in M_W was evident.

3.2
Cellulose in Lyocell Dopes

The cellulose solvent NMMO (*N*-methylmorpholine-*N*-oxide) monohydrate is a white crystalline solid at room temperature with a melting point of 84 °C. Dissolution of cellulose and further processing of the spinning dope is thus carried out at temperatures of about 100 °C. As NMMO is a relatively strong oxidant, which is also frequently used as oxidizing agent in organic synthesis [166], it exerts pronounced oxidative stress on the dissolved pulp material, which is additionally intensified by the elevated process temperatures. Oxidizing solvent and pulp interact sufficiently long to cause possible severe effects, e.g., decreased fiber properties, discoloration of the resulting fibers, thermal instabilities of the dope and even explosions and uncontrolled degradation reactions, sometimes called "exothermic events" [167]. The oxidative action of the solvent NMMO on the solute cellulose was thus a topic of highest importance.

By means of the CCOA method the carbonyl profiles of pulps upon Lyocell processing were studied. It was shown that the overall carbonyl content of the pulp decreased continuously upon dissolution in NMMO at 120 °C. This overall decrease was relatively small, but significant. A detailed evaluation, as given in Fig. 11, revealed that this overall decrease was a superposition of counteracting processes with regard to regions of different molecular weight. Up to a DP of 50, already existing carbonyl groups were consumed, which was mainly attributed to changes of the hemicellulose parts [168]. The same decrease in carbonyls, although less pronounced, was found for a molecular weight region up to DP 200. In contrast to these MW regions, the content of carbonyl groups increased moderately in high-molecular weight material above DP 200.

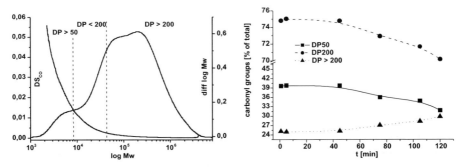

Fig. 11 Cellulose (beech sulfite pulp) dissolved in NMMO (Lyocell dope). *Left*: DS_{CO} and differential MWD of the starting pulp, molecular weights of DP = 50 and DP = 200 are indicated by *vertical dashed lines*. *Right*: Time course of the overall carbonyl content for three selected molecular weight ranges. Reprinted with permission from Biomacromolecules (2002) 4:743. Copyright (2002) American Chemical Society

In the short chain material with a high relative content of reducing end groups (e.g., one per 50 or 100 anhydroglucose units) a major reaction pathway was the oxidation of these structures to carboxylic (gluconic) acid residues, paralleling the behavior of model compounds. At the same time, it was likely that a minor amount of new carbonyl groups was generated along the chain in unselective oxidation at C-2, C-3 and C-6, which was outweighed by the oxidation of reducing ends, so that the observed overall decrease in carbonyl groups resulted. In high-molecular weight parts, in contrast, the relative amount of reducing ends was rather low (e.g., one per 500 or 1000 anhydroglucose units), so that the generation of new carbonyl groups became dominant. The introduction of carbonyl groups, especially into longer chains, was an important result as carbonyl groups have been demonstrated to be directly correlated with chromophore formation in the Lyocell system [169, 170].

In summary, processing of a beech sulfite pulp sample in NMMO at elevated process temperatures caused a net decrease of the carbonyl content. The conversion of reducing ends to carboxylic acids, being the dominant process in lower-molecular weight material, was counteracted by an unselective introduction of keto groups along the chain, which was prevailing in the high-molecular weight region. The latter process will also play a role in subsequent discoloration reactions.

3.3
Alkalization of Cellulose and Aging of Alkali Cellulose

Alkalization ("steeping") is a crucial process step in the production of cellulose derivatives. It is employed prior to certain derivatization reactions, such as xanthation in the viscose process or etherification in the production of carboxymethyl cellulose [171]. Especially in the viscose process the alkalization is not only used for activating the hydroxyl groups, but also to free the pulp from impurities, such as hemicelluloses. The steeping step involves treatment of bleached pulp with strong alkali hydroxides, mostly 18% NaOH, to convert cellulose into sodium cellulose I. Excess sodium hydroxide is pressed off and the press cake, consisting of approx. 34–35% cellulose and 15–16% NaOH, is left in the presence of air for several hours, which is referred to as aging or "ripening". In this process stage the appropriate pulp viscosity, i.e., the cellulose DP, is adjusted to the needs of further processing to viscose. The changes in the molecular weight distribution are brought about by oxidative processes [96, 173–175] which involve both introduction and conversion of oxidized functions, such as keto and aldehyde groups, and chain cleavage by subsequent alkali-induced reactions, such as β-alkoxy-eliminations. The CCOA method was employed to examine these oxidative changes: for the first time the introduction and conversion of oxidized functionalities were monitored in dependence of both time course and molecular weight distribution.

In Fig. 12, the molecular weight distribution and the carbonyl-DS curves of the genuine pulp sample, the pulp after steeping and removal of excess alkali ("pressing"), and the samples after different times of aging are shown. Both the decrease in molecular weight and the intended narrowing of the distribution over time were evident. The overall carbonyl content decreased significantly during the alkali treatment, approx. by about 50%. In contrast, the carbonyl-ΔDS plot in Fig. 13, depicting the difference curve between starting pulp and alkali pulp after steeping, indicated no significant change in carbonyls in the remaining pulp, except for the lowest molecular weight part. The overall decrease was consequently due to a removal of low-molecular weight hemicelluloses and celluloses, being no longer contained in the pulp sample so that the carbonyl functions were thus no longer detectable.

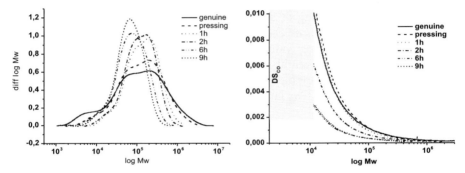

Fig. 12 Alkalization of cellulose and aging of alkali cellulose (beech sulfite pulp). DS_{CO} (*right*) and differential MWD (*left*) after steeping and different aging times. Reprinted with permission from Biomacromolecules (2002) 4:743. Copyright (2002) American Chemical Society

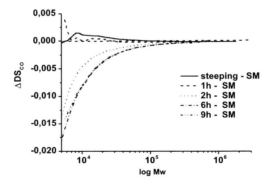

Fig. 13 Alkalization of cellulose and aging of alkali cellulose (beech sulfite pulp). ΔDS_{CO} plots for pulps after steeping and different aging times, SM: starting material. Reprinted with permission from Biomacromolecules (2002) 4:743. Copyright (2002) American Chemical Society

With increasing aging times, the carbonyl content in the pulp decreased further, as shown by the carbonyl-ΔDS plots after 2, 6 and 9 hours (Fig. 13). The main reason for this loss in CO groups was a progressing oxidation of reducing end groups to carboxyl functions, and a removal of keto groups, e.g., by β-elimination and rearrangements. After an aging time of about 4 hours the decline in carbonyls slowed down, and eventually leveled off at a low content of 4 μmol/g (Fig. 14, left). A detailed evaluation revealed once more pronounced differences in the reactivity of different molecular weight regions with regard to the carbonyl groups. In very high-molecular material with a DP > 2000, the amount of carbonyls stayed nearly unchanged (Fig. 14, right) as only very few carbonyl groups (reducing end groups) were contained. The relatively large amount of carbonyl groups and reducing end groups in lowest molecular and low-molecular weight ranges of DP < 50 and DP < 200 caused a continuous decrease during the aging procedure due to the oxidative consumption processes discussed above. Interestingly, the number of carbonyls in the mid-range of 200 < DP < 2000 was found to increase. Thus, only in this region the oxidative introduction of keto groups overcompensated the oxidative and alkali-promoted consumption of carbonyls, which dominated in the other MW ranges.

In summary, the application of the CCOA method disclosed two different processes in the alkalization and aging of cellulose, the initial removal of low-molecular weight material with a high content of carbonyls by dissolution leaving behind a carbonyl-depleted purified pulp, and the subsequent, slower process of oxidatively converting remaining carbonyl groups to carboxyl groups in the pulp. This confirms the traditional concept of alkalization [172, 173] as constructed from the results of conventional analytical methods. In the lower-molecular weight regions, large amounts of carbonyls

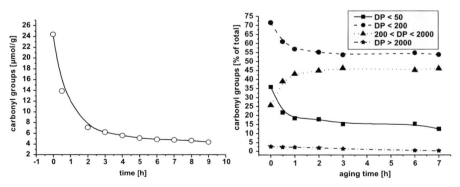

Fig. 14 Alkalization of cellulose and aging of alkali cellulose (beech sulfite pulp). Time course of the total carbonyl content (*left*) and the carbonyl content in different molecular weight ranges (*right*). Reprinted with permission from Biomacromolecules (2002) 4:743. Copyright (2002) American Chemical Society

were oxidatively removed, while in the mid-range generation of new keto groups outweighed this consumption.

3.4
Irradiation of Cellulose

Carbonyl group profiles of samples having undergone high-energy irradiation exhibited significant differences from chemically oxidized samples. Figure 15 shows ΔDS_{CO} plots of a UV-irradiation treatment with differing irradiation times. The cellulose sample used was a bleached prehydrolysis kraft pulp with the hemicellulose part being significantly lowered by alkaline extraction. This cellulose, free of lignin and most hemicelluloses, proved to be very stable under the conditions of UV radiation of up to 10 d. No significant reduction in the molecular weight was observed, although a largely uniform increase in carbonyls for the mid- and high-molecular weight areas was found—in addition to a stronger increase in the low-molecular weight range (Fig. 15, left).

How UV irradiation eventually changed the integrity of the cellulose was highly dependent on the origin of the cellulose (pulping) as well as structure and amount of the accompanying polymers (lignin and hemicellulose). While lignin acted as an antioxidant preserving cellulose in TMP pulp upon irradiation [174, 175] and unbleached pulp upon accelerated aging [176], residual lignin in prehydrolysis kraft pulp was shown to have the opposite effect. Radiation with UV light decreased the molecular weight significantly. A similar amount of residual lignin in a beech sulfite pulp, however, did not show such an effect on the DP of cellulose.

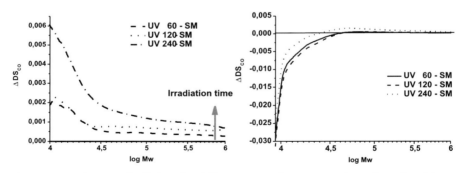

Fig. 15 UV radiation of cellulose for different times (h). ΔDS_{CO} plots of a pulp without hemicellulose (*left*) and with hemicellulose (*right*). The ΔDS_{CO} plots show the MW-dependent differences in the carbonyl content between the respective irradiated pulp samples and the non-irradiated starting material (SM). Reprinted with permission from Holzforschung (2004) 58:597. Copyright (2002) Walter de Gruyter

Also the presence of hemicellulose influenced the oxidation of the cellulose by UV light. After removal of the hemicelluloses, the UV irradiation caused significant introduction of carbonyls into the low-molecular weight part. A strong decrease in carbonyls of the low-molecular weight cellulose was found when hemicellulose was present, see the ΔDS_{CO} plot in Fig. 15 (right). Thus, the hemicelluloses responded differently, somehow acting as a sacrificial substrate which is further oxidized to carboxyls, saving the cellulose from being oxidized itself.

β-Irradiation, usually called "electron beaming" or "e-beaming", was found to induce severe changes due to the much higher energy input as compared to UV. The cellulose was significantly degraded, and this loss in DP was accompanied by the introduction of carbonyl groups over the whole molecular weight range (Fig. 16). In comparison to UV radiation, the amount of carbonyls introduced by e-beaming was about threefold.

In summary, from the studies on different types of oxidative modifications of celluloses it can be generalized that chemical means affect the low-molecular weight areas primarily. Higher oxidant dosages or more severe oxidation conditions increased the effect also on the high-molecular weight parts. By contrast, irradiation by UV light or by electron beams caused a constant increase in carbonyl groups also in the mid- and high-molecular weight areas. It was concluded that chemical oxidations, carried out under heterogeneous conditions, mostly influenced the readily accessible less ordered regions in cellulose, while radiation treatments were also able to penetrate into well ordered "crystalline" regions, causing a more uniform oxidation of the material over all MW areas.

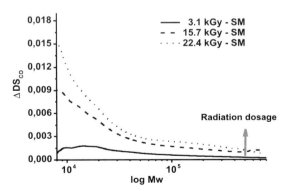

Fig. 16 ΔDS plots for β-irradiation (electron beaming) of cellulose with different dosages. The ΔDS_{CO} plots show the MW-dependent differences in the carbonyl content between the respective irradiated pulp samples and the non-irradiated starting material (SM). Reprinted with permission from Biomacromolecules (2002) 4:743. Copyright (2002) American Chemical Society

3.5
Aging of Cellulose in Paper

Two major processes are involved in aging of paper, hydrolysis induced by the presence of water and acids, and oxidative processes triggered by different factors such as metal ions, air contaminants, or simply the presence of air over long periods of time. Both pathways cause an increase in carbonyl and carboxyl groups. However, aging is also influenced by endogenic factors of the paper itself. Origin of the cellulosic material, pulping and papermaking procedures and especially papermaking additives, such as sizing agents, play a crucial role in this respect. Storage, mainly under ill-defined conditions, sometimes over centuries in the case of historical papers and documents, is an additional unknown variable. A straightforward assessment of the mechanisms of paper aging is thus rather problematic. The following section will illustrate how CCOA and FDAM method can assist in conservation and aging studies.

The molecular weight distribution of rag papers from four centuries is shown in Fig. 17. The highest M_w in this example was observed for the oldest paper. Endurance and stability of rag paper is highly influenced by the provenience of the material, the papermaking procedure and the storage conditions. Thus, no conclusion as to the age can be drawn from the molecular weight distributions measured.

A common feature observed at naturally aged rag paper samples is the relatively high amount of carbonyl groups at the low- and mid-range of MW, approx. below 10^5 g/mol or DP 6000. Roughly 50% of all carbonyls fall below DP 300. From the amount of reducing ends as estimated from M_n it was

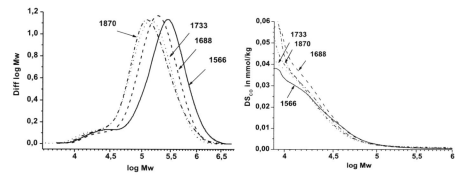

Fig. 17 Molecular weight distribution (*left*) and DS$_{CO}$ plots (*right*) of rag papers from different centuries. Data given in the graphs correspond to the approximate year of production. The total amounts of carbonyl groups and the REG estimated from M_n (in parenthesis) were as follows: 1566 – 17.7(7.5) µmol/g; 1688 – 21.8(8.1) µmol/g; 1733 – 24.2(11.8) µmol/g; 1870 – 25.9(10.7) µmol/g

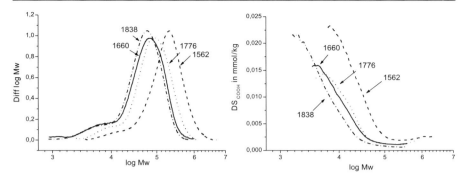

Fig. 18 Molecular weight distribution (*left*) and DS$_{COOH}$ (*right*) of rag papers from different centuries. Data given in the graphs correspond to the approximate year of production. The total amounts of carboxyl groups were as follows: 1562 – 23.6 μmol/g; 1660 – 20.5 μmol/g; 1776 – 17.3 μmol/g; 1838 – 15.44 μmol/g

calculated that the amount of keto functions, which are related to oxidative processes, is about 50% of the total carbonyls, while the remaining half is attributed to reducing ends that are naturally present or a result of hydrolysis. Rag paper samples having experienced accelerated aging exhibited carbonyl profiles comparable to naturally aged material.

The MWD and carboxyl group distributions for different rag papers are given in Fig. 18. The amount of carboxyl groups also increased in the low MW region. Hence, it was concluded that during natural aging oxidation preferentially occurred in the low MW region, or that the progressing degradation of rag paper produced higher amounts of acidic low-molecular weight material with consequently increased amounts of carboxyl groups in this region.

3.6
Visualization of Oxidized Groups on Paper Surfaces

Analysis of historic paper materials and chemical processes occurring at specific locations of paper, such as ink or pigment lines, often calls for a visualization method. Since CCOA and FDAM emit fluorescence in the UV region, these labels cannot be used for direct imaging. Hence, other chromophores were applied, but the labeling principle and the underlying chemistry was maintained [163]. A representative example is given in Fig. 19, showing a model paper with a grid of copper pigments subjected to accelerated aging. Next to the pigment lines hydrophobic areas developed which became visible upon wetting with water (Fig. 19, left). These hydrophobic areas coincided with locations of higher oxidative damage in terms of a higher concentration of carbonyl groups. This was caused by migration of copper ions into the surrounding paper parts as shown by laser ablation MS techniques. The oxidative injure was visualized using a fluorescence label

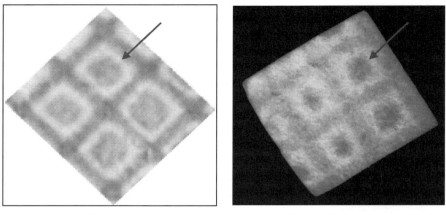

Fig. 19 Simulated copper corrosion in paper. Wetting (*left*) and fluorescence labeling (*right*) to visualize areas of pronounced oxidative damage

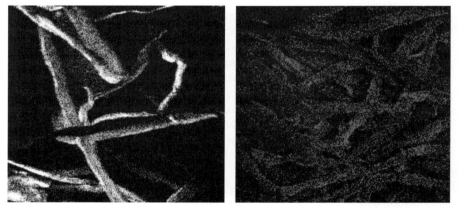

Fig. 20 Group selective labeling agents for chemical microscopy by TOF-SIMS. Analysis of a CCOA-labeled (*right*) and an FDAM-labeled beech sulfite pulp

emitting in the visible region, showing a stronger fluorescence emission for the hydrophobic, stronger oxidized areas (Fig. 19, right). Hence, also relatively minor spatial differences in oxidation states of paper surfaces can be made visible.

CCOA and FDAM are also suitable labels for surface analysis by TOF-SIMS. Both labels exhibit a distinct mass spectrum which can be used to visualize the surface distribution of carbonyl or carboxyl groups across single fibers by chemical microscopy (Fig. 20) (courtesy of Fardim P, Abo Academy Turku, Finland).

3.7
Outlook

The application of fluorescence labels in combination with GPC can be considered a step forward in the analysis of oxidized functionalities in cellulosics. However, a large number of questions still remain to be addressed in the future. If oxidized functionalities are considered as "substituents" along the polymer chain of cellulose, then a thorough analysis of the substituent distribution within the cellulose chains and per anhydroglucose unit should provide many new insights. The differentiation of aldehyde and keto functions will be a next step. Also the exact position of carbonyls (keto or aldehyde) within the AGU needs to be resolved, and differences in their reactivity determined. Furthermore, it is an open question whether oxidation occurs statistically within cellulose chains or forms clusters of highly oxidized areas.

With selective labels available, visualization of oxidative damage in all spatial dimensions of paper materials will become a major issue, especially with regard to conservation issues and restoration of historical paper documents. For the determination of carboxyl groups improvements in the reliable quantification of lactones can be considered a future challenge.

The analysis of oxidized groups in celluloses will remain a hot topic in cellulose chemistry and analytics, as we are only beginning to understand the enormous importance of these groups in governing chemical behavior and macroscopic properties of cellulosic material.

Acknowledgements The financial support of the work by the Austrian Christian Doppler Research Society, by Lenzing AG, Austria, and the Austrian Fonds zur Förderung der Wissenschaftlichen Forschung (project L188-N11), are gratefully acknowledged. We are grateful to all students, co-workers and cooperation partners that were involved in the studies described.

References

1. Nord S, Samuelson O, Simonson R (1962) Svensk Papperstidn 65:767
2. Larsson K, Samuelson O (1969) Svensk Papperstidn 72:97
3. Alfredsson B, Samuelson O, Sandstig B (1963) Svensk Papperstidn 18:703
4. Lemeune S, Jameel H, Chang HM, Kadla JF (2004) J Appl Polym Sci 93:1219
5. Lewin M, Epstein JA (1962) J Polym Sci 58:1023
6. Green JW (1980) In: Pigman W, Horton D, Wander JD (eds) The Carbohydrates, Chemistry and Biochemistry, vol. 1B, chap. 24, 2nd edn. Academic, New York
7. Hollemann AF, Wiberg N (1985) Lehrbuch der anorganischen Chemie, 91–100 Aufl. De Gruyter, Berlin
8. Gierer J (1997) Holzforschung 51:34
9. Gratzl JS (1992) Das Papier 10A:V1
10. Chirat C, Lachenal D (1997) Holzforschung 51:147
11. Kishimoto T, Nakatsubo F, Murakami K, Umezawa TJ (1995) Wood Chem Technol 15:453

12. Godsay MP, Pearce EM (1984) TAPPI Oxygen Delignification Conference Proceedings, p 55
13. Pan G, Chen CL, Chang HM, Gratzl JS (1981) The Ekman Days Proceedings 2, p 132
14. Kishimoto T, Nakatsubo F, Murakami K (1993) Mokuzai Gakkaishi 39:1049
15. Katai A, Schuerch C (1966) J Polym Sci 4:2683
16. Olkkonen C, Tylli H, Forsskahl I, Fuhrmann A, Hausalo T, Tamminen T, Hortling B, Janson J (2000) Holzforschung 54:397
17. Magara K, Ikeda T, Tomimura Y, Hosoya S (1998) J Pulp Paper Sci 24:264
18. Lewin M, Mark HF (1997) Macromol Symp 118:715
19. Lewin M, Ettinger A (1969) Cell Chem Technol 3:9
20. Garves K (1997) Holzforschung 51:526
21. Fischer K, Goldberg W, Wilke M (1985) Lenzinger Ber 59:32
22. Schleicher H, Lang H (1994) Das Papier 12:765
23. Takacs E, Wojnarovits L, Földvary C, Borsa J, Sajo I (2000) Radiat Phys Chem 57:399
24. Gilbert BC, King MD, Thomas CB (1984) Carbohydr Res 125:217
25. Arney JS, Jacobs AJ (1979) Tappi J 62:89
26. Marraccini LM, Kleinert TN (1962) Svensk Papperstidn 65:126
27. Kleinert TN, Marracchini LM (1963) Svensk Papperstidn 66:189
28. Tylli H, Forsskahl I, Olkkonen C (1995) J Photochem Photobiol A Chem 87:181
29. Heitner C, Scaiano JC (eds) (1993) Photochemistry of lignocellulosic materials. Advances in Chemistry Series 531. ACS, Pointe Clair, Canada, p 192
30. Forsskahl I (1994) Photochem Photobiol 3:503
31. Potthast A, Schiehser S, Rosenau T, Sixta H, Kosma P (2004) Holzforschung 58:597
32. Feller RL, Lee SB, Bogaard J (1986) The kinetics of cellulose deterioration. In: Needles HL, Zeronian SH (eds) Historic Textile and Paper Materials: Conservation and Characterization. Advances in Chemistry Series 212. ACS, Washington DC, p 329
33. Bouchard J, Overend RP, Chornet E (1992) J Wood Chem Technol 12:335
34. Calvini P (2005) Cellulose 12:445
35. Whitmore P, Bogaard J (1994) Restaurator 15:26
36. Kolar J (1997) Restaurator 18:163
37. Whitmore P, Bogaard J (1995) Restaurator 16:10
38. Margutti G, Conio G, Calvini P, Pedemont E (2001) Restaurator 22:67
39. Strlic M, Kolar J (2005) Ageing and stabilisation of paper. National and University Library, Ljubljana
40. Hofenk de Graaff (1994) In: Verrschoor H, Mosk J (eds) Contributions of the Central Research Laboratory to the Field of Conservation and Restoration. Central Research Laboratory for Objects of Art and Science, Amsterdam, p 21
41. Malesic J, Kolar J, Strlic M (2002) Restaurator 23:145
42. Lee SB, Feller RL, Bogaard J (1985) J Imaging Sci 29:61
43. Klemm D, Schumann D, Udhardt U, Marsch S (2001) Prog Polym Sci 26:1561
44. Zhu Y, Zajicek J, Serianni AS (2001) J Org Chem 66:6244
45. Malaprade L (1928) Bull Soc Chim France 43:683
46. Calvini P, Gorassini A (2002) Restaurator 23:48
47. Varma AJ, Chavan VB (1995) Polym Degrad Stab 49:245
48. Varma AJ, Chavan VB, Rajmohanan PR, Ganapathy S (1997) Polym Degrad Stab 58:257
49. Kim UJ, Kuga S, Wada M, Okano T, Kondo T (2000) Biomacromolecules 1:488
50. Röhrling J, Potthast A, Rosenau T, Sixta H, Kosma P (2002) In: Finn Carbohydrate Soc, 7th European Workshop on Lignocellulosics and Pulp, Book of Abstracts, 26–29 August 2002, Turku, Finland, p 23–26

51. Clode DM, Horton D (1971) Carbohydr Res 19:329
52. Luetzow AE, Theander O (1974) Svensk Papperstidn 77:312
53. Calvini P, Gorassi A (2002) Restaurator 23:205
54. Rowen JW, Forziati FH, Reeves RE (1951) J Am Chem Soc 73:4484
55. Fan QG, Lewis DM, Tapley KN (2001) J Appl Polym Sci 82:1195
56. Ant-Wuorinen O, Visapaa A (1963) Paperi ja Puu 45:81
57. Morooka T, Norimoto M, Yamada T (1989) J Appl Polym Sci 38:849
58. Rosenau T, Potthast A, Kosma P, Saariaho AM, Vuorinen T, Sixta H (2005) Cellulose 12:43
59. Sixta H (2006) Handbook of Pulp. Wiley, Weinheim, p 315
60. Sjöström E (1977) Tappi J 60:151
61. Ebringerova A (2006) Macromol Symp 232:1
62. Isogai A, Kato Y (1998) Cellulose 5:153
63. Painter TJ (1977) Carbohydr Res 55:95
64. Painter TJ, Cesaro A, Delben F, Paoletti S (1985) Carbohydr Res 61:140
65. Johansson M, Samuelson O (1977) Carbohydr Res 54:295
66. Thompson NS, Kaustinen OA, Ross R (1963) Tappi J 46:490
67. Ross R, Thompson NS (1963) Tappi J 46:376
68. Buchert J, Telemann A, Harjunpaa V, Tenkanen M, Viikari L, Vuorinen T (1995) Tappi J 78:125
69. Vuorinen T, Fagerström P, Buchert J, Tenkanen M, Teleman A (1999) J Pulp Paper Sci 5:155
70. Combs BS, Carper WR, Stewart JJP (1992) J Mol Struct (Theochem) 258:235
71. Diehl HW, Pokorny M, Zissis E, Ness RK, Fletcher HG (1974) Carbohydr Res 38:364
72. Huang CH, Lai WL, Lee MH, Chen CJ, Vasella A, Tsai Y-C, Liaw SH (2005) J Biol Chem 280:38831
73. Ayers AR, Ayers SB, Eriksson KE (1978) Eur J Biochem 90:171
74. Hallberg BM, Henriksson G, Pettersson G, Vasella A, Divne C (2003) J Biol Chem 278:7160
75. Higham CW, Gordon-Smith D, Dempsey CE, Wood PM (1994) FEBS Lett 351:128
76. Moe ST, Holen AK, Schult T (2002) J Carbohydr Chem 21:513
77. Achwal WB, Shanker G (1972) J Appl Polym Sci 16:1873
78. Bohrn R, Potthast A, Rosenau T, Kosma P, Sixta H (2005) Synlett 20:3087
79. Kato L, Cameron RE (1999) Cellulose 6:23
80. Diniz FJMB, Gil MH, Castro JAAM (2004) Wood Sci Technol 37:489
81. Bohrn R (2005) Dissertation, University of Natural Resources and Applied Life Sciences, Vienna
82. TAPPI Method T-430 om-94 (1994)(Braidy); Zellcheming method IV/8/70 (Schwalbe-Sieber)
83. Cyrot J (1957) Chimie Analytique 39:449
84. Rehder W, Philipp B, Lang H (1965) Das Papier 19:502
85. Lewin L (1972) In: Whistler RL, BeMiller JN (eds) Methods in Carbohydrate Chemistry, vol. 6. Academic, New York, p 76
86. Ströle U (1956) Makromol Chem 20:19
87. Tihlarik K, Pasteka M (1991) Starch/Stärke 43:83
88. Geiger E, Wissler A (1945) Helv Chim Acta 28:1648
89. Rehder W, Philipp B, Lang H (1965) Das Papier 19:502
90. Szabolcs O (1961) Das Papier 15:41
91. Strlic M, Pihlar B (1997) J Anal Chem 357:670
92. Pommering K, Rein H, Bertram D, Müller R (1992) Carbohydr Res 233:219

93. Calvini P, Conio G, Lorenzoni M, Pedemonte E (2004) Cellulose 11:99
94. Kongruang S, Penner MH (2004) Carbohydr Polym 58:131
95. Horn SJ, Eijsink VGH (2004) Carbohydr Polym 56:35
96. Sihtola H, Neimo L (1963) Tappi J 46:730
97. Blaha J, Cerny P, Jahn K (1984) Angew Macromol Chem 128:99
98. Nevell TP (1963) In: Whistler RL, Green JW, BeMiller JN, Wolfrom ML (eds) Methods in Carbohydrate Chemistry, vol. 3. Academic, New York, p 164
99. Siggia S, Maxcy W (1947) Ind Eng Chem Anal Ed 19:1023
100. Ellington AC, Purves CB (1953) Can J Chem 31:801
101. Houdier S, Legrand M, Boturyn D, Croze S, Defrancq E, Lhomme J (1999) Anal Chim Act 382:253
102. Röhrling J (2002) Dissertation, University of Technology, Vienna
103. Kostic M, Potthast A, Rosenau T, Kosma P, Sixta H (2006) Cellulose DOI 0.1007/s10570-005-9040-1
104. Chai XS, Hou QX, Zhu JY, Chen SL, Wang SF, Lucia L (2003) Ind Eng Chem Res 42:5440
105. Chai XS, Hou QX, Zhu JY (2003) Ind Eng Chem Res 42:5445
106. Wilson K (1966) Svensk Papperstidn 69:386
107. Ant-Wuorinen O (1951) Paperi ja Puu 33 B:105
108. Ant-Wuorinen O (1951) Paperi ja Puu 33 B:174
109. Sjöström E, Haglund P (1961) Svensk Papperstidn 64:438
110. Samuelson O, Törnell B (1961) Svensk Papperstidn 64:155
111. Sihtola H (1954) Paperi ja Puu 36:149
112. Nabar GM, Shenai VA (1970) J Appl Polym Sci 14:1215
113. Husemann E, Weber OH (1942) J Prakt Chem 159:334
114. Weber OH (1955) Das Papier 9:16
115. Phillipp B, Rehder W, Lang H (1965) Das Papier 19:1
116. Fardim P, Holmbom B (2002) Tappi J 2:28
117. Wilson K (1948) Svensk Papperstidn 51:45
118. TAPPI Method T237 cm-98 (1998) TAPPI standard 1998
119. Doering H (1956) Das Papier 10:140
120. Rebek M, Kirnbauer A, Semlitsch MFK (1960) Das Papier 14:510
121. Katz S, Beatson RP, Scallan AM (1984) Svensk Papperstidn 87:R48
122. Saake B (1992) Dissertation, University of Hamburg
123. Fardim P, Holmbom B, Ivaska A, Karhu J, Mortha G, Laine J (2002) J Nordic Pulp Paper Res J 17:346
124. Sawatari A, Nakamura H (1993) Sen'i Gakkaishi 49:279
125. Nishiyama S, Funato N, Sawatari A (1993) Sen'i Gakkaishi 49:357
126. Jayme G, Rohmann EM (1965) Das Papier 19:719
127. Stübchen-Kirchner H (1962) Chem Ztg 63:319
128. Ant-Wuorinen O, Visapää A (1963) Paperi ja Puu 45:35
129. Fontaine T, Fournet B, Karamanos Y (1994) J Microbiol Methods 20:149
130. Selvendran RR, March JF, Ring SG (1979) Anal Biochem 96:282
131. Shatalov AA, Pereira H (2004) Cellulose 11:109
132. Gailing MF, Guibert A, Combes D (1998) Biotechnol Tech 12:165
133. Slavik I, Pasteka M, Kucerova M (1967) Svensk Papperstidn 70:229
134. Slavik I, Pasteka M, Kucerova M (1967) Svensk Papperstidn 70:365
135. Achwal WB, Murali R (1986) J Appl Polym Sci 32:3913
136. Tenkanen M, Gellerstedt G, Vuorinen T, Teleman A, Perttula M, Li J, Buchert J (1999) J Pulp Paper Sci 25:306

137. Ragnar M (2001) Nordic Pulp Paper Res J 16:68
138. Bjarnestad S, Dahlman O (2002) Anal Chem 74:5851
139. Tenkanen M, Hausolo T, Siikaho M, Buchert J, Viikari L (1995) Proceedings 8th ISWPC, Helsinki, Finnland, III:189
140. Rydlund A, Dahlman O (1997) J High Resol Chromatogr 20:72
141. Gellerstedt G, Li J (1996) Carbohydr Res 294:41
142. Chai XS, Zhu JY, Li J (2001) J Pulp Paper Sci 27:165
143. Jääskeläinen AS, Saariaho AM, Vuorinen T (2005) J Wood Chem Technol 25:51
144. Saariaho AM, Hortling B, Jääskeläinen AS, Tamminen T, Vuorinen T (2003) J Pulp Paper Sci 29:363
145. Evtuguin DV, Daniel AID, Pascoal Neto C (2002) J Pulp Paper Sci 28:189
146. Hruby VJ, Meyer JP (1998) Chemical synthesis of peptides. In: Hecht SM (ed) Bioorganic Chemistry: Peptides and Proteins. Oxford University Press, Oxford, p 27
147. Kunishima M, Kawachi C, Morita J, Terao K, Iwasaki F, Tani S (1999) Tetrahedron 55:13159
148. Dünges W (1997) Anal Chem 49:442
149. Toyo'oka T (2002) Anal Chim Acta 465:111
150. Lam S, Grushka E (1978) J Chromatogr 158:207
151. Nakabayashi S, Kudo I, Kuma K, Matsunaga K, Hasebe K (1993) Anal Chim Acta 271:25
152. Nimura N, Kinoshita T (1980) Anal Lett 13:191
153. Barker SA, Monti JA, Christian ST, Benington F, Morin RD (1980) Anal Biochem 107:116
154. Yamauchi Y, Tomita T, Senda M, Hirai A, Terano T, Tamura Y, Yoshida S (1986) J Chromatogr 357:199
155. Nimura N, Kinoshita T, Yoshida T, Uetake A, Nakai C (1988) Anal Chem 60:2067
156. Schneede J, Ueland PM (1992) Anal Chem 64:315
157. Takadate A, Tahara T, Fujino H, Goya S (1982) Chem Pharm Bull 30:4120
158. Bohrn R, Potthast A, Schiehser S, Rosenau T, Sixta H, Kosma P (2006) Biomacromol 7:1743
159. Röhrling J, Potthast A, Rosenau T, Lange T, Ebner G, Sixta H, Kosma P (2002) Biomacromol 3:959
160. Röhrling J, Potthast A, Rosenau T, Lange T, Borgoards A, Sixta H, Kosma P (2002) Biomacromol 3:969
161. Chayes R, Dvir R, Gould S, Harell A (1971) Anal Biochem 42:283
162. Lattova E, Perreaul H (2003) J Chromatogr B 793:167
163. Henniges U, Prohaska T, Banik G, Potthast A (2006) Cellulose DOI 10.1007/s10570-005-9030-3
164. Potthast A, Röhrling J, Rosenau T, Sixta H, Kosma P (2003) Biomacromol 4:743
165. Chirat C, Lachenal D (1994) Holzforschung 48:(Suppl)133
166. Albini A (1993) Synthesis, p 263
167. Rosenau T, Potthast A, Sixta H, Kosma P (2001) Progr Polym Sci 26:1763
168. Adorjan I, Potthast A, Rosenau T, Sixta H, Kosma P (2005) Cellulose 12:51
169. Rosenau T, Potthast A, Adorjan I, Hofinger A, Sixta H, Firgo H, Kosma P (2002) Cellulose 9:283
170. Rosenau T, Potthast A, Kosma P (2002) Preprints ICC 2002, 1st International Cellulose Conference, Kyoto, Japan, p 29
171. Fengel D (1980) Das Papier 34:428
172. Entwistle D, Cole EH, Wooding NS (1949) Textile Res J, p 527
173. Entwistle D, Cole EH, Wooding NS (1949) Textile Res J, p 609

174. Barthel P, Philipp B (1967) Faserforsch Textiltech 18:266
175. Schmidt JA, Rye CS, Gurnagul N (1995) Polym Degrad Stab 49:291
176. Zou X, Gurnagul N, Uesaka T (1993) J Pulp Pap Sci 19:J235
177. Nagel G, Potthast A, Rosenau T, Kosma P, Sixta H (2005) Lenzinger Ber 84:27–35

Adv Polym Sci (2006) 205: 49–96
DOI 10.1007/12_097
© Springer-Verlag Berlin Heidelberg 2006
Published online: 30 August 2006

Nanocelluloses as Innovative Polymers in Research and Application

Dieter Klemm[1,3] (✉) · Dieter Schumann[2,3] · Friederike Kramer[1] ·
Nadine Heßler[1] · Michael Hornung[4] · Hans-Peter Schmauder[4] ·
Silvia Marsch[2,3]

[1]Institut für Organische Chemie und Makromolekulare Chemie,
Friedrich-Schiller-Universität Jena, Humboldtstraße 10, 07743 Jena, Germany
dieter.klemm@uni-jena.de

[2]Klinik für Mund-, Kiefer- und Gesichtschirurgie/Plastische Chirurgie,
Friedrich-Schiller-Universität Jena, Erlanger Allee 101, 07747 Jena, Germany

[3]POLYMET Jena e.V., Lessingstraße 8, 07743 Jena, Germany

[4]Forschungszentrum für Medizintechnik und Biotechnologie e.V., fzmb,
Geranienweg 7, 99947 Bad Langensalza, Germany

Abstract Cellulose is a fascinating and almost inexhaustible and sustainable natural polymeric raw material characterized by exciting properties such as hydrophilicity, chirality, biodegradability, broad chemical-modifying capacity, and the formation of different semicrystalline fiber morphologies. If cellulosics such as bacterial cellulose or strongly disintegrated wood cellulose are composed of nanosized fibers and the nanofiber structuring determines the product properties, these polymers are described as nanocelluloses.

Because of the extraordinary supramolecular structure and exceptional product characteristics as high-molecular and high-crystalline cellulosics with a water content up to 99%, nanocelluloses require increasing attention. This review assembles the current knowledge in research, development, and application in the field of nanocelluloses through examples. The topics combine selected results on nanocelluloses from bacteria and wood as well as their use as technical membranes and composites with the first long-time study of cellulosics in the animal body for the development of medical devices such as artificial blood vessels, and the application of bacterial nanocellulose as animal wound dressings and cosmetic tissues.

Keywords Nanocelluloses · Membranes · Composites · Medical devices · Animal wound dressings

Abbreviations

B	BASYC®
BASYC®	bacterial-synthesized cellulose
BC	bacterial cellulose
ca	carotid artery
CD-31	cluster of differentiation – 31
CMC	carboxymethyl cellulose
DN	double network
DSM	Deutsche Sammlung von Mikroorganismen und Zellkulturen
E. coli	*Escherichia coli*
ec	endothelial cell
EDL	musculus extensor digitorum longus
EtOH	ethanol
fb	fibroblast
fbb	fibroblast branches
GlcNAc	*N*-acetyl glucosamine
HS	Hestrin–Schramm
l	lumen
MC	methyl cellulose
MFC	micro-fibrilated cellulose
NMR	nuclear magnetic resonance spectroscopy
PET	poly(ethylene terephthalate)
PTFE	poly(tetrafluoroethylene)
SEM	scanning electron microscopy
TEM	transmission electron microscopy
UV	ultraviolet
vr	vitalized region

1
Introduction

A material consisting of 99% water and 1% of a high-molecular and high-crystalline polymer with well-defined molecular and supramolecular structure represents an unusual but exciting matter in polymer research and application. The description applies perfectly to specific cellulosics characterized by a nanosized architecture and presented in this contribution as nanocelluloses.

It is well known that polysaccharide cellulose is – in general – a very important and fascinating biopolymer and an almost inexhaustible and sustainable polymeric raw material. The trend towards renewable resources and the tailoring of innovative products for science, medicine, and technology has led to a global renaissance of interdisciplinary cellulose research and the use of this abundant organic polymer over the last decade [1].

The polymer cellulose is of particular importance owing its unique structure, which is quite different from the common synthetic polymers. Formed by repeated connection of glucose building blocks, the highly functionalized, linear stiff-chain homopolymer is characterized by its hydrophilicity, chirality, biodegradability, and broad chemical-modifying capacity. This molecular structure is also the basis for extensive hydrogen-bond networks forming semicrystalline fiber morphologies. The properties of cellulosics are therefore considerably determined by the supramolecular order and specific assembling, these again being controlled by the origin and treatment of the cellulose [1].

Against this background and in view of broadening the range of cellulose use, including the development of novel materials with ground-breaking new features, it is important to have access to various types of celluloses of miscellaneous provenance, supramolecular structure, exceptional properties, different availability, and expanded application areas.

The dominant pathway to cellulose material is its production from plants. In the seed hairs of cotton, cellulose is available in an almost pure state. In contrast, wood cellulose forms a native composite with lignin and other carbohydrates (hemicelluloses) from which it is isolated by large-scale chemical pulping, separation, and purification processes. Wood pulp [2] remains the most important industrial feedstock for the production of paper and cardboard, for cellulose regenerate fibers and films as well as for the industrial-scale synthesis of the palette of cellulose esters and ethers. These derivatives are used as very important and well-known active components in coatings, optical films, sorption media, and additives in building materials, drilling techniques, pharmaceutics, foodstuffs, and cosmetics. Numerous novel applications of cellulose also take advantage of its biocompatibility and chirality.

Apart from plants, certain bacteria, algae, and fungi produce cellulose as well. Amongst the cellulose-forming bacteria, *Acetobacter* strains – reclassified as the genus *Gluconacetobacter* – are especially suitable for the formation of cellulose. They are not pathogenic, are commonly found on fruits and fruit products, and can be cultivated under laboratory conditions.

One of the first applications of the natural product bacterial cellulose (BC) was the use as a calorie-free dessert called Nata de Coco, today a common Asian food.

In terms of the molecular formula, BC is identical to cellulose of plant origin – apart from alien groups such as carbonyl and carboxyl units in the latter as a result of the plant cellulose processing (see also the contribution of A. Potthast et al. in this volume). But important structural features and properties significant for practical application of BC are quite different from wood cellulose: high purity, high degree of polymerization (up to 8000), high crystallinity (of 70 to 80%), high water content to 99%, and high mechanical stability.

These specific parameters are caused by the subsequently outlined biosynthetic formation of the BC and the resulting particular supramolecular structure as a network of nanofibers formed during self-assembly of the cellulose molecules in the aqueous culture medium, free from potential composite partners as in wood biosynthesis, reviewed in [1].

If cellulosics such as BC are composed of nanosized fibers and the nanofiber structuring determines the product properties, these polymers are described as nanocelluloses.

The similar term *nanosized cellulose* is used in case of isolated crystallites and whiskers formed by acid-catalyzed degradation of cellulosics. This field and the application of that nanosized cellulose, e.g. in composites, have been intensively investigated. Typical examples have been presented in current papers [3, 4] and at the 231st American Chemical Society (ACS) national meeting in Atlanta.

Moreover, the same term was recently used for small cellulose patches formed as parts of open spin-coat films produced from low-concentration trimethylsilylcellulose solutions and by subsequent heterogeneous desilylation of the primary formed silylcellulose layers under cellulose regeneration [5].

In the last years, growing worldwide activity can be observed regarding extensive scientific investigation and increasing efforts for the practical use of the nanocelluloses. An overview of the increase of annually papers on BC since 2000 is presented in Fig. 1.

It is the aim of this contribution to present advances in research, development, and application in the field of nanocelluloses. The topics combine selected results on nanocellulosics from bacteria and wood as well as their use as technical membranes and composites with the first long-time study of cellulose in the animal body for the development of medical devices such as artificial blood vessels, and the application of bacterial nanocellulose as animal wound dressings and cosmetic tissues. Therefore, the review has brought together colleagues from chemistry, medicine, and biotechnology.

Important further fields of BC research and development are already the subject of numerous original papers and review articles. Therefore, these topics are not part of this overview. This applies predominantly to recent knowledge on the biosynthesis of BC [6], its application as wound dressings in human medicine [7], the design of electronic paper [8] and the

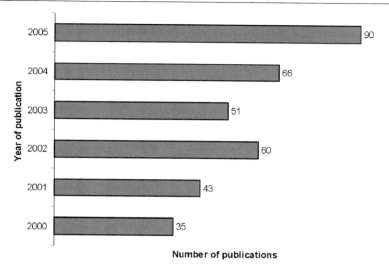

Fig. 1 Increase of annual publications on BC since 2000 (SciFinderScholar2006, bacterial cellulose as entered)

development of BC-additives in paper-making [9] as well as the extensive structure characterization by solid-state nuclear magnetic resonance (NMR) spectroscopy [10].

2
Types of Nanocelluloses

As described before, one type of nanocellulose is formed directly as the result of biosynthesis of special bacteria. A very pure product with subsequently reported important properties is formed that necessitates challenging biosynthesis/biotechnological handling and the development of large-scale production.

Another kind of nanocellulose can be prepared from the nearly inexhaustible source of feedstock wood using controlled mechanical disintegration steps to produce the favored product properties.

2.1
Nanocellulose from Bacteria

In 1886, A. J. Brown [11] discovered BC as a biosynthetic product of *Gluconacetobacter xylinus* strains. He identified a gelatinous mass, formed on the solution during the vinegar fermentation as cellulose. In the middle of the 20th century, Hestrin and Schramm developed a special culture medium for

Gluconacetobacter xylinus to optimize cellulose formation on the laboratory scale [12]. As a result of systematic and comprehensive research over the last 10–15 years, broad knowledge of the formation and structure of BC has been acquired. This work is an important part of the integration of biotechnological methods into polysaccharide chemistry and the development of cellulose products with new properties and application potential.

The biosynthesis of BC occurs at a cellulose-synthesizing complex in the bacterial cell. It starts from uridine diphosphate glucose and proceeds via the addition of this intermediate to the end of the growing cellulose molecule. The chain exits the cell as a so-called *elementary fibril* through pores at the bacterium surface. In the presence of an aqueous culture medium and in the absence of composite components (such as in wood, Sect. 2.2) the self-assembly of the cellulose molecules leads to a highly swollen three-dimensional (3D) network with a distinct tunnel and pore structure and a water content of up to 99%. Using the common complex medium from Hestrin and Schramm [12] in a static culture, the BC is shaped as pellicles at the air/liquid interface (Fig. 2). Biosynthesis leads with a high yield of about 40% – in relation to the bacterial strain – to nanocellulose of high purity. Residual bacteria and components of the culture medium can be removed by boiling with weak basic media and subsequent washing with water.

A recent model of the BC structure in the never-dried state was given by Fink et al. [13]. Anhydrous nano-fibrils in the range 7×13 nm appear hydrated as a whole and are aggregated to flat microfibrils with a width of 70–150 nm. This means that the water is outside of the crystalline cellulose nano-units and between these elements. A shell of noncrystalline cellulose chains passes around neighboring microfibrils to produce a microfibril band

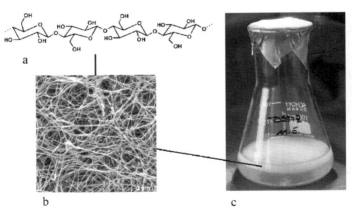

a

b   c

Fig. 2 Shape and structure of BC. **a** molecular cellulose chain, **b** scanning electron microscopy (SEM) of freeze-dried nanofiber network (magnification 10 000), **c** pellicle of bacterial nanocellulose from common static culture

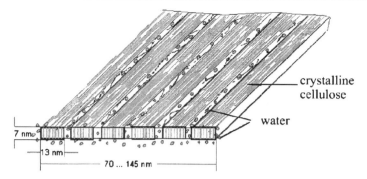

Fig. 3 Model of initially hydrated BC fibrils. Reprinted with permission from [1]

(ribbon) with a width of about 0.5 μm. The arrangement of these microfibrils and ribbons is shown in Fig. 3 and also demonstrated in Fig. 8.

Both bacterial and native plant cellulose (so called cellulose I) coexist in two crystal modifications I_α (triclinic) and I_β (monoclinic). The difference consists in the H-bonding systems and in the conformation of neighboring cellulose chains. The I_α/I_β ratio depends on the origin of the cellulose [13].

As mentioned before, bacterial nanocellulose is characterized by specific and extraordinary properties based on the accessible hydrated nanofibrillated network. Besides the high degree of polymerization and crystallinity, this nanocellulose is especially distinguished by its biocompatibility and moldability during cultivation (see Sect. 4 for examples) and excellent mechanical properties (Sect. 3). BC fibers possess a small thermal expansion coefficient (similar to that of glass), a very high Young's modulus (134 GPa), and tensile strength (2 GPa). The latter are in the range of aramid fibres (e.g. Kevlar) [14, 15].

2.2
Nanocellulose from Wood

In contrast to BC, cellulose from wood is composed of fibers that are about one hundred times thicker [17] (Fig. 4a). Because of the complex and expensive cultivation of BC (sophisticated medium and long cultivation time), it is also a challenge to produce nano-fibrillated celluloses from wood. The substructures of wood are only accessible by chemical treatment [16] and mechanical disintegration procedures.

In the last 25 years, there have been efforts to reduce wood fibers in size. As a first step, in the early 1980s Turbak et al. [18] developed a micro-fibrillated cellulose (MFC). Today, there are different ways to produce materials with controlled fiber diameters.

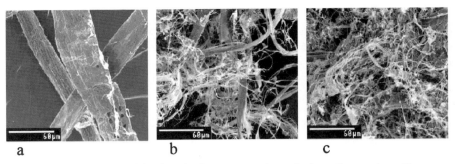

a   b   c

Fig. 4 SEM of **a** untreated kraft pulp, **b** 16 passes through the refiner pulp, **c** 30 passes through the refiner pulp. Reprinted with permission from [20]

At first, a water suspension of pulp has to go through a mechanical treatment that consists of a spring-loaded valve assembly (refiner), where the slurry is pumped at high pressure. The formed MFC is moderately degraded and extremely expanded in surface area. In recent years, cellulose with a nanoscale web-like structure (Fig. 4c) has been made. The fiber diameters are in the range 10–100 nm [19, 20]. The degree of fibrillation depends on the number of passes through the refiner (Fig. 4b, c).

Another technique to prepare wood MFC/nanocellulose is described by Takahashi et al. [21]. The aim was the creation of strong composites in tension using hot-pressed fibers without synthetic polymers but with the original wood components hemicelluloses and lignin as binders. The starting material was bamboo because of its high cellulose content. Bamboo fiber bundles and monofilaments were ground under high-speed conditions using stone disks. A combination of thermal and alkali pre-treatments, given the appropriate ratio of cellulose, hemicelluloses and lignin in the monofilaments, led to strong adhesion between the fibers under the hot-press conditions.

Suzuki and Hattori [22] treated a pulp with a solid concentration of 1–6% with a disk refiner more than 10 times. The fibers obtained had a length of less than 0.2 mm.

There have also been some investigations into the properties of nanocellulose from wood, which has an amazing water-storage capacity, similar to BC. A dispersion of these cellulose fibers in water with a solid content of only 2% leads to a mechanically stable transparent gel. Further properties of this nanocellulose will be discussed – in relation to BC – in Sect. 3.2. The wood nanocellulose fibers are suitable for solidification of emulsion paints and filter aids, useful for both primary rough filtration and precision filtration [23, 24]. Furthermore, nanocellulose from wood is used in paper-making as a coating and dye carrier in paper tinting [25, 26]. Moreover, it can be utilized in the food industry as a thickening agent [27], a gas-barrier and in moisture-resistant paper laminate for packaging [28]. In cosmetics, wood nanocellulose is suitable as an additive in skin-cleansing cloths [29], and as part of disposal diapers, sanitary napkins and incontinence pads [30]. Possible medical applications are directed to excipients such as binders, fillers, and/or disintegrants in the development of solid dosage forms [31].

Besides application in its pure form, it is possible to use nanocellulose from wood in polymer composites (Sect. 3.2). In embedding tests, the tensile strength of such composites was five times higher than of the original polymers. This result, as well as its natural origin, make this nanocellulose attractive for combination with different (bio)polymers. Possible applications for such reinforced (bio)polymers could arise in areas such as medicine, the food industry, and gardening [19, 20, 32, 33]. In these sectors, properties such as biodegradability, high mechanical strength, and – where required – optical transparency are important.

It should also be mentioned that the application of wood nanocellulose prepared by the described techniques – where the cell wall is further disintegrated by mechanical treatment – leads to lower-strength cellulose fiber-reinforced composites than in the corresponding BC materials [34].

3
Nanocellulose Membranes and Composites in Technical Applications

Membranes and composites from cellulose and cellulose esters are important domains in the development and application of these polymer materials. The most important segment by volume in the chemical processing of cellulose contains regenerated cellulose fibers, films, and membranes. In the case of the cellulose esters mainly cellulose nitrate and cellulose acetate as well as novel high-performance materials created therefrom are widely used as laminates, composites, optical/photographic films and membranes, or other separation media, as reviewed in [1]. The previously specified nanocelluloses from bacteria and wood tie in with these important potentials and open novel fields of application.

The direct formation of stable and manageable BC fleeces as the result of bacterial biosynthesis in the common static culture is significant. This and their exciting properties described in Sect. 2.1 have led to increasing use of BC as a membrane material and composite component. Contaminations incorporated from the culture medium and bacterial cells can be removed from the BC by smooth purification methods depending on the application area.

One recent example of the formation and application of foils/membranes of unmodified bacterial nanocellulose is described by George and co-workers [35]. The processed membrane seems to be of great relevance as a packaging material in the food industry, where continuous moisture removal and minimal-oxygen-transmission properties play a vital role. The purity, controllable water capacity, good mechanical stability, and gas-barrier properties of BC are important parameters for this application. Therefore, BC may be an option for plant cellulose foils.

For the purification of the membranes, the common treatment with 0.1–2.0 N aqueous sodium/potassium hydroxide solutions has been compared with the purification of raw BC membranes using aqueous sodium/potassium carbonate. Whereas hydroxide purification leads to a decrease of the tensile strength and an elongation of the material this effect is lower in the case of carbonate treatment. Moreover, the oxygen transmission rate is higher after carbonate washing.

To open up new application fields, the structure and properties of BC membranes/composites can be modified by low-molecular organic and inorganic compounds, including monomers and metals/metal oxides, via carbohydrates and polysaccharides, up to different types of other natural and synthetic polymers.

Compared to nanocellulose from wood, BC has the major advantage of modifiability during biosynthesis by simple addition of water-soluble compounds to the culture medium (in situ modification).

Moreover, the nanosized fibers of the swollen nanocelluloses can be coated with different components and the pore system can be loaded with agents (post-modification). Further methods in this field use well-known procedures

Table 1 Nanocelluloses as membranes in a technical application

Entry	Authors	Citation	Title
1	Amano Y, Kanda T, Nozaki K (2005)	JP 2005320657 A2 20051117	Manufacture of bacterial cellulose with desired shape
2	Slezak A, Jarzynska M (2005)	Polim Med 35:15	Development of Kedem–Katchalsky equations of the transmembrane transport for binary nonhomogeneous non-electrolyte solutions
3	Uraki Y (2005)	Cell Commun 12:170	Honeycomb patterned bacterial cellulose
4	Levy NLF, Kurokawa EC, Podlech PAS (2004)	WO 2004050986 A1 20040617	A process of obtaining a cellulosic wet sheet and a membrane and equipment for their manufacture
5	Nishimura T (2004)	Kobunshi 53:790	Development of high performance leukocyte reduction filters based on ultra-fine non-woven fabric
6	Oshima T, Kawasaki M, Ohto K, Inoue K (2004)	JAERI-Conf 2004-014:55	Development of a novel adsorbent based on bacterial cellulose for adsorption of rare earth metals
7	Evans BR, O'Neill HM, Jansen VM, Woodward J (2003)	US 2003113610 A1 20030619	Metallization of bacterial cellulose for electrical and electronic device manufacture

for polymer modification such as reinforcement, grafting, and interpenetrating networks.

A survey of the large variety of technical uses of further BC membranes is shown in Table 1. A more detailed discussion of typical examples of these nanocelluloses applications is given in the following.

3.1
In Situ Modification

The addition of water-soluble compounds affects the biosynthesis of cellulose chains (polymer formation), and they may also be adsorbed by single fibers as the nanofiber network is built (crystallization). Moreover, the bacterial cells themselves can be modified.

The cultivation of BC in the presence of N-acetyl glucosamine (GlcNAc) causes a variation of the polymer formation by the insertion of GlcNAc units. It is possible to produce very thin membranes of the nanocellulose-chitin hybrid formed in this way [36].

Yamanaka et al. [37] described the influence of bioactive organic agents such as nalidixic acid and chloramphenicol (antibiotics) as well as dithiothreitol (a reducing reagent) as additives to the BC culture medium. In this case, not only the crystallization of the fibers and the material properties are influenced but the *Gluconacetobacter* cells are themselves also changed.

Using antibiotics in a concentration of 0.1 mM, a 2–5 times elongation of the cell length was observed due to inhibition of cell division. The fibers became 1–2 times wider compared to common BC. Because of the better uniplanar orientation of the wider fibers after pressing of the membranes, the Young's modulus of this material increases by up to 34–42%.

In contrast, adding reducing reagents (1.0 mM) leads to a 0.6 times cell shortening. The fibers will be thinner and – because of the shorter cells – the number of terminal complexes is also reduced. No variation of properties was observed compared unmodified BC.

The additive effects on the bacterial cells can be seen in Fig. 5. The modification of the BC fibers is shown in Fig. 6.

The addition of carboxymethyl cellulose (CMC) and methyl cellulose (MC) to the nutrient solution (0.5–2.0 m/v %) effects a change of the fiber and network architecture [38] (Fig. 7). In the first case, the crystallization of the BC is affected by agglomeration of the CMC onto the fibers during self-assembly. The additive seems to compete with the BC for hydrogen-binding sites during ribbon construction [39] (Fig. 7a). In the second case, the adsorption takes place mainly on the planar ribbons (Fig. 7c).

The water-holding capacity of such CMC and MC composites can be 100 times higher than that of unmodified BC. Besides controlling the water content, these composites can be stored in the dried state and reswollen before use, reaching higher water absorption than pure never-dried BC [40].

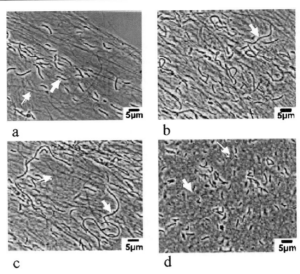

Fig. 5 Typical optical micrographs showing the effect of chemical reagents on cell morphology: **a** control, **b** nalidixic acid (0.1 mM), **c** chloramphenicol (0.3 mM), **d** dithiothreitol (1.0 mM). *Bold arrow* bacterial cell, *arrow* cellulose fibers. Reprinted with permission from [37]

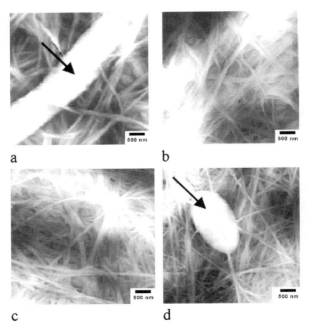

Fig. 6 Typical atomic force microscopy images showing the effect of chemical reagents on BC fibers. **a–d** the same as in Fig. 5; → bacterial cell. Reprinted with permission from [37]

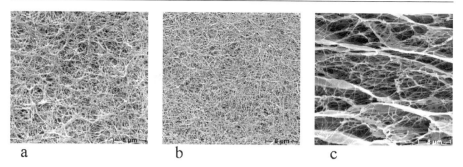

a   b   c

Fig. 7 SEM (magnification 3000) of freeze-dried BC. **a** CMC-modified (thicker fibers), **b** unmodified, **c** MC-modified (strong widening of the fiber network under amplification of the planar layers)

Moreover, these in situ modified CMC/BC composites exhibit good ion-exchange capacities with specific adsorption ability for lead and uranyl ions. Their application to ion exchangers has been described [41].

When chitosan is added to the culture medium the BC/chitosan composites formed include the properties of both components, mainly an extraordinary bactericidal and barrier effect against microorganisms. Therefore, this modified BC is well suited for wound dressings [42].

By combining BC with hemicelluloses such as xyloglucan, pectin, xylan, and acetyl glucomannan models for the primary and secondary plant cell wall could be created [39, 43–45]. When grown in the presence of xyloglucan and acetyl glucomannan, the ratio between the BC allomorphs I_α/I_β (Sect. 2.1) is decreased and the cellulose ribbons also seem to become disrupted and smaller. During biosynthesis, loose bundles of cellulose microfibrils were formed. The decrease of crystallite size indicates that, in the case of acetyl glucomannan, the tight aggregation of microfibrils to form a ribbon by adhering to the surface of nascent microfibrils is prevented. Therefore, it may influence the second and third stages of cellulose ribbon formation (Fig. 8). BC and xyloglucan were uniformly strained along the sample length, too. Here, in contrast, the presence of pectin did not lead to any change in the I_α/I_β ratio or ribbon width. These latter results are in agreement with those from onion epidermis [46].

Ishiada et al. [47, 48] investigated a special *Gluconacetobacter xylinus* strain directly producing composites of BC and acetan. Acetan is a mixture of glucose, mannose, glucuronic acid, and rhamnose (in the ratio 4/1/1/1). Their thesis was that the production of acetan reduces the yield of BC. To increase the cellulose rate they used a mutant *Gluconacetobacter xylinus* strain that had no acetan-producing ability. Surprisingly, BC production did not increase but actually decreased. When acetan was added to the culture medium of the mutated strain, the BC crop was almost the same as that from the acetan-producing strain. An important result of these experiments was that the no

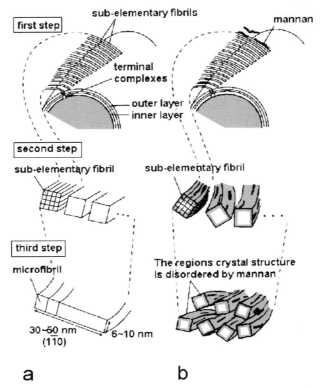

Fig. 8 Schematic comparison of cellulose structure formation of BC synthesized in **a** Hestrin–Schramm (HS) medium (Fig. 3), **b** glucomannan-modified HS medium. Reprinted with permission from [44]

acetan producer needs less time to generate the same amount of BC under these conditions. Experiments with other soluble polysaccharides like xanthan and agar showed the same results. This indicates that the formation of BC is not influenced specifically by acetan; an adequately viscous water solution of other polysaccharides has the same effect. A further consequence of this study is that the amplification of acetan-synthesis genes in such *Gluconacetobacter xylinus* strains can shorten culture times [49].

3.2
Post-Modification

There have been numerous investigations into the subsequent modification of bacterial and wood nanocelluloses. The additives range from other polysaccharides, albuminoids such as gelatine, different types of monomers and synthetic polymers, to metals, metal oxides, and inorganic fibers. On the other hand, if membranes and composites are modified with these nanocelluloses, materials with drastically different properties compared to ordinary fiber-based products can be obtained. For example, materials reinforced with nanocellulose may find practical application where properties such as biodegradability are desired.

Yano and Nakahara [15] used accessory polysaccharides to form composites with wood MFC/nanocelluloses. The disintegrated wood celluloses were mixed with starch as a binder and then hot-pressed between porous metal plates. Using a starch content of 2 wt %, the bending strength reached 310 MPa, compared to 250 MPa for unmodified fibers. Concurrently, the Young's modulus decreased from 16 to 12.5 GPa. When the starch content was 20 wt %, the bending strength decreased to 270 MPa. This indicates that added starch may act not only as a binder but also as a plasticizer.

Dubey et al. [50, 51] and Pandey et al. [52] demonstrated by means of impregnation of BC surfaces with chitosan that these modified membranes can be used for pervaporative separations of EtOH/H_2O azeotropes. After drying the BC (one week under vacuum at 60 °C), the coating takes place by dipping the membrane into 1% acetic acid solution of chitosan for 48 h. After removing, washing several times with distilled water, and drying for 24 h, the membranes were heated at 60 °C under vacuum for about 30 h. These modified membranes contain 8 ± 2 wt % of chitosan. They are biodegradable and resistant to alcohols, ketones, aldehydes, hydrocarbons, ethers, and aprotic solvents as well as dilute acids/bases.

Because of its hydrophilic nature even unmodified BC shows great potential to separate azeotropes such as EtOH/H_2O. It adsorbs seven times more water than ethanol. This selectivity and a reasonable flux increase with growing temperature and thinning of the membrane. In addition, the BC membranes also show a high water affinity in aqueous binary mixtures of organic solvents.

Using two natural polymers, BC and gelatine, Nakayama et al. [53] created high-mechanical-strength double-network (DN) hydrogels. The BC was immersed in an aqueous solution of gelatine that was crosslinked by N-(3-dimethylaminopropyl)-N'-ethylcarbodiimide hydrochloride. This composite formation led to a substantial improvement of the mechanical properties. Despite its high water content (70 wt %) a fracture strength of 5 MPa and an elastic modulus of 4 MPa could be reached, which is similar to articular cartilage. The disadvantages of both components, i.e. the brittleness of the gelatine gel and the lack of mechanical toughness of the swollen BC, could be compensated (Fig. 9). Combinations of BC with polysaccharides, such as sodium alginate, gellan gum, and t-carrageenan, showed a similar enhancement in the mechanical strength.

Nakagaito and co-workers [20, 54] produced composites from wood and BC nanocelluloses with phenolic resins. The cellulose were used in the form of sheets that were first impregnated with the phenolic resin, then stacked in

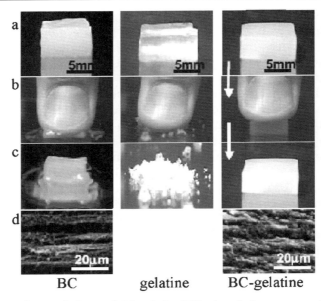

BC   gelatine   BC-gelatine

Fig. 9 Pictures of BC, gelatine, and BC-gelatine DN gels. **a** before compression, **b** during compression, **c** after 10 min compression, **d** SEM images of the stratified structure. Reprinted with permission from [53]

layers and compressed. The bending strength of the wood-based composites was remarkably high (370 MPa), while in comparison the Young's modulus was very low (19 GPa). The BC materials reached bending strengths of up to 425 MPa and a Young's modulus of 28 GPa. Stress–strain curves of unmodified sheets of both types showed a significantly higher modulus and strength in the case of BC. Therefore, additional experiments were made with disintegrated BC, which is more similar to the wood product (Fig. 10). These tests showed bending strengths and Young's moduli similar to wood-based composites. This means that the high values of the BC-based materials are really due to the extraordinary network structure of BC, up to now only producible by nature.

Because elements with diameters less than one tenth of the visible-light wavelength range are free from light scattering, nanosized fibers such as BC are considered to be ideal reinforcement agents in optically transparent polymers. Yano et al. [15] and Nogi et al. [55] have developed such BC-reinforced composites by impregnation of dried BC sheets with acrylic or epoxy resins and subsequent UV curing. Even with a high fiber content of 60–70 wt % the composites showed a high degree of transparency, up to 80% (Fig. 11).

Choi and co-workers [56] modified BC with cation-exchangeable acrylic acid units by UV graft polymerization. For this purpose, the BC was dried at 80 °C, immersed in methanol with benzophenone as a photoinitiator for

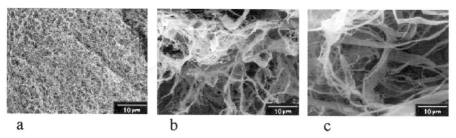

a   b   c

Fig. 10 SEM of the supramolecular structure of **a** BC pellicle, **b** disintegrated BC, **c** MFC. Reprinted with permission from [54]

Fig. 11 High transparency of a BC nanocomposite demonstrated by the view on a flower through a corresponding foil. Reprinted with permission from [55]

3 h and air-dried for 30 min. To activate the BC membrane it was irradiated with UV light for 3 min under a nitrogen atmosphere. After the addition of an acrylic acid solution, it was treated with UV light for 5–20 min, purified and stored in NaCl solution. The resulting composite membranes show a rising density with increasing UV irradiation time because of the anchorage of the grafted poly(acrylic acid) (Fig. 12). This is combined with a growing ion-exchange capacity. In addition, the modified membranes exhibit excellent tensile strength due to the high crystallinity of BC.

To improve the thermal, mechanical, and viscoelastic properties of cellulose acetate butyrate, it was reinforced with nanocellulose crystals prepared from BC by acid hydrolysis. Using this *nanosized cellulose* (Sect. 1) a significant improvement in the properties of the composites was demonstrated [57].

Besides impregnation with organic agents and polymers, BC can be coated with metals such as Pd, Au, and Ag. For this purpose, never-dried BC has been loaded with the respective metal salts by storage in the corresponding aqueous solutions for 18–24 h at 38 °C. After heating the mixture for 2 h and drying, thin, flexible and thermally stable metal-impregnated membranes were ob-

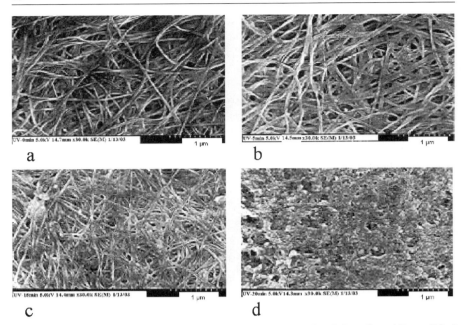

Fig. 12 SEM of membrane surfaces. **a** unmodified BC, **b** to **d** poly(acrylic acid)-modified BC membranes formed by time-dependent UV irradiation (**b** 5 min. **c** 10 min. **d** 20 min). Reprinted with permission from [56]

tained. The authors assume that their BC possesses reducing activity initiating the formation of Pd, Au, and Ag from the aqueous solutions of the metal salts. Dried thin membranes of this type are suitable for the construction of electrode assemblies. Using Pd-modified membranes, the generation of hydrogen after incubation with sodium dithionite could be realized.

In addition, metal BC can be combined with enzymes immobilized in the membranes for the design of biosensors and biofuel cells [58, 59].

In a similar way, a well-adhered surface modification of BC fibers can be achieved with TiO_2 nanoparticles (with a diameter of about 10 nm) by the hydrolysis of titanium tetraisopropanolate adsorbed onto the fibers. It was observed that the titania-coated surface appears to be dense and have low porosity and to consist of near-spherical grains. By washing with sodium carbonate solution, the TiO_2 films were not removed during neutralization. It seems that the particles have formed strong interactions with BC. The coated membranes showed substantial bactericidal properties under UV radiation and white light (containing a small fraction of UV) conditions, too. This effect is caused by the photocatalytic destruction of the bacterial cells.

Even in the absence of light a decrease in cell growth was observed compared to unmodified BC. These materials are recommended for use as bactericidal surfaces in medical, healthcare, and hygienic applications [60].

In addition to sterilizing surfaces, it is possible to produce mesoporous titania networks consisting of interconnected anatase nanowires by using BC membranes as natural biotemplates. The coating process is carried out by dipping the purified BC after solution exchange into a mixture of titanium tetrabutanolate, acetyl acetone, and isopropanol. After removing the reaction medium the hydrolysis was finished by storage on air. The obtained BC–titania hybrid was heated at 500 °C, resulting in crystalline titania thin films under loss of the BC template. The novel titania nanowire networks may find applications, e.g., as photocatalysts, photovoltaics, and support in bone-tissue engineering [61].

Katagiri [62] investigated the combination of silica–alumina fibers and wood MFC/nanocellulose (95/5 wt %). A further development is oil-retaining sheets with good durability for cleaning rolls [63], which are made from 100 parts inorganic fibers to 5 parts wood MFC/nanocellulose. These sheets are characterized by a weight of $36.3\ \mathrm{g/m^2}$, thickness of 0.21 mm, void volume of 92%, and silicone and oil retention capacity of $0.74\ \mathrm{g/cm^3}$.

4
Development of Medical Devices

Supplementary to the technical use of nanocelluloses reviewed in the previous sections, BC in particular has great potential as a natural biomaterial for the development of medical devices and applications in healthcare and veterinary medicine.

Well-known products include wound dressing, temporary skin, and connective-tissue replacement (BioFill®, Gengiflex®) [7]. Numerous groups are working on these applications and investigating the healing effect of this external BC material (Table 2).

As well as being used as a scaffold for tissue engineering, Hutchens et al. [64] described the creation of a calcium-deficient hydroxyapatite, the main mineral component of bone. Calcium phosphate particles were precipitated in BC by consecutive incubation of calcium chloride and sodium phosphate solutions. Initial tests with osteoblasts in the in vitro evaluation showed that solid fusion between the material and the bone tissue is possible. Hence, this material is a good candidate for use as a therapeutic implant to regenerate bone and heal osseous damage.

In contrast, in experimental and clinical medicine only a few groups have been active in the research and development of shaped BC as implant biomaterial [65–76]. Therefore – from our viewpoint – it is necessary to specify BC design and handling for medical applications in detail.

The following subsections describe techniques for implantation, biological reactions in animal body as well as histological and ultrastructural investigations of incorporated BC. The first long-time study of nanocellulose in the

Table 2 Nanocelluloses as membranes in medical application

Entry	Authors	Citation	Title
1	Slezak A, Kucharzewski M, Jasik-Slezak J (2005)	Polim Med 35:23	Medical properties of membrane dressings made from bacterial cellulose
2	Slezak A, Jasik-Slezak J, Kucharzewski M (2005)	Polim Med 35:15	Biophysical properties of membrane dressings made from bacterial cellulose
3	Schoenfelder U, Abel M, Wiegand C, Klemm D, Elsner P, Hipler U-C (2005)	Biomaterials 26:6664	Influence of selected wound dressings on PMN elastase in chronic wound fluid and their antioxidative potential in vitro
4	Czaja W, Kawecki M, Krystynowicz A, Wysota K, Sakiel S, Wroblewski P, Glik J, Bielecki S (2004)	227th ACS National Meeting, Anaheim, CA, USA	Application of bacterial cellulose in treatment of second and third degree burns
5	Legeza VI, Galenko-Yaroshevskii VP, Zinov'ev EV, Paramonov BA, Kreichman GS, Turkovskii II, Gumenyuk ES, Karnovich AG, Khripunov AK (2004)	Bull Exp Biol Med 138:311	Effects of new wound dressings on healing of thermal burns of the skin in acute radiation disease
6	Serafica GC (2004)	227th ACS National Meeting, Anaheim, CA, USA	Development of new bioengineered cellulose wound dressing with a unique dual fluid handling capability
7	Krystynowicz A, Czaja W, Pomorski L, Kolodziejczyk M, Bielecki S (2000)	Med Fac Landbouww Univ Gent 65:213	The evaluation of usefulness of microbial cellulose as a wound dressing material

animal body for the development of artificial blood vessels and other medical devices of BC will be presented.

4.1
Artificial Blood Vessels

Our investigations on BASYC® (**BA**cterial-**SY**nthesized **C**ellulose) as artificial blood vessel and cuff for nerve suturing [65] are reviewed below. Because of the essential shaping and structure design as well as the application of this innovative biomaterial for experimental surgery we have taken advantages of the *Gluconacetobacter xylinus* strain DSM 14666 as a particularly suitable *cellulose factory* in the laboratory.

Thus it was demonstrated that BC can be suitably shaped for application during biosynthesis [77] (Fig. 13). With a patented matrix reservoir technology developed for this purpose, it is possible to synthesize cellulose in the shape of very regularly hollow bodies of different length, wall thickness and inner diameter (Fig. 14) directly in the culture medium.

The arrow-marked BASYC® tube in Fig. 14 has an inner diameter of 1–3 mm, length of about 5 mm, and wall thickness of 0.7 mm. These parameters are sufficient for experimental microsurgical requirements. Figure 15 shows a schematic picture of the cultivation vessel, in which the glass matrix is immersed in a vertical or horizontal position in a larger volume of nutrient solution. The tube-shaped BC is produced in the nutrient medium, which has entered between the outer and inner matrices, and is supplied with oxygen by a second opening to the air space. After removal of the tubes from the culture medium a purification method including washing with water is necessary. By exchanging the swelling agent water with physiological salt solution, the BASYC® tubes can be stored under cooling.

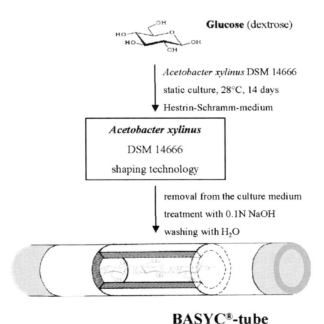

BASYC®-tube

Fig. 13 Scheme of BASYC® tube formation by *Gluconacetobacter xylinus* (*former Acetobacter xylinus*) (DSM 14666) starting from glucose [65]

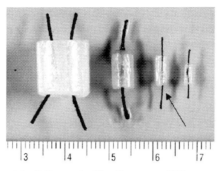

Fig. 14 BASYC® tubes with different inside diameter, different wall thickness, and different length. *Black lines* symbolize the blood stream. → BASYC® tube sufficient for experimental microsurgical applications. Reprinted with permission from [65]

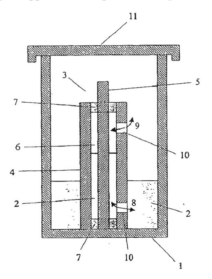

Fig. 15 Patented matrix-reservoir culture technique producing shaped BC. 1 – reservoir, 2 – culture medium (reservoir and matrix part), 3 – air, 4 – outer matrix, 5 – inner matrix, 6 – space between outer and inner matrix, 7 – spacer, 8 – interaction with reservoir, 9 – interaction with air, 10 – passage, 11 – lid [78]

These biotechnologically formed artificial pipelines with an internal diameter of less than 3 mm (Fig. 14) were investigated for their application as a new type of biomaterial for surgery on microvessels (microsurgery). This work is characterized by an intensive cooperative effort between physicians, chemists and biologists.

The application is derived from new microsurgical techniques that can repair small blood vessels and nerves by sutures with optical equipment. The well-known synthetic implant materials for conduits of larger vessels (inner diameter 3–6 mm) made of poly(tetrafluoroethylene), poly(ethylene) and poly(urethane) have proved insufficient, often resulting in thrombosis.

The wall of the BASYC® tubes consists of BC loaded with water in the nanofiber network up to 99%. The hollow space of the material transports water, monovalent ions and small molecules, but not biopolymers or corpuscular blood constituents. The stored water not only stabilizes the cellulose network, but also contributes to the tissue- and hemocompatibility of the nanocellulose device.

For BASYC® vessel implants, low roughness of the inner surface (feature size ∼ 15 nm) is especially significant and can be obtained with the matrix reservoir technique. This degree of roughness is within an order of magnitude of that of typical blood vessels in rats. BASYC® bioartificial pipelines also meet other significant demands for small-vessel replacement: they have a constant shape, are sufficiently stable against internal and external pressure, are flexible and elastic, and are capable of handling a tight microsurgical suture.

The application of BASYC® as a microvessel endoprosthesis was investigated, e.g., in the case of a typical end-to-end anastomosis using the carotid artery of the rat. To demonstrate the implantation area of BASYC® a part of the arterial blood-vessel system of the rat is shown in Fig. 16.

Figure 17 shows an example of microsurgical work with the shaped cellulose material as an interposition implant (end-to-end) immediately after operation. Early results of investigations of implantation as small vessels of the rat as an experimental model are described as follows.

Four weeks postoperatively, the experimental animal was anaesthetized again and re-operated. The inspection of the treated vessel showed that the carotid-artery BASYC® complex was wrapped with connective tissue, and

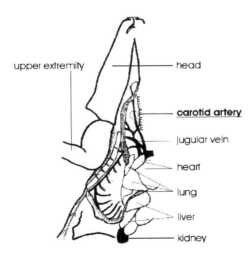

Fig. 16 Scheme of arterial blood-vessel system of the rat. The carotid artery at the animal model is marked. Reprinted with permission from [65]

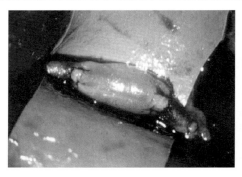

Fig. 17 BASYC® interposition in the carotid artery of the rat immediately after microsurgical operation. The implant is placed on the tissue and connected with the vessel on both sides by sutures. Reprinted with permission from [65]

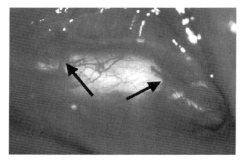

Fig. 18 Carotid artery–BASYC® complex four weeks postoperatively. → anastomosis areas. Reprinted with permission from [65]

pervaded with small vessels like *vasa vasorum*. As shown in Fig. 18 the BASYC® interposition was completely incorporated into the body without any rejection reaction. The arrows indicate the two anastomosis areas. During the postoperative period the animals were not treated with anticoagulation agents. All interpositioned BASYC® tubes had a patency rate of 100% over the entire investigation time provided that the anastomosis was sufficient. There were no signs of coagulation or proliferation.

Upon reconnection of a dissected carotid artery with a BASYC® tube, the internal surface of the BASYC® material becomes completely covered by an endothelial cell layer after a residence time of four weeks, as determined by histological examination of the preparation along with a specific test for endothelial cells (Fig. 19). Blood remnants can be found in the lumen. As observed by scanning electron microscopy (SEM), the complete colonization of the BASYC® region with endothelial cells covering both parts of the suture (Fig. 20) can be substantiated. Apparently, the BASYC® material is a good substrate (matrix) for the anchoring of autogenous cells.

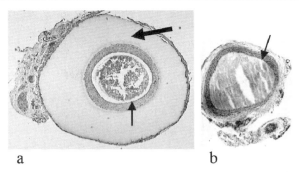

a   b

Fig. 19 Histological preparation (haematoxylin and eosin staining) of square sections. **a** BASYC® in the middle region of the interposition four weeks after implantation in the carotid artery of the rat, **b** untreated carotid artery of the rat. → endothelial cell layer. ▶ BASYC® layer

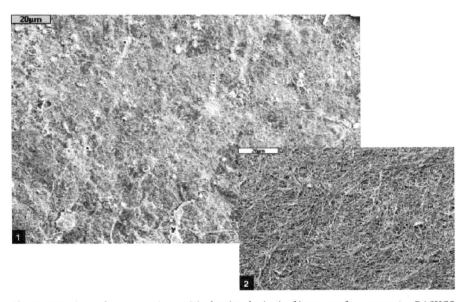

Fig. 20 SEM (sample preparation: critical point drying) of inner surface areas. 1 – BASYC® in the middle region of the interposition four weeks after implantation in the carotid artery of the rat with endothelial cells, 2 – BASYC® before incorporation. Reprinted with permission from [65]

Upon coverage of longitudinally cleaved tubes with bovine endothelial cells in cell culture tests, a distinct sprouting of the initially spherical cells takes places within 24 h. SEM shows that the resulting filaments barely differ from the fibers of BC (Fig. 21). This structure also benefits the rapid endothelial colonization of BASYC®.

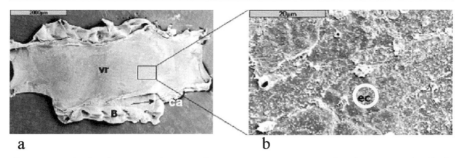

a   b

Fig. 21 Longitudinal section of a carotid artery-BASYC®-preparation (SEM examination). B – BASYC®, ca – carotid artery, vr – vitalized region, ec – endothelial cells, **a** region of anastomosis between BASYC® and carotid artery (→ suture material is visible), **b** dense endothelial cover on the entire inner surface

The latest results of BASYC® relate to long-term incorporation (1 year) in the rat. Special interest was directed to investigation of the contact region between the artificial prosthesis and the newly formed blood vessel. Therefore, longitudinal sections of the preparation were examined.

The results of the immunohistochemical labeling (CD-31) document a complete lining with endothelial cells of the inner surface including the sensitive region of anastomosis (Fig. 22).

SEM investigations confirmed this result. In the top view the complete endothelial cover is visible (Fig. 21). More detailed information was obtained with the help of transmission electron microscopy (TEM) shown in Fig. 29.

The TEMs revealed a change of material structure in the contact region with the newly built blood vessel. Beside cellulose we discovered still different structures in region 1 (Fig. 24a). Branches of fibroblasts, cellular compounds of the newly built vessel, have penetrated the cellulosic network and in this way anchored themselves in the artificial vessel prosthesis. Sections of the fibroblast branches and the contact region are clearly visible (Fig. 24b).

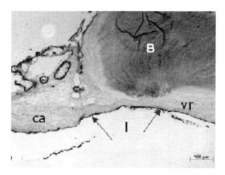

Fig. 22 Longitudinal section of a carotid artery–BASYC® preparation. → CD-31 labeling for endothelial cells. B – BASYC®, ca – carotid artery, vr – vitalized region, l – lumen

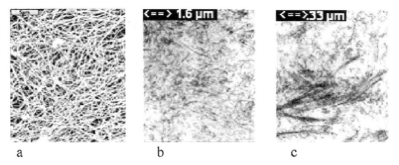

a   b   c

Fig. 23 SEM and TEM examination of longitudinal sections. **a** SEM of the ultrafine network structure of BASYC®, **b** and **c** TEM pictures of the BASYC® layer

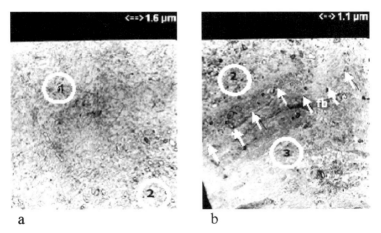

a   b

Fig. 24 TEM of BASYC® layer in contact with blood flow. BASYC® layer **a** with hints to fibroblast cells, **b** in contact with blood shows fibroblast body. → contact region to the blood, fb – fibroblast, 1 – BASYC®, 2 – BASYC® with newly built structures (fibroblast and collagen), 3 – newly formed vessel, 1 to 2 to 3 – increase of fibroblast detection

Under higher magnification (Fig. 25) it can be proved that the fibroblasts are active. Active fibroblasts produce collagen. In contrast to the smooth cellulose fibrils the collagen fibrils appear striped.

To complete the study, the inner surface of the vitalized artificial prosthesis was examined. In the TEM a complete endothelial cover facing the blood stream can be seen (Fig. 26).

The results of the one-year experimental investigations confirm the results of the four-week studies. The integration processes are in progress. The homogeneity and the neointima anchorage capability correspond with the recipient blood vessel, the carotid artery of the rat. The detected collagen confirms vital fibroblasts and these induce an increased integration of BASYC®.

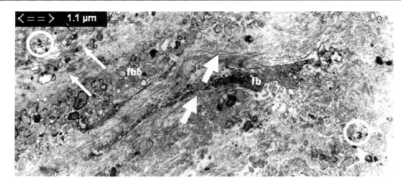

Fig. 25 TEM of the contact region. 2 – BASYC® with newly built structures (fibroblast and collagen), 3 – newly formed vessel. → cellulose fibers (without striped structures), ▶ collagen fibers (*striped structures*), fb – fibroblast, fbb – fibroblast branches (*sections*)

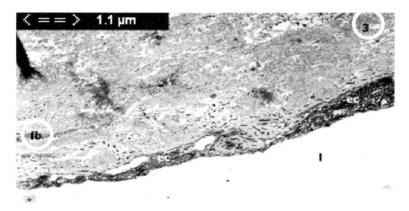

Fig. 26 TEM of the region of the newly built blood vessel. 3 – newly formed vessel, fb – fibroblast, ec – endothelial cell, l – lumen

The endothelial cells assume their entire function in the neointima. In all one-year animals ($N = 5$) no thrombosis was recognized.

The high patency rate of the small vessels is apparently favored because of the nature-orientated morphology of BASYC®. These findings are evidence for a process of integration without degradation (vitalization).

4.2
Cuffs for Nerve Surgery

Upon dissection of the *nervus ischiadicus* of the rat and subsequent reconnection by a typical microsurgical suture, a protective cover (cuff) of BASYC® prevents connective tissue from growing into the nerve gap and favors the adhesion of the fascicles, which facilitates early regeneration of the nerve and rapid return of the muscle function. The good incorporation of BC-forming connective tissue on the surface of the protective cover is particularly noteworthy (Fig. 27).

After the observation time from 4 to 26 weeks postoperatively the BASYC® tube was covered with connective tissue and small vessels within (Fig. 27b). Neither an inflammation reaction nor an encapsulation of the implant was observed. The regeneration of nerve function improved after 10 weeks, compared to an uncovered anastomosed nerve. As an indicator of the regeneration of nerve function the increasing muscle weight of the *musculus extensor digitorum longus* (EDL) of the rat was determined. Another indicator was the reappearance of acetylcholine as the transmitter of nerve impulses to the executive organ. This agent was indirectly detected by determination of the activity of acetylcholinesterase using Gomori's method of acetylthiocholineiodid staining [79].

Using the aforementioned animal model (sciatic nerve/EDL), in another experiment BASYC® served as a drug depot for a substance with neuroregenerative properties (experiments were carried out in cooperation with the Max Planck Research Unit, Enzymology of Protein Folding, Halle/Saale, Germany) [80]. Immediately after nerve reconstruction the substance was applied to the anastomosis and stored in the BASYC® cover. For comparison a placebo was examined likewise. At the end of the observation time (8 and 10 weeks) the EDLs were removed and the muscle weight was determined. In comparison to the placebo the tested substance caused quicker return of innervation, measured by the muscle weight. In addition to the method of muscle-weight determination, the functional recuperation of the paralyzed legs was evaluated by the observation of the walking behavior. These results correspond to those of the muscle-weight determination.

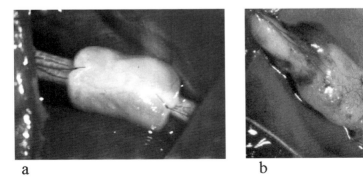

a   b

Fig. 27 Application of a BASYC® tube in micronerve surgery. **a** sciatic nerve of the rat with a BASYC® tube as prospective cover immediately after operation, **b** sciatic nerve of the rat with the BASYC® cover 10 weeks after operation. Reprinted with permission from [65]

4.3
Microsurgical Model for Training

The successful application of the microsurgical operation technique (manipulation under microscope with 20–30 times magnification of the anatomical structure) is impossible without intensive microsurgical training. For reasons of animal protection it is necessary to use simple models consisting of rubber membranes or plastic tubes in the first step. These models are very easy to handle but they do not however permit a close-to-reality training of microsurgical anastomosis techniques. Besides the preparation of sufficient anastomosis, careful handling with living tissue is an important requirement for the success of microsurgical operations. Bruising and stretching of the microvessel are traumatizing and result in damage to the vessel wall [81]. These are limiting factors for successful anastomosis together with lesions caused by temporary vessel occlusion with microclips [82–86].

Under this criterion we developed a new model that enables training of all microsurgical techniques by copying conditions in the living organism.

The BASYC® tubes are integrated components of the model system. The practice unit permits optimal and close-to-reality training, i.e. manipulation and anastomosis in a wet milieu. Incorrect handling with microsurgical in-

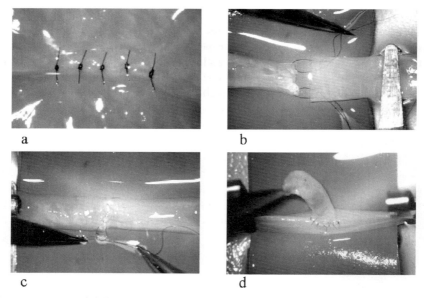

a   b

c   d

Fig. 28 Training of different microsurgical suture techniques with the help of BASYC®.
a end-to-end anastomosis (single knot suture), **b** end-in-end anastomosis, **c** end-to-end anastomosis (continuous suture), **d** end-to-side anastomosis. Reprinted with permission from [65]

struments results in visible changes of the shape of the BASYC® tube, as in a natural vessel.

Consequently, the model leads to improve discipline in terms of careful preparation and operation.

In Fig. 28 some techniques are shown. The training model was used successfully in practical microsurgical courses and reduces the number of experimental animals necessary.

4.4
Present State and Trends

Implantation materials, which are in direct contact with blood, have to meet a particularly large range of requirements: bio- and blood compatibility, mechanical strength against blood pressure, impermeability to the blood and its constituents, and sterilizability. In addition, the healing process that takes place on the inner and the outer surface of the artificial vessel is very different. The inner surface of the biomaterial should not stimulate adhesion of cellular blood components but should be covered with endothelial cells, whereas the outer surface of the prosthesis should be wrapped with connective tissue.

Commercial human prosthesis materials such as poly(tetrafluoroethylene)/ PTFE (Teflon®, Gore-Tex®), poly(ethylene terephthalate)/PET (Dacron®), poly(urethane) or silicone, which are applied in heart, vessel and thorax surgery with an inner vessel diameter of greater than 3 mm are unsuitable for microsurgical requirements. According to the definition microvessels have an inner diameter smaller than 3 mm. In a lot of publications experimental studies of the short- and long-time behavior of synthetic small-caliber prosthetic grafts (3–6 mm) have been reported. Because of the high risk of thrombosis these grafts have not, to date, been used in clinical applications [87–96]. A recent solution is the autogenous transfer of veins for small-vessel reconstruction. Reasons for failures in synthetic implants are seen in a hypoplasia of the intima in the region of anastomosis, an extensive perivascular fibrosis or high thrombogenity of the implanted material [97, 98]. Greisler and co-workers described that fabrication of poly(propylene) into an arterial substitute might result in an efficacious prostheses because of the physical properties (e.g. high tensile strength) and relatively inert behavior of the material. After implantation in the abdominal aorta of dogs, the authors determined better late patency for poly(propylene) grafts (4 mm I.D. ×50 mm in length) than for Dacron® and ePTFE. In one month, a confluent endothelialized surface was seen in all explants with a homogenous surface [99].

Expensive attempts were made to optimize the inner surface of the implant materials. The scientific work is concentrated on the development of suitable coatings based on proteins such as collagen [100, 101], gelatine [101–103] and albumin [101, 104, 105] as well as chitosan–poly(vinylalcohol) blends [106], heparin [97], or hyaluronic acid [107]. Further investigations included the transplantation of endothelial cells onto the surface of the endoprostheses [108–116]. In addition, the fixation of cell receptors or growth factors [117–119] onto coating materials is an interesting development in the field of biomaterial research.

Our animal experiments demonstrate that the structure and properties of the cellulosic microvessel endoprosthesis material cause rapid adhesion and optimal coating with autogenous cells in the living body [66–74, 120–125].

The results of initial experimental applications in microsurgery and microsurgical training substantiate the remarkable advantages of this nanostructured polysaccharide: handling in an easy and practice-orientated way, blood and tissue compatibility, consistency comparable to that of the blood vessel, and vitalization in the living body.

The test of BASYC® tubes as small-vessel prosthesis for other artificial pipelines (arteries, veins, lymph, and urine leading vessels) is under progress. Initial results of the implantation of BASYC® material into jugular veins of 10 white rats showed no complications such as wall-proliferation, stenosis or thrombosis. A normal blood flow through the interposition as well as the formation of connective tissue around it were detectable 4 and 12 weeks, respectively, after the operation [124].

Besides the promising use of BASYC® in experimental vascular surgery we see the possibility for the application of BC as a soft-tissue substitution material in various medical fields because of its extraordinary properties.

Internal medicine, urology, gynecology, otolaryngology, maxillofacial and plastic surgery could be potential users of this designed nanocellulose. For this reason, hollow cellulose tubes with different inside diameter and different wall thickness have been prepared in micro-dimensions. Figure 14 shows a collection of BASYC® tubes. Foils, patches, and other shapes are possible, too.

The immobilization of biologically active substances within the cellulosic network structure to optimize healing and regeneration processes as well as to test the potential pharmaceutical suitability of natural or synthetic substances are also study points for the future.

5
Bacterial Nanocellulose in Veterinary Medicine and in Cosmetics

5.1
Animal Wound Dressing

In veterinary practice, wounded animals often appear and their treatment is complicated because of bacterial infections. This section describes clinical experiences with wounds to horses and sheep. In the case of horses, the wounds are larger, usually contaminated and have a tendency to hypergranulation. A further significant problem in this kind of animals is stable fixation of the treatment material, e.g., on the legs, as the horses are very restless.

- Wound characteristics
  Because animal wounds are usually extensively damaged and contaminated a primary wound dressing is often not successful. The damaged skin will become necrotic in a few days and stitches are not stable. Furthermore, in distal limbs with lower blood circulation, higher tension at the edges, lower temperatures, and hyperplastic granulation, delayed healing occurs.
  These kinds of wounds will be closed in the secondary healing processes, where the formation of granulation tissue first takes place. Covering with epithelium is the final stage of wound healing. Often such secondary healing processes take a long time and require a lot of material [126].
- Wound treatment
  One aim of modern wound treatment with new materials is the reduction of infection and the normalization of healing. In veterinary medicine, experience with such optimized materials are limited and conclusions are contradictory. Some investigations have been carried out on the leg of horses, but – in comparison with other animals or kinds of wounds – these show different healing processes characteristics [127–133]. The majority of clinical trials introduce modern wound-treatment materials applied in human medicine to veterinary practice. These studies show positive results [134–136].
  In principle, there are two different forms of wound treatment, under dry conditions and in a damp atmosphere. In the latter case, the rate of epithelial proliferation is increased. Around the wound, a specific microclimate will develop that is important for a more natural course of the healing process [137–139]. Such wound-treating materials allow gas exchange, the absorption of exudate from the wounds, and a long-time stable damp atmosphere [140, 141]. In these wound dressings semi-occlusive and occlusive wound bandages are used. Because of the lower knowledge in animal wound dressing, applications in veterinary medicine need currently more detailed work than in highly developed human medical treatment [7, 142].
  From this viewpoint, larger experiments were realized at the fzmb using BC wound dressings, mainly with horses and sheep.

Application to horses was carried out in veterinary hospitals within the bounds of possibility in the course of normal treatment (Fig. 29).

The randomized study with sheep was the subject of a special animal experiment (Fig. 30). Comparable artificial wounds were created and the

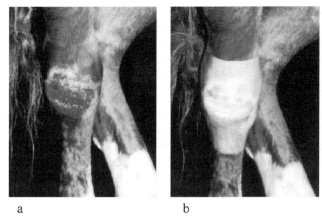

a   b

Fig. 29 Wound treatment of horses. **a** wound after accident, **b** wound with the BC dressing

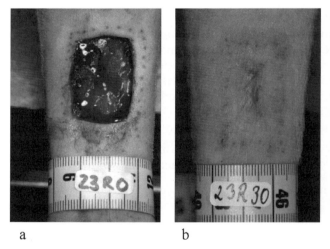

a   b

Fig. 30 Wound healing of sheep. **a** fresh artificial wound (day 0), **b** wound area after treatment with BC dressing (day 30)

process of healing was controlled. In this way, the permeability to microorganisms, the extent of wound bleeding, and the characteristics (duration and degree) of the healing process were investigated. Furthermore, the contribution of epithelial proliferation rate, microbial contamination of the wounds, wound contraction, area of the epithelium formed as well as the area of the granulation tissue, wound swelling, and mathematically derived parameters for statistical evaluation were studied [126, 127, 129, 131, 133, 143–149].

The healing process of BC-treated wounds was compared with sheep wounds treated with the registered products Comfeel plus® [135] and Vetisept®, the latter product containing a combination of poly(vinylpyrrolidone) and iodine. It is known that the bactericidal activity of this ointment is intensive and has a long-time effectiveness.

During clinical practice with horses as well as the experiment with sheep the dressings were changed each third day. The main results of the experiments with sheep are presented in Tables 3–5 as well as in [150–152]. The data obtained are summarized as follows (Table 6):

- Wounds covered with hydrocolloid Comfeel plus® healed significantly faster than with BC or the bactericidal ointment (Table 3).
- Wounds treated with BC did not show any retraction but a significant contraction with a clearly slower rate of epithelial proliferation.
- Removing BC layers that, due to desiccation, had adhered to the wound margin resulted in an often painful dressing change for the sheep. Furthermore, the newly grown epithelium of the skin seemed to be irritated by the removal, thus prolonging the healing of the wounds.
  By preventing the desiccation of the BC during the treatment, sticking to the wound margin and the subsequent prolongation of the healing process could be avoided. Using thicker layers of BC or covering the dressed wound with a poly(urethane) foil seems the most suitable method to achieve this effect.
- Unlike the wounds treated with BC and bactericidal ointment the wounds covered with the hydrocolloid Comfeel plus® were highly contaminated with several bacteria (Tables 4 and 5). The growing microorganisms were members of both pathogenic and non-pathogenic groups.

The low number of microorganisms under the Vetispet® is due to the application of iodine as a strong bactericidal component. Particularly conspicuous is the extremely low number of microorganisms growing under the BC and the small number of pathogenic bacterial strains as well as of gram-negative species (Table 6).

This is understandable from the viewpoint of the special biopolymer structure of BC. Its high humidity and water content is important for the development of a specific atmosphere at the wounds.

For other polysaccharides such as chitosan a high antimicrobial activity has been described – mainly against *Escherichia coli* and *Staphylococcus aureus* [153, 154]. This activity of chitosan is the basis for the development of materials for wound healing, e.g., in combination with *N,O*-(carboxymethyl)chitosan and collagen [155].

Contrary to the results of the sheep experiments, in the case of horses no covering between the BC and the wound surface was detectable. During dressing changes a yellowish creamy liquid appears (wound exudation, with only low numbers of microorganisms). Rapid formation of healthy granulations was detectable so that the wound healing process was faster. No more than 45–50% of the horses showed hypergranulations. This had to be removed sur-

Table 3 Course of wound contraction with respect to the wound area at day 0 using different dressing materials for induction of the wound healing in sheep [150]

Dressing material		Wound contraction (%)					
	Day	0	3	6	9	12	15
Vetisept®		0	12	10	2	23	33
BC		0	4	10	21	38	50
Comfeel plus®		0	12	– 5	18	30	43

Table 4 Bacterial growth under different dressing materials for induction of wound healing in contaminated sheep wounds of sheep (semiquantitative analysis). Growth of bacteria per day [150]

Dressing material		Bacterial growth							
	Day	0		3		6		9	
		%	Mean [a]	%	Mean [a]	%	Mean [a]	%	Mean [a]
Vetisept®		4	(+)	29	(+)	29	+	29	+++
BC		17	(+)	46	(+)	29	+	29	+
Comfeel plus®		21	(+)	100	+++	100	++++	100	++++

[a] – no growth, (+) absolutely low growth, + low growth, ++ growth, +++ strong growth, ++++ very strong growth, R Spread over the whole Petri dishes

gically only in some cases. These clinical experiences show conspicuously few disorders in wound healing in the case of horses.

BC is a good material not only for wound treatment and other fields of veterinary medicine, but also as a scaffold material for cell cultivation in tissue engineering [156, 157]. On such scaffolds the fzmb has cultivated the following cell types successfully: human osteoblasts, human osteogenic sarcoma cells (SAOS-2), equine osteoblast lines and chondrocytes, and mesenchymal stem cells.

It could be demonstrated that BC is biocompatible and a stable basis for further research in tissue engineering and also the development of new technological variants for veterinary medicine.

In practice, further important aspects of BC are the focus of interest, concerning cooling of overtaxed muscles and particularly wound treatment of animals such as horses, sheep, cows, cats, and dogs. Extremely highly infected wounds are frequent in dogs after car crashes or similar accidents [143]. Furthermore, treatment of badly healing and permanent wounds, e.g., ulcers, and in the clinical and home-care sector both for human and veterinary medicine, as well as specific applications in tissue engineering will be major future developments.

Table 5 Semi-quantitative investigation of the bacterial species using different dressing materials

Dressing material	Species [a]	Concentration (mean values [b])
Vetisept® [c]	Several species of unproblematic *Cocci* (not haemolytic)	(+)– +
	Sporulating bacteria (unproblematic)	+
	E. coli	(+)– +++
	Sporulating bacteria, haemolytic	+++
	Staphylococcus intermedius	+
	Micrococcus sedentarius	++ –R
	Actinomyces pyogenes (gram-positive rods, haemolytic, in only one case)	+++
	Staphylococcus xylosus	
BC [d]	Several species of unproblematic *Cocci* (not haemolytic)	(+)
	Aerococcus viridans	(+)
	Sporulating bacteria (unproblematic)	(+)
	Enterobacter gesgoriae	++
	Staphylococcus felis	+++
	E. coli	++
	Actinomyces pyogenes (gram-positive rods, haemolytic, in only one case)	+++ (R)
Comfeel plus® [e]	Gram-negative *Cocci*	(+)–R
	Sporulating bacteria (unproblematic)	(+)– ++++
	Staphylococcus intermedius	R
	Streptococcus uberis	+++
	E. coli	+++ –R
	Cocci, haemolytic	++++
	Sporulating bacteria, haemolytic	+ – ++
	Enterobacteriaceae	++
	Enterococcus faecium	+++
	Streptococcus parasanguis	++++
	Actinomyces pyogenes (gram-positive rods, haemolytic)	+++ –R

[a] A list of all the kinds of microorganisms detected under the dressing material. The main species are in bold
[b] – no growth, (+) very low growth, + low growth, ++ growth, +++ strong growth, ++++ very strong growth, R: spread over the whole Petri dish
[c] The species were not found in all samples and in some cases no significant contamination was detected
[d] The species were only found in a small number of samples and in most cases no significant contamination was detected
[e] After the third day of the experiment these microorganisms were detectable in all tested samples. The concentrations found were high in all cases

Table 6 Summarized results of the experiment with sheep and a comparison of the effects of all dressing materials for induction of wound healing [150]

Entry	BC	Vetisept®	Comfeel plus®
1	Adhesion to the wound margin	Adhesion to the wound bottom	Removal of the dressing material without alteration of the wound surface
2	Painfull dressing change	Painfull dressing change	
3	Tearing by removal		
4	Mild or medium bleeding	Medium bleeding	Very slight bleeding
5	Wound secretion not completely absorbed	Wound secretion not completely absorbed	Wound secretion completely absorbed
6		Retraction	
7	Slight bacterial contamination (not in all tested samples)	Slight bacterial contamination	Strong bacterial contamination
8		Strong swelling of the wound	
9			Hypergranulations
10		Incrustation	
11			Short-time epithelization. Fast wound healing
12	Healing response by wound contraction	Healing response by epithelial proliferation	Equal participation of contraction and epithelial proliferation

5.2
Cosmetic Tissues

Because of its extraordinary properties such as mechanical stability, high wetness with a water content of over 95%, and high purity, BC is an important material for applications in the cosmetic sector. For instance, two cosmetic tissue products are successful on the market: a series of masks based on BioCellulose and the mask basis material NanoMasque®. Both tissues are produced from pure BC and alternatively impregnated with active substances applied in cosmetics such as plant extracts, extracts from algae, essential oils, and panthenol. The additives are bound inside the cellulose matrix, e.g., by hydrogen bonds. These bonds are selective enough to localize the substances in the matrix as well as to allow their migration into the skin during application (Fig. 31).

For the preparation and use of BC as a carrier in cosmetics the following main conditions have to be considered:

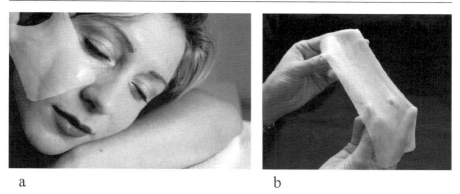

a   b

Fig. 31 Example of cosmetic application of NanoMasque®. **a** treatment of the skin, **b** NanoMasque® material

- Washing and cleaning
  For the purification of this biopolymer no treatment with bases (e.g., sodium or potassium hydroxide) should be carried out. Traces of alkaline hydroxides can accumulate inside the polymer and can induce irritation of the skin. Other possibilities for successful washing and cleaning of BC are described in a patent. The authors have used solutions with an alkaline reaction – but free of bases – and list medical application [158].
- Endotoxin test
  After purification the homogeneity of BC is proved by an endotoxin test using an E-TOXATE®-Kit (Sigma–Aldrich) [158]. The results demonstrate that the biopolymer is free of endotoxins.
- Biocompatibility
  Investigations of the biocompatibility of BC are carried out using two methods: cell culture and clinical examination.
  Using human keratinocytes (HaCaT cells) to determine the direct interaction of purified BC with cells in culture, good biocompatibility of BC with cells was shown, with no cytotoxicity effects.
  Several randomized studies with humans have been realized: a human patch test for 72 h and a repetitive epicutaneous test for six weeks, as well as tests on the influence of BC on the moistness of the skin after short-term treatment (20 min, measuring of moistness up to 12 h after treatment). The results demonstrate that the applied BC will not cause any unwanted skin reactions due to irritating, sensitizing or early allergic effects and that the biopolymer is safe for this indication. An additional finding was that the moistness of the treated skin was significantly higher than in untreated areas [159–161].
  Based on these results, an application of materials based on purified BC, e.g., as NanoMasque®, in the field of cosmetic treatment is possible.

6
Further Examples of Nanocellulose Application

In the previous sections, the application potential and the use of nanocelluloses as technical materials as devices in human and veterinary medicine as well as cosmetics have been described. Because of the exceptional properties of these innovative polymers more widespread utilization has been observed in recent years. In this section some additional and quite different examples will be reported. Further information can bee found in Table 7.

A classical and growing field is the usage of BC in the food industry. Lin and Lin [162], for example, have developed Chinese-style meatballs containing BC. These balls (also known as kung-wan) are a very typical emulsified meat product. After cooking in water they are ready to eat directly. Because of its unique suspending, thickening, water-holding, stabilizing, bulking, and fluid properties bacterial nanocellulose is a very suitable and versatile ingredient in this case, and for processing other foods. The addition of BC occurs in a vacuum emulsifier when mixing all components. Important features of this application are: a reduction in food retention in the intestine and an increase in bile acid secretion as well as a lowering of cholesterol and triacylglycerol levels in animal experiments.

BC added to ice cream can serve as a stabilizer. Zhou et al. [163] found that BC was the best ingredient for ice cream to resist meltdown and heat stock.

A completely different application of BC is the manufacturing of biosensors. For successful employment in biosensor use, biocatalysts have to be immobilized in biocompatible matrices that stabilize structure and preserve activity while providing electronic connections. Photosynthetic protein complexes from higher plants and green algae are membrane-bound complexes that may be useful as nanoscale light sensors. These photo-systems show particular problems such as loss of activity due to sensitivity to chemicals and light. Therefore, a medium that allows transmission of visible light in the required range is needed. BC, synthesized by the bacterium *Gluconacetobacter hansenii*, was used as a matrix for the immobilization of the described components. This system can be doped with metals and ceramics or used as coatings on electrode surfaces [164].

7
Conclusion

The aim of the paper was to demonstrate the current state of research and development in the field of nanocelluloses on the basis of selected examples. The extraordinary supramolecular nanofiber network structure and the resulting valuable properties have led to a real challenge and extensive global activity. It is the intention of this work to broaden knowledge in this subject area

Table 7 Some special applications of nanocelluloses

Entry	Authors	Citation	Publication
1	Helenius G, Backdahl H, Bodin A, Nannmark U, Gatenholm P, Risberg B (2006)	J Biomed Mater Res (Part A) 76:431	In vivo biocompatibility of bacterial cellulose
2	Ono H, Nakamura M, Hayashi M (2006)	WO 2006004012 A1 20060112	Cellulose nonwoven fabrics with high porosity
3	Bielecki S, Krystynowicz A, Czaja W (2005)	WO 2005003366 A1 20050113	A method for the production of bacterial cellulose
4	Bodin A, Backdahl H, Helenius G, Gustafsson L, Risberg B, Gatenholm P (2005)	229th ACS National Meeting, San Diego, CA, United States	Bacterial cellulose as scaffold for tissue engineering
5	Dai Y (2005)	CN 1664027 A 20050907	Digital jet-printing ink composition and its production
6	Ek R, Mihranyan A, Andersson S-B, Bosson B, Lindberg N-O (2005)	US 2005053665 A1 20050310	Nicotine formulations containing cellulose
7	Kiyota Y, Konishi A, Uematsu T (2005)	JP 2005279074 A2 20051013	Hemostatic patches containing blood cell-impermeable fine fiber structure
8	Sato K (2005)	JP 2005283567 A2 20051013	Protein chip using bacteria cellulose and lipid bilayer
9	Tabuchi M, Baba Y (2005)	Anal Chem 77:7090	Design for DNA separation medium using bacterial cellulose fibrils
10	Tabuchi M, Kobayashi K, Fujimoto M, Baba Y (2005)	Lab Chip 5:1412	Bio-sensing on a chip with compact discs and nanofibers
11	Basler EA, Dadang S, Kondo K, Tanito M (2004)	JP 2004201590 A2 20040722	Microbial cellulose gels, their manufacture, and use for food materials
12	Ng C-C, Shyu Y-T (2004)	World J Microbiol Biotechnol 20:875	Development and production of cholesterol-lowering *Monascus*-nata complex
13	Sato K (2004)	JP 2004271540 A2 20040930	Protein chip

Table 7 (continued)

Entry	Authors	Citation	Publication
14	Tsukada T, Midorikawa M (2004)	JP 2004014592 A2 20040115	Separator for double layer capacitor
15	Martin M, Evans B, O'Neill H, Woodward J (2003)	Appl Opt 42:6174	Laser-induced breakdown spectroscopy used to detect palladium and silver metal dispersed in bacterial cellulose membranes
16	Tsukada T, Midorikawa M (2003)	JP 2003347166 A2 20031205	Capacitor separators with suppressed internal resistivity and long life
17	Butterfield DA, Colvin J, Liu J, Wang J, Bachas L, Bhattacharrya D (2002)	Anal Chim Acta 470:29	Electron paramagnetic resonance spin label titration: a novel method to investigate random and site-specific immobilization of enzymes onto polymeric membranes with different properties
18	Jung H, Wilson DB, Walker LP (2002)	Biotechnol Bioeng 80:380	Binding mechanisms for *Thermobifida fusca* Cel5A, Cel6B, and Cel48A cellulose-binding modules on bacterial microcrystalline cellulose
19	Serafica G, Mormino R, Bungay H (2002)	Appl Microbiol Biotechnol 58:756	Inclusion of solid particles in bacterial cellulose
20	Shao W, Tang M, Li S, Xiong Z (2002)	Shipin Kexue 23:167	Study on application of bacterial cellulose in fermented soy milk ice cream
21	Herbert W, Chanzy HD, Ernst S, Schuelein M, Husum TL, Kongsbak L (2001)	WO 2001025470 A1 20010412	Cellulose films for screening
22	Krystynowicz A, Bielecki S, Czaja W, Rzyska M (2000)	Prog Biotechnol 17 (Food Biotechnology): 323	Application of bacterial cellulose for clarification of fruit juices
23	Sheu F, Wang CL, Shyu YT (2000)	J Food Sci 65:342	Fermentation of *Monascus purpureus* on bacterial cellulose-nata and the color stability of *Monascus*-nata complex

and stimulate the practical application of nanocelluloses. From the scientific and economic viewpoint, these innovative polymers, which are exciting examples of the large and significant biopolymer family of cellulosics, are on the threshold of a breakthrough that is also being driven by recent extraordinary activities in the field of nanosized materials.

Moreover, if cellulose-forming bacteria could be cultivated on a large industrial scale, the requirements for cellulose could be satisfied entirely from this source in the future.

Acknowledgements The financial support of the work by the Friedrich-Schiller-Universität Jena, the Deutsche Forschungsgemeinschaft (priority programm 1011), and the Thuringian Ministry of Science, Research, and Culture as well as of the Federal Ministry of Economics and Technology, the Thuringian Ministry of Technology and Labour and the European Union (LEONARDO programme) is gratefully acknowledged. We thank all coworkers, cooperation partners, and students that were involved in the results described.

References

1. Klemm D, Heublein B, Fink H-P, Bohn A (2005) Angew Chem Int Ed 44:3358
2. Sixta H (ed) (2006) Handbook of Pulp. Wiley, Weinheim
3. Ljungberg N, Bonini C, Bortolussi F, Boisson C, Heux L, Cavaille JY (2005) Biomacromolecules 6:2732
4. Samir MASA, Alloin F, Dufresne A (2005) Biomacromolecules 6:612
5. Kontturi E, Thüne PC, Alexeev A, Niemantsverdriet JW (2005) Polymer 46:3307
6. Brown RM Jr, Saxena IM (2004) Cellulose 11:491
7. Czaja W, Krystynovicz A, Bielecki S, Brown RM Jr (2006) Biomaterials 27:145
8. Shah J, Brown RM Jr (2005) Appl Microbiol Biotech 66:352
9. Song H, Zhang Y, Guo H (2004) Ziran Kexueban 29:73
10. Horii F, Erata T (2006) 231st ACS National Meeting, Atlanta, GA, USA
11. Brown AJ (1886) J Chem Soc 49:432
12. Hestrin S, Schramm M (1954) Biochem J 58:345
13. Fink H-P, Purz HJ, Bohn A, Kunze J (1997) Macromol Symp 120:207
14. Yano H, Sugiyama J, Nakagaito AN, Nogi M, Matsuura T, Hikita M, Handa K (2005) Adv Mater 17:153
15. Yano H, Nakahara S (2004) J Mater Sci 39:1635
16. Klemm D, Philipp B, Heinze T, Heinze U, Wagenknecht W (1998) The accessibility of cellulose. In: Comprehensive Cellulose Chemistry. Wiley, Weinheim, p 29
17. http://www.empa.ch/plugin/template/empa/*/27303, 16.09.2005
18. Turbak AF, Snyder FW, Sandberg KR (1983) J Appl Polym Sci: Appl Polym Symp 37:815
19. Nakagaito AN, Yano H (2005) Appl Phys A 80:155
20. Nakagaito AN, Yano H (2004) Appl Phys A 78:547
21. Takahashi N, Okubo K, Fujii T (2005) Bamboo J 22:81
22. Suzuki M, Hattori Y (2004) WO Patent 2004009902
23. http://www.empa.ch/plugin/template/empa/*/39683, 16.09.2005
24. Miyagawa A, Fujikawa A (1992) Adv Filtr Sep Tech 6:314

25. Matsuda Y, Hirose M, Ueno K (2001) US Patent 6183596
26. Matsuda Y (2000) Sen'i Gakkaishi 56:192
27. Ioelovich M, Leykin A (2004) Sci Israel – Technol Adv 6:17
28. Koga S (2000) JP Patent 2000303386
29. Kyomori K, Tadokoro I, Matsumura T (2005) JP Patent 2005126867
30. Fukuda S, Takahashi M, Yuyama M, Oka N (2001) JP Patent 2001293033
31. Kumar V (2002) WO Patent 2002022172
32. http://www.empa.ch/plugin/template/empa/729/22027, 16.09.2005
33. http://www.ofek.org.il/eng/public.php, 25.07.2005
34. Gindl W, Keckes J (2004) Comp Sci Technol 64:2407
35. George J, Ramana KV, Sabapathy SN, Bawa AS (2005) World J Microbiol Biotechnol 21:1323
36. Tokura S, Tamura H (2005) 229th ACS National Meeting, San Diego, CA, USA
37. Yamanaka S, Ishihara M, Sugiyama J (2000) Cellulose 7:213
38. Haigler CH, White AR, Brown AR Jr, Cooper KM (1982) J Cell Biology 94:64
39. Astley OM, Chanliaud E, Donald AM, Gidley MJ (2001) Int J Biol Macromol 29:193
40. Seifert M, Hesse S, Kabrelian V, Klemm D (2004) J Polym Sci A1 42:463
41. Sakairi N, Suzuki S, Ueno K, Han SM, Nishi N, Tokura S (1998) Carbohydr Polym 37:409
42. Ciechanska D (2004) Fibres Textil Eastern Eur 12:69
43. Weimer PJ, Hackney JM, Jung HJG, Hatfield RD (2000) J Agric Food Chem 48:1727
44. Tokoh C, Takabe K, Fujita M, Saiki H (1998) Cellulose 5:249
45. Tokoh C, Takabe K, Sugiyama J, Fujita M (2002) Cellulose 9:351
46. Kacurakova M, Smith AC, Gidley MJ, Wilson RH (2002) Carbohydr Res 337:1145
47. Ishida T, Sugano Y, Nakai T, Shoda M (2002) Biosci Biotechnol Biochem 66:1677
48. Ishiada T, Mitari M, Sugano Y, Shoda M (2003) Biotechnol Bioeng 83:474
49. Chao Y, Mitarai M, Sugano Y, Shoda M (2001) Biotechnol Prog 17:781
50. Dubey V, Pandey LK, Saxena C (2005) J Memb Sci 251:131
51. Dubey V, Saxena C, Singh L, Ramana KV, Chauhan RS (2002) Sep Purif Technol 27:163
52. Pandey LK, Saxena C, Dubey V (2005) Sep Purif Technol 42:213
53. Nakayama A, Kakugo A, Gong JP, Osada Y, Takai M, Erata T, Kawano S (2004) Adv Funct Mater 14:1124
54. Nakagaito AN, Iwamoto S, Yano H (2005) Appl Phys A 80:93
55. Nogi M, Handa K, Nakagaito AN, Yano H (2005) Appl Phys Lett 87:243110/1
56. Choi Y-J, Ahn Y, Kang M-S, Jun H-K, Kim IS, Moon S-H (2004) J Chem Technol Biotechnol 79:79
57. Grunert M, Winter WT (2002) J Polym Environ 10:27
58. Evans BR, O'Neil H, Howe JY, Grennbaum E (2005) 230th ACS National Meeting, Washington, DC, USA
59. Evans BR, O'Neil H, Hugh M, Malyvanh VP, Lee I, Woodward J (2003) Biosens Bioelectron 18:917
60. Daoud WA, Xin JH, Zhang Y-H (2005) Surf Sci 599:69
61. Zhang D, Qi L (2005) Chem Commun 21:2735
62. Katagiri Y, Kakizaki Y, Endo H, Mizuno T (2004) JP Patent 2004277904
63. Katagiri Y (2004) JP Patent 2004198981
64. Hutchens SA, Evans BR, O'Neill HM (2004) Abstracts, 56th Southeast Regional Meeting of the American Chemical Society, Research Triangle Park, NC, USA
65. Klemm D, Schumann D, Udhardt U, Marsch S (2001) Prog Polym Sci 26:1561

66. Klemm D, Udhardt U, Marsch S, Schumann D (1999) Polym News 24:373
67. Schumann D, Marsch S, Geyer U, Stein A, Heinze Th, Klemm D, Schmauder HP (1996) Jubilee Congress of the European Association for Cranio-Maxillofacial Surgery, Zurich, Switzerland
68. Schumann D, Marsch S, Geyer U, Klemm D, Schmauder HP (1997) 47. Jahreskongreß der Deutschen Gesellschaft für Mund-, Kiefer- und Gesichtschirurgie, Hamburg, Germany
69. Marsch S, Udhardt U, Schumann D (1999) VII Congress of Polish Society of Plastic, Reconstructive and Esthetic Surgery, Polanica Zdroj, Poland
70. Udhardt URE, Marsch SE, Klemm DO, Schumann DA (1999) The Third International Conference on Advanced Polymers via Macromolecular Engineering, Colonial Williamsburg, USA
71. Schumann DA, Marsch SE (1999) Congress, Surgery Room of the 21st Century, Glasgow, UK
72. Udhardt URE, Klemm DO, Marsch SE, Schumann DA (2000) 219th ACS National Meeting, San Francisco, USA
73. Marsch S, Udhardt U, Wurdinger J, Schumann D (2000) Fifth EFSM Congress of the European Federation for Societies of Microsurgery, Coimbra, Portugal
74. Schumann D, Marsch S, Udhardt U, Klemm D (2000) 11th Annual Meeting of EURAPS, Berlin, Germany
75. Svensson A, Nicklasson E, Harrah T, Panilaitis B, Kaplan DL, Brittberg M, Gatenholm P (2005) Biomaterials 26:419
76. Bäckdahl H, Helenius G, Bodin A, Nannmark U, Johansson BR, Risberg B, Gatenholm P (2006) Biomaterials 27:2141
77. White DG, Brown RM Jr (1989) Prospects for the commercialization of the biosynthesis of microbial cellulose. In: Schuerch C (ed) Cellulose and wood – chemistry and technology. Wiley, New York, p 573
78. Klemm D, Marsch S, Schumann D, Udhardt U (2001) WO Patent 2001/61026 A1
79. Romeis B (1968) Mikroskopische Technik. Oldenburg, München Wien, p 303
80. Christner C, Wyrwa R, Marsch S, Küllertz G, Thiericke R, Grabley S, Schumann D, Fischer G (1999) J Med Chem 42:3515
81. Grünberg S, Krause A (1996) PhD thesis, Friedrich-Schiller-Universität Jena
82. Schädlich J (1978) Diploma thesis, Friedrich-Schiller-Universität Jena
83. Schädlich J (1998) PhD thesis, Friedrich-Schiller-Universität Jena
84. Tilgner A, Heiner H (1981) Zentralbl Chir 106:961
85. Tilgner A (1986) Z Exp Chir Transplant Künstliche Organe 19:100
86. Schumann D, Marsch S (1995) H + G Zeitschrift Hautkrankheiten 70:391
87. Bschorer R, Koveker GB, Gehrke G, Jeschke M, Hermanutz V (1994) Int J Oral Maxillofac Surg 23:389
88. Hess F, Jerusalem C, Braun B, Grande P (1985) Microsurgery 6:155
89. Joos KM, Sandra A (1990) Microsurgery 11:268
90. Matsumoto H, Hasagawa T, Fuse K, Yamamoto M, Saigusa M (1973) Surgery 74:519
91. Samuels RM, McGeaschie JK, Prendergast FJ, Storrie EA (1989) Microsurgery 10:274
92. Bravo C, Tichú F, Misek I, Horký D, Stastyná M, Válka J, Veselýet J (1997) J Acta Vet Brno 66:183
93. Maeda M, Fukui A, Nakamura T, Inada Y, Tamai S, Haga S, Tatsumi K, Yamamoto H, Ogata S, Iwata H, Ikada Y (1998) J Reconstr Microsurg 14:491
94. Yoneyama T, Ishihara K, Nakabayashi N, Ito M, Mishima Y (1998) J Biomed Mater Res 43:15

95. Doi K, Matsuda T (1997) J Biomed Mater Res 37:573
96. Lanzetta M, Owen E (1992) Microsurgery 13:76
97. Ritter EF, Kim YB, Reischl HP, Serafin D, Rudner AM, Kiltzmann B (1997) Surgery 122:888
98. Viskardi PF, Page EA, Clark HG, Serafin D, Klitzmann B (1997) J Reconstruct Microsurgery 13:303
99. Greisler HP, Tattersall CW, Henderson SC, Cabusao EA, Garfield JD, Kim DU (1992) J Biomed Mater Res 26:1383
100. Guidoin R, Marceau D, Couture J, Rao TJ, Merhi Y, Roy PE, De la Faye D (1989) Biomaterials 10:156
101. Hirt SW, Aoki M, Demertzis S, Siclari F, Haverich A, Borst HG (1993) J Vasc Surg 17:538
102. Jonas RA, Ziemer G, Schoen FJ, Britton L, Castaneda AR (1988) J Vasc Surg 7:414
103. Sasajima T, Inaba M, Azuma N, Goh K, Koshiko S, Kubo Y, Miyamoto K, Tokita M, Komai T (1997) Artif Organs 21:287
104. Sigot-Luizard MF, Domurado D, Sigot M, Guidoin R, Gosselin C, Marois M, Girard JF, King M, Badour B (1984) J Biomed Mater Res 18:895
105. Cziperle DJ, Joyce KA, Tattersall CW, Henderson SC, Cabusao EB, Garfield JD, Kim DU, Duhamel RC, Greisler HP (1992) J Cardiovasc Surg 33:407
106. Haimovich B, Difazio L, Katz D, Zhang L, Greco RS, Dror Y, Freeman A (1997) J Appl Polym Sci 63:1393
107. Kito H, Matsuda T (1996) J Biomed Mater Res 30:321
108. Bhat VD, Klitzman B, Koger K, Truskey GA, Reichert WM (1998) J Biomater Sci Polym 9:1117
109. Meinhart J, Deutsch M, Zilla P (1997) ASAIO J 43:M515
110. Carr HM, Vohra R, Sharma H, Smyth JV, Rooney OB, Dodd PD, Walker MG (1996) Ann Vasc Surg 10:469
111. Vohra R, Thomson GJ, Carr HM, Sharma H, Walker MG (1991) Br J Surg 78:417
112. Vohra RK, Thompson GJ, Sharma H, Carr HM, Walker MG (1990) Artif Organs 14:41
113. Curti T, Pasquinelli G, Preda P, Freyrie A, Laschi R, D'Addato M (1989) Ann Vasc Surg 3:351
114. Herring MB, Compton RS, LeGrand DR, Gardner AL, Madison DL, Glover JL (1987) J Vasc Surg 6:114
115. Lindblad B, Burkel WE, Wakefield TW, Graham LM, Stanley JC (1986) Acta Chir Scand 152:653
116. Herring MB, Gardner AL, Glover JA (1978) Surgery 84:498
117. Hubbel JA, Massia SP, Desai NP, Drumheller PD (1991) Biotechnology 9:568
118. Zarge JI, Gosselin C, Huang P, Vorp DA, Severyn DA, Greisler HP (1997) J Surg Res 67:4
119. Masuda S, Doi K, Satoh S, Oka T, Matsuda T (1997) ASAIO J 43:M530
120. Schumann D, Marsch S (1998) Suppl 1, XIV Congress of the European Association for Cranio-Maxillofacial Surgery, EACMFS, Helsinki, Finland
121. Marsch S, Schumann D (1998) International Workshop for Microsurgery at the European Surgical Institute, Norderstedt, Germany
122. Schumann D, Marsch S, Udhardt-Geyer U, Wolf H (1998) Workshop, European Surgical Institute, Hamburg, Germany
123. Klemm DO, Udhardt URE (1999) Congress, Surgery Room of the 21st Century, Glasgow, UK

124. Wurdinger J, Marsch S, Udhardt U, Schumann D (2000) Fifth E.F.S.M. Congress (European Federation for Societies of Microsurgery) Coimbra, Portugal, 20:268
125. Klemm D, Udhardt U, Marsch S, Schumann D (2000) Fourth International Symposium Alternative Cellulose – Manufacturing, Forming, Properies, Rudolstadt, Germany
126. Bonath K (1998) Kleintierkrankheiten Band 3. Orthopädische Chirurgie und Traumatologie
127. Morgan PW, Binnington AG, Nmiller CW, Smith DA, Valliant A, Prescott JF (1994) Vet Surg 23:494
128. Ramsey DT, Pope ER, Wagner-Mann C, Berg JN, Swaim SG (1995) Am J Vet Res 56:941
129. Agren MS, Mertz PM, Franzen L (1997) J Am Acad Dermatol 36:53
130. Shaw M, Longhofer K (1999) Veterinary Record 19:374, 408
131. Howard RD, Stashak TS, Baxter GM (1993) Am J Vet Res 54:2150
132. Knottenbelt DC (1997) Veterinary Dermatology 1997:273
133. Goodrich LR, Moll HD, Crisman MV, Lessard P, Bigbie RB (2000) Am J Vet Res 61:326
134. Cockbill SM, Turner TD (1995) Vet Rec 136:362
135. Cockbill SM (1998) Veterinary Dermatology 1998:87
136. Swaim SF (1997) Veterinary Dermatology 1997:249
137. Turner TD (1979) Pharm J 31:222
138. Lippert H, Piatek S (1996) In: Sedlarik KM (ed) Wundheilung und Wundauflagen, 1. Int. Hartmann-Wundkongress, WVG, Stuttgart, p 50
139. Anonymous (2001) Am Assoc Equine Pract 35:107
140. Swaim SF, Wilhalf D (1985) Compend Contin Educ Pract Vet 7:146
141. Chvapil M, Chvapil TA, Owen JA (1987) J Trauma 27:278
142. Seaman S (2002) J Am Podiatr Med Assoc 92:24
143. Dlouhy U, Frank C (2002) Kleintiermedizin 2:53
144. Lee AH, Swaim SF, McGuire JA, Hughes KS (1987) J Am Vet Med Assoc 190:416
145. Madison JB, Hamir AN, Ehrlich HP, Haberman J, Topkis V, Villasin JV (1991) Am J Vet Res 52:1128
146. Madison JB, Gronwall RR (1992) Am J Vet Res 53:1575
147. Ramsey DT, Pope ER, Wagner-Mann C, Berg JN, Swaim SF (1995) Am J Vet Res 56:941
148. Wilmink JM, Stolk PW, Van Weeren PR, Barnefeld A (1999) Equine Vet J 31:53
149. Wilmink JM, Van Weeren PR, Stolk PW, Van Mil FN, Barnefeld A (1999) Equine Vet J 31:61
150. Otten V (2005) PhD thesis Universität Gießen
151. Litzke LF, Otten V, Ezer J, Schmauder HP (2005) Proc Equitana Essen, p 5
152. Schmauder, H-P (2005) Proceedings MEDTEC Germany, Session 102 (PowerPoint presentation)
153. Qin C, Li H, Xiao Q, Liu Y, Zhu J, Di Y (2006) Carbohydr Pol 63:367
154. Liu H, Du Y, Wang X, Sun L (2004) Int J Food Microbiol 95:147
155. Chen RN, Wang GM, Chen CH, Ho HO, Sheu MT (2006) Biomacromol, DOI 10.1021/bm050754b
156. Hornung M, Ludwig M, Schmauder HP, Wilke I, Ponomarev I (2005) DE-OS 103 61 898 A1
157. fzmb e.V. (2005) DE 203 21 115 U1
158. Frankenfeld K, Hornung M, Lindner B, Ludwig M, Mülverstedt A, Schmauder H-P (2000) DE Patent 100 22 751 C2

159. Voss W, Dermatest GmbH, Münster, Germany (2001) Report on the testing of moistness
160. Voss W, Dermatest GmbH, Münster, Germany (2004) Report on a Human Patch Test
161. Voss W, Dermatest GmbH, Münster, Germany (2004) Special dermatological Report on a Sensitization Test According to Magnussen (Repetitive Epicutaneous Test)
162. Lin KW, Lin HY (2004) J Food Sci 69: SNQ107
163. Zhou J, Dong M, Jiang H (2003) Shipin Gongye Keji 24:25
164. Evans BR, O'Neill HM, Greenbaum A (2005) 57th Southeast/61st Southwest Joint Regional Meeting of the American Chemical Society, Memphis, TN, USA

Adv Polym Sci (2006) 205: 97–151
DOI 10.1007/12_095
© Springer-Verlag Berlin Heidelberg 2006
Published online: 8 June 2006

Material Functionalization of Cellulose and Related Polysaccharides via Diverse Microcompositions

Yoshiyuki Nishio

Division of Forest and Biomaterials Science, Graduate School of Agriculture,
Kyoto University, 606-8502 Kyoto, Japan
ynishio@kais.kyoto-u.ac.jp

Abstract This review surveys advances in current studies of a variety of polymer compositional materials based on cellulose and related polysaccharides, the major studies being directed toward elaborate designs of environmentally conformable and/or biocompatible materials or highly functionalized material systems for many-faceted prospective applications. The scope covers graft copolymerization, miscible or compatible polymer blending and interpenetration, nanohybridization with inorganic minerals, surface modification of matrices, and mesoscopic self-assembling such as the formation of liquid crystals. In each individual case, it will be shown that cellulosic polysaccharides are microscopically incorporated and cooperatively interact with other adequate polymeric or inorganic ingredients to attain some functional objective. However, essential factors are the inherent characteristics of cellulosic polymers, such as the reactivity allowing hydroxyl substitution, hydrogen-bonding formability, complexing ability, semi-rigidity, chirality, etc. Material functionalities arousing interest include highly controllable biodegradability coupled with ease of processing for relatively common use as consumables, and further extensions for special uses, from mechanical and adsorptive performance to use in novel optical, electro-optical, and ion-conductive devices, etc.

Keywords Cellulose · Polysaccharides · Functionalization · Microcomposition · Graft copolymers · Polymer blends · Naonocomposites · Liquid crystals

Abbreviations

AFM	Atomic force microscopy
ATR-FTIR	Attenuated total-reflection Fourier-transform infrared spectroscopy
ATRP	Atom-transfer radical polymerization
BC	Bacterial cellulose
CA	Cellulose acetate
CAB	Cellulose acetate butyrate
CAP	Cellulose acetate propionate
CAPh	Cellulose acetate hydrogen phthalate
CB	Cellulose butyrate
CELL	Cellulose
CE	Cellulose ester
CL	ε-Caprolactone
CMC	Carboxymethyl cellulose
CP	Cellulose propionate
CTC	Cellulose tricarbanilate
CV	Cellulose valerate
DCA	Dichloroacetic acid
DD	Degree of deacetylation
ΔH_f	Enthalpy of fusion
DMAc-LiCl	N,N-Dimethylacetamide-lithium chloride
DMSO	Dimethyl sulfoxide
DS	Degree of substitution
DSC	Differential scanning calorimetry
EC	Ethyl cellulose
(E-CE)C	Ethyl-cyanoethyl cellulose
ETBE	Ethyl $tert$-butyl ether
H	Applied magnetic field
HAp	Hydroxyapatite
H_c	Coercive force
HECA	Hydroxyethyl cellulose acetate
HPC	Hydroxypropyl cellulose

IPN	Interpenetrating network
LA	D,L-lactic acid
LCST	Lower critical solution temperature
LDH	Layered double hydroxide
M	Magnetization
MA	Methyl acrylate
MMA	Methyl methacrylate
M_r	Remanent magnetization
MRI	Magnetic resonance imaging
MS	Molar substitution
M_s	Saturation magnetization
NCA	N-Carboxy α-amino acid anhydride
NMMO	N-Methylmorpholine N-oxide
NMP	N-Methyl-2-pyrrolidone
P	Cholesteric (or chiral nematic) pitch
PA66	Aliphatic polyamide 66
PAA	Poly(acrylic acid)
PACMO	Poly(acryloyl morpholine)
PAN	Poly(acrylonitrile)
PCL	Poly(ε-caprolactone)
P26ClPTA	Poly(2,6-dichloro-1,4-phenylene terephthalamide)
PDMAm	Poly(N,N-dimethylacrylamide)
PEG	Poly(ethylene glycol)
PEO	Poly(ethylene oxide)
PF	Paraformaldehyde
PHA	Poly(hydroxyalkanoate)
PHB	Poly(3-hydroxybutyrate)
PHBV	Poly(3-hydroxybutyrate-co-3-hydroxyvalerate)
PLA	Poly(lactic acid)
PLLA	Poly(L-lactic acid) or poly(L-lactide)
PMMA	Poly(methyl methacrylate)
PVA	Poly(vinyl alcohol)
PVAc	Poly(vinyl acetate)
PVP	Poly(N-vinyl pyrrolidone)
P(VP-co-GMA)	Poly(N-vinyl pyrrolidone-co-glycidyl methacrylate)
P(VP-co-VAc)	Poly(N-vinyl pyrrolidone-co-vinyl acetate)
PVPh	Poly(vinyl phenol)
P4VPy	Poly(4-vinylpyridine)
SnOct$_2$	Tin octoate (or tin(II) 2-ethylhexanoate)
SPM	Superparamagnetic
$T_{1\rho}^{H}$	Proton spin-lattice relaxation time in the rotating frame
t-BuOK	Potassium $tert$-butoxide
T_c	Cloud point
TEC	Triethyl citrate
TEOS	Tetraethylorthosilicate
T_g	Glass-transition temperature
THF	Tetrahydrofuran
T_m	Melting temperature
TMOS	Tetramethoxyorthosilicate
TMP	Thermomechanical wood pulp

TMS	Trimethylsilyl
TPEOCELL	6-O-Trityl-2,3-O-diPEOcellulose
VA	Vinyl alcohol
WISE	Wide line separation
λ_M	Wavelength of maximum light reflectance

1
Introduction

Progressively since the middle of the 1980s, naturally occurring polysaccharides, represented by cellulose and its relatives, have been reevaluated not only as sustainable resources but also as fascinating chemicals with various uses as materials. It is now better recognized that these are environmentally conformable substances and that they possess great potential to be developed for new industrial applications in themselves or in combination with supplementary ingredients.

Natural polysaccharides from plant, or sometimes animal, sources exhibit a range of characteristics at the molecular and supramolecular levels, in addition to their obvious availability and renewability. Their characteristics at the structural level are associated with their hydrogen-bonding ability, side-group reactivity, which can be modified covalently or by ionic bonds, enzymatic degradability, chirality, semi-rigidity, etc. Examples of the higher-order structure involve their ability to form fibrous crystalline entities, chelate complexes, lyo-gels, liquid crystals, etc., which are mostly enabled by innate molecular characteristics. On the basis of accurate understanding of these characteristics, advanced use of polysaccharides beyond their traditional role as commodities in clothing, paper, and foodstuff is desirable.

Designing of highly functionalized materials based on carbohydrate polymers via *multicomposition* is a useful approach. The dimension of the multi-composing objects can range from the gross bulk level ($> 10^{-3}$ m) to a single molecule ($< 10^{-8}$ m), but in view of the hierarchy of biological resources, the derivatization or modification of materials will be significant at several specific structure levels.

For convenience, multicomponent polymeric materials based on cellulosics may be grouped into three classes, namely: (a) combinations of wood with plastics (WPC), (b) mechanical mixtures in the form of fibers, such as cotton/polyester staple-mixed fibers and cellulosic fiber-filled polymer sheets, and (c) incorporations of cellulosics at a hyperfine structural level. The latter can be further ramified, for instance as follows:

- novel graft copolymers such as cellulose-*graft*-poly(hydroxyalkanoate)s;
- miscible or compatible blends of cellulose/synthetic polymers;

- interpenetrating cellulose/polymer networks;
- biodegradable or biocompatible polyester blends with cellulosics;
- fibrillar composites of bacterial cellulose;
- cellulose microfibril (or microcrystallite)-reinforced polymer matrices.

Concerning the classes (a) and (b), which fall into a traditional research field, available reviews [1–4] cover recent progress. Advances in class (c) have accelerated over the past two decades to form a part of the lively research fields covering the utilization and functional development of cellulose and other polysaccharides. Actually, this kind of microscopic approach is also useful for enhancing the performance of various gross composites, for example, through modification of their bulk surfaces or the adhesion of ingredient interfaces.

A more recent trend in polymer materials research is the hybridization of cellulosic polysaccharides with inorganic compounds; natural and synthetic layered clays, silica, zeolites, metal oxides, and apatites are employable as nanoscale components. In addition, if mesoscopic assemblies such as liquid-crystalline ordering are used in the construction of new compositional systems, the variety of functionalized cellulosic materials will be further expanded.

The present article reviews work over the last decade on a variety of microcompositional materials based on cellulose and related polysaccharides, especially centering around graft copolymers, miscible or compatible blends and networks, nanohybrids with inorganic minerals, and mesomorphic molecular assemblies. The general theme divides into the elaborate design of *environmentally conformable and/or biocompatible materials* and that of *multifunctionalized material systems for many-faceted prospective applications* for the use of cellulosics. One of the functionalities that will be spotlighted is temporal or spatiotemporal controllable biodegradability coupled with processing into dimensioned materials for use as consumables. Other areas will include improved mechanical and adsorptive performance and novel optical, electro-optical, magnetic, and ion-conductive functions for specialized use. Regarding the synthesis of cellulosic polymer molecules, which is to be overviewed in this issue by other researchers, the author will solely refer to molecular derivatizations or decorations in connection with the design of the functional materials mentioned above.

2
Graft Copolymers

Graft copolymerization has been a practical way to modify cellulosic molecules or to alter the surface properties of cotton fibers and woody materials for more than half a century [1, 5–7]. It is still a useful method, not only to improve some of the original properties of polysaccharides but also to introduce new functionality into copolymers products.

2.1
General Trend: Polysaccharide-*graft*-poly(hydroxyalkanoate)s

A current advanced use of the grafting technique is in the design of biodegradation-controllable graft copolymers of cellulose and related polysaccharides. In particular, the employment of aliphatic hydroxy acids or cyclic esters as monomer ingredients for the graft reaction is a viable approach, since the resulting graft chains, categorized as poly(hydroxyalkanoate)s (PHAs) in a wide sense, are well known as biodegradable and biocompatible polymers [8–11]. The difference in degradability between the carbohydrate backbone and the PHA graft chains makes it possible to regulate the overall degradation rate of the polymer material.

For selectiveness of initiation of graft reactions, the use of a tin octoate (SnOct$_2$) (or tin(II) 2-ethylhexanoate) catalyst is convenient for preparing this type of copolymers, polysaccharide-*graft*-PHAs, because the ring-opening polymerization of cyclic esters such as lactides and ε-caprolactone (CL) can be initiated efficiently by hydroxo-initiators (i.e. based on hydroxyl groups) in the presence of SnOct$_2$ [12–16], as exemplified in Scheme 1a.

The hydroxo-initiator R OH may be a polysaccharide or may be modified by a certain degree of substitution by hydroxyl groups (Scheme 1b). Following this scheme, many graft copolymers have been synthesized and characterized:

- dextran-*graft*-polylactide (**g1a**) and dextran-*graft*-poly(lactide-co-glycolide) (**g1b**) [17];
- pullulan-*graft*-poly(ε-caprolactone) (**g2a**) and pullulan-*graft*-poly(L-lactide) (**g2b**) [18];
- starch-*graft*-poly(ε-caprolactone) (**g3**) [19, 20];
- cellulose diacetate-*graft*-poly(ε-caprolactone-co-L-lactide) (**g4**) [21];
- partially deacetylated chitin-*graft*-poly(ε-caprolactone) (**g5**) [22];
- cellulose acetate-*graft*-poly(lactic acid) (**g6a**) [23], cellulose acetate-*graft*-poly(δ-valerolactone) (**g6b**), etc. [24].

Some graft products are particularly significant for medical and pharmaceutical applications. For example, the use **g1** and **g2** as a drug-delivery matrix offers the possibility of manipulating the release and biocompatibility of hydrophilic drug substances, since the water uptake and degradation of the copolymers can be regulated by varying the polysaccharide/(co)polyester composition. Poly(ε-caprolactone) (PCL)-grafted polysaccharide-type substances, e.g., **g2a** and **g3**, can also serve as compatibilizers in composites with various matrix resins, wherein the matrix of the composites is a component that is miscible or mechanically compatible with PCL. This is because this

(a) By using butanol (BuOH) as co-initiator

$$SnOct_2 + Bu-OH \rightleftharpoons OctSn-OBu + OctH$$

OctSn-OBu + ε-caprolactone (CL) $\longrightarrow$ OctSn-O—$\overset{O}{\underset{(CH_2)_5}{C}}$OBu $\overset{nCL}{\rightleftharpoons}$ PCL

(b) By using polysaccharides as hydroxo-initiator

R-OH + CL or Lactide $\overset{SnOct_2}{\longrightarrow}$ R-O-$[CO(CH_2)_5]_n$OH

R-O-$[COCH(CH_3)]_n$OH

Polysaccharide-*graft*-aliphatic polyester

$\left(\begin{array}{l} \text{R-OH = pullulan;} \\ \text{cellulose acetate;} \\ \text{starch, etc.} \end{array} \right)$

Scheme 1 **a** Ring-opening polymerization of cyclic esters in the presence of SnOct$_2$ and hydroxo-initiator, and **b** its application to the synthesis of polysaccharide-*graft*-aliphatic polyesters (or poly(hydroxyalkanoate)s)

polyester has the rare property of being compatible with a wide variety of polymers [19].

For the control of graft modification, the water-mediated preparation of **g3** [20] and **g5** [22] is a useful approach to obtain target products. The total yield of the graft copolymers or the density of polyester grafts onto the trunk backbone can be controlled by changing the content of water, which acts as a swelling agent or OH-protector for the polysaccharide substrate, leading instead to a large amount of homo-polyesters that can be produced by the ring-opening initiation of the input monomer with the hydroxyl groups of the water during the graft reaction. In the synthesis of the chitin-based graft copolymer **g5**, it is interesting that selective introduction of PCL to the deacetylated C-2 position, i.e., onto the amino group, was achieved using SnOct$_2$ in the presence of water [22], although the reaction mechanism still appears to be ambiguous.

As a matter of course, grafting using stannous octoate is possible for celluloses. In contrast to the case for unmodified cellulose, however, the relatively good solubility of cellulose derivatives in some organic solvents, which varies according to the degree of substitution (DS), makes this reaction possible in an appropriate homogeneous system.

Cellulose acetate (CA) is an important cellulose derivative, widespread as an industrially established product [25]. This cellulose ester of DS ≤ 2.5 has been reported to be degraded by microorganisms [26, 27], and more recently, it has been demonstrated that even conventional CA goods such as cigarette filters are decomposable in mature compost under aerobic conditions as well as in sludge under anaerobic conditions [28]. Therefore, further use of CA beyond present applications may be expected, e.g., in agroindustrial, sanitary, and bio-related fields. However, this cellulosic material shows a high glass-transition temperature (T_g), which results in limited thermal processibility. In practice, low-molecular-weight plasticizers are usually employed for thermal molding of CA. A problem with this convenient plasticization method is bleeding out of the plasticizers with time. The synthesis of thermoplastic derivatives of CA by graft reactions with a suitable substituent, exemplified by **g4** and **g6** in the above list, may be a way to address this issue in an environmentally friendly way. Such copolymerization of aliphatic esters onto rigid cellulosic polymers can also be significant when creating widely controllable properties for PHA family, some of which exhibit undesirable mechanical performance in their solid materials.

2.2
Highly Functionalized Cellulosic Graft Copolymers

Synthesis of wide ranges of compositions of CA-*graft*-poly(lactic acid)s (**g6a**, CA-*g*-PLAs) has been accomplished by Teramoto et al. [23] by combining three graft polymerization methods: (1) co-polycondensation of lactic acid in diphenyl ether, (2) ring-opening copolymerization of L-lactide in dimethyl sulfoxide (DMSO), and (3) ring-opening copolymerization similar to the second method, but in bulk without DMSO, each initiated at the residual hydroxyl positions of CA (acetyl DS = 2.15) with the aid of SnOct$_2$. In their subsequent work [24], additional series of CA-*graft*-PHAs (CA-*g*-PHAs) were synthesized by ring-opening copolymerization of (R,S) – β-butyrolactone, δ-valerolactone, and ε-caprolactone onto CA, by virtue of a suitable catalyst, solvent, and procedure for each individual case. To achieve a diversity of molecular architectures of the respective graft series, the acetyl DS of the starting CA material was also varied, resulting in different levels of intramolecular density of grafts. CA-*g*-PHAs were thus prepared over a wide range of molar substitution (MS) values to cover almost the whole range of CA/PHA weight ratios, where MS is defined as the average number of introduced oxyalkanoyl units per anhydroglucose residue.

The CA-*g*-PLA products (acetyl DS = 2.15) exhibit a composition-dependent, single T_g in the respective differential scanning calorimetry (DSC) thermograms [23]. As shown in Fig. 1, the T_g decreases from 202 °C of the original CA to $\sim$ 50 °C with increasing PLA content up to $w_{PLA} \approx 0.79$ (MS $\approx$ 14) and converges a little under the T_g (62 °C) of the PLA homopolymer with a further increase of the lactyl content. The T_g–composition data obeys a simple rule of mixing of both ingredients, CA and PLA, if the molecular-weight dependence of the side-chain-component T_g is taken into consideration [24].

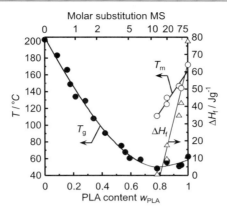

Fig. 1 Composition dependence of the thermal transition parameters, T_g, T_m, and ΔH_f, estimated by DSC for CA-*g*-PLA samples (acetyl DS = 2.15). Data for T_m (melting temperature) and ΔH_f (enthalpy of fusion) were obtained in the first run, while T_g was evaluated in the second. (Quoted from [23] with an adequate modification)

Grafting a larger amount of lactyl units (MS $\geq$ 14) on the CA renders the PLA side-chains of the copolymers crystallizable.

A similar evaluation was made for the other series of CA-*g*-PHAs to establish a general relationship between their molecular architecture and thermal transition behavior [24]. Of particular interest is the finding that the composition dependence of the T_g of the cellulosic graft copolymers was represented well in terms of a formulation based on a comb-like polymer model [29], when CAs of acetyl DS $\approx$ 2 were employed as a trunk polymer.

The cellulose derivative–aliphatic polyester grafts, CA-*g*-PHAs, may be a promising group of biodegradable materials. The degradation rate would be widely variable not only by altering the copolymer composition but also by controlling the phase structure involving crystalline morphology. Figure 2 illustrates a result of enzymatic hydrolysis experiments [30] conducted with proteinase K for two selected compositions (MS = 4.7 and 22) of CA-*graft*-poly(L-lactide) (CA-*g*-PLLA; equivalent to CA-*g*-PLA copolymerized with L-lactide, covering the compositions of MS > 3 in Fig. 1). The variations of weight loss of the film specimens exposed to the enzyme in a Tris-HCl buffer are plotted as a function of elapsed time, together with the corresponding data for a PLLA homopolymer sample. The copolymer films employed were, in advance, subjected to different thermal treatments, viz., they were solely quenched from the molten state (A) or, further, annealed at temperatures below (B) or above (C) their T_g. The notations A, B, and C distinguish the three treatments for the respective compositions. The heat treatment B gives rise to so-called physical aging of the two samples of MS = 4.9 and 22, both maintaining a totally vitrified state [31]. Treatment C allows the copolymer film of MS = 22 to produce a crystalline phase, while the other of MS = 4.9 is

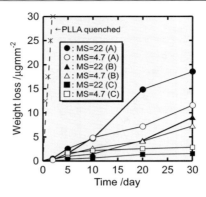

Fig. 2 Weight loss for PLLA and CA-g-PLLA films exposed to proteinase K in a Tris-HCl buffer (37 °C, pH 8.6), plotted as a function of elapsing time. See text for a description of the notations used. (Reproduced from [30])

still completely amorphous and only undergoes plastic flow during treatment. It can be seen from Fig. 2 that the hydrolysis rate is decreased by the graft modification itself and by selecting the lower PLLA content unless the PLLA crystallinity develops. It is also demonstrated that the temporal control of the enzymatic degradation of CA-g-PLLA is possible through supramolecular rearrangement such as a tighter packing of PLLA grafts during the isothermal treatment, attended by the physical aging or crystallization phenomenon.

Regarding the spatial aspects of the enzymatic degradation of CA-g-PLLA, a surface characterization [30] was carried out for melt-molded films by atomic force microscopy (AFM) and attenuated total-reflection Fourier-transform infrared spectroscopy (ATR-FTIR) before and after the hydrolysis test with proteinase K. As exemplified in Fig. 3 for a copolymer of MS = 22, the AFM study showed that hydrolysis for a few weeks caused a transformation of the original smooth surface of the test specimen (Fig. 3a) into a more undulated surface with a number of protuberances of 50–300 nm in height and less than a few micrometers in width (Fig. 3b). The ATR-FTIR measurements proved a selective release of lactyl units in the surface region of the hydrolyzed films, and the absorption intensity data monitored as a function of time was explicable in accordance with the AFM result.

The degraded film specimens often impart an iridescent color. This effect may be attributed to the interference of visible light due to diffused reflection between the protuberances formed on the surface of the films, which is interpretable in terms of a thin-layered gradation of the refractive index near the film surface.

Complementary observations of the aftereffects of enzymatic hydrolysis on the graft copolymers show *spatiotemporally controlled degradation*, leading to a new method for fine surface abrasion as well as for degradation-rate regulation of polymeric materials.

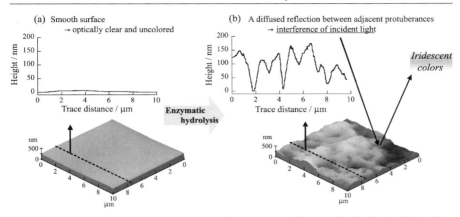

Fig. 3 AFM images of a CA-*g*-PLLA film of MS = 22 (melt-quenched), obtained **a** before and **b** after the enzymatic hydrolysis for 30 days. The *line profiles* of the sections were taken along the *dotted lines* in the respective AFM images. (Rearranged by using data from [30])

2.3
Other Examples of Tailor-Made Graft Copolymers

Ohya et al. [32] proposed a trimethylsilyl (TMS) protection method to achieve graft polymerization of cyclic esters on polysaccharide. Through this protection, the polysaccharide becomes soluble in organic solvents and the number of grafted chains should also be controllable. As shown in Scheme 2, for example, poly(lactic acid) (PLA)-grafted amylose was obtained via homogeneous graft polymerization of L-lactide on mostly trimethylsilylated amylose (TMSAm) in tetrahydrofuran (THF) by using potassium *tert*-butoxide (*t*-BuOK) as an initiator and subsequent removal of the TMS groups with methanol. The graft product showed a higher hydrolytic degradability compared with PLA alone due to reduced crystallinity, and the possibility of showing blood compatibility was suggested for biomedical and pharmaceutical applications. Biodegradable pullulan-*graft*-PLA was also synthesized by the relevant method using TMS-protected pullulan [33].

Chitin derivatives having monodisperse poly(2-alkyl-2-oxazoline) side-chains were synthesized by the reaction of amino groups of about 50% deacetylated chitin with living poly(2-alkyl-2-oxazoline) (alkyl: methyl or ethyl) (Scheme 3) and examined for miscibility with commodity polymers such as poly(vinyl alcohol) (PVA) [34, 35]. Soil burial degradation behavior of the blend system of PVA/chitin-*graft*-poly(2-alkyl-2-oxazoline) was also investigated with aqueous solution-cast films [36]. The burial test indicated that the blending accelerated a process of enzymatic biodegradation of PVA, possibly due to an increase in segmental mobility and a remarkable reduction in

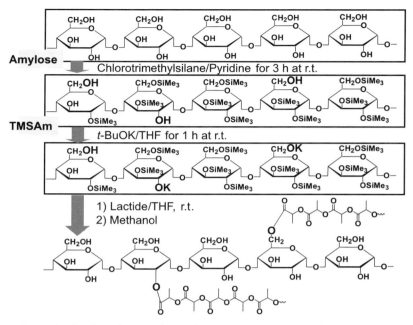

Scheme 2 Synthesis of amylose-*graft*-PLA by a TMS protection method [32]

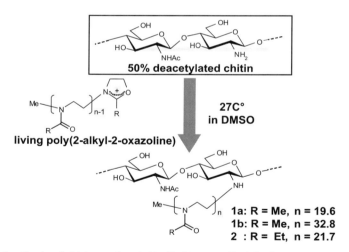

Scheme 3 Synthesis of chitin-*graft*-poly(2-alkyl-2-oxazoline) [35]

crystallinity of the PVA component, originating from the partial miscibility with the chitin graft copolymers.

A graft copolymer of chitosan (degree of deacetylation, DD = 88%) was prepared by grafting D,L-lactic acid (LA) onto the primary amino groups in the polysaccharide without any specific catalyst, according to the route shown in Scheme 4 [37]. First, chitosan is dissolved in an aqueous LA solution to give a homogeneous viscous polyelectrolyte solution. By heating the solution, dehydration of the chitosan amino lactate salt will occur to form amide linkages between the chitosan and LA, along with polycondensation of LA, which then acts as its own catalyst. In the final treatment, methanol is an efficient extractant to remove unreacted monomers, unbound oligomers, and salt-linked LA side-chains. The graft copolymers thus obtained form pH-sensitive, physically crosslinked gels in aqueous solutions. The hydrogels are of general interest for biomedical applications, possibly useful as artificial actuators or switches in biochemical separation systems, and in controlled-release systems for drug delivery.

Heinze et al. found that DMSO in combination with tetrabutylammonium fluoride trihydrate dissolved cellulose (degree of polymerization ≤ 650) within 15 min at room temperature [38]. They also demonstrated that homogeneous esterification of cellulose is possible in this solvent system. The applicability of this new solvent system to cellulose grafting has recently been proved by adoption of cyclic compounds such as lactones and N-carboxy α-amino acid anhydrides (NCAs) [39]. ε-Caprolactone was facilely graft-polymerized on cellulose at a graft rate of $\sim 65\%$ (per trunk weight of 100), and NCAs at over 100%, in the respective homogeneous reaction systems at $< 60\,^{\circ}\mathrm{C}$.

Worthy of special remark is surface grafting via a novel technique of polymer synthesis, atom-transfer radical polymerization (ATRP), the usefulness of which has received much recognition due to the control of the molecular

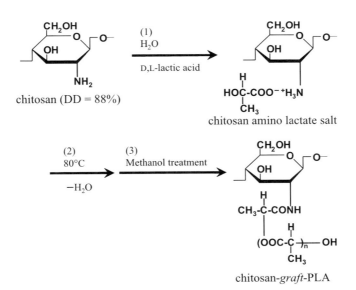

Scheme 4 Preparation of a hydrogel-forming chitosan-*graft*-PLA [37]

weight and the distribution of the polymerized product. Tsujii et al. [40] applied ATRP to the preparation of *polymer brushes* tethered to silicon wafers with great success. This technique should also be widely applicable to surface modifications of polysaccharide materials [41].

3
Polymer Blends and Microcomposites

As is well established, polymer/polymer blending is an important method to improve the original physical properties of one or both of the components, or to obtain new polymeric materials showing widely variable properties without parallel in homopolymers. There have been numerous blend studies for various polymer pairs from both the fundamental and practical viewpoints. A few reviews [7, 42] have described a general scheme for preparation and characterization of the blends and microcomposites of unmodified cellulose with synthetic polymers, mainly based on works performed until 1994. The present review will cover the articles published on this topic since the mid 1990s, with extensions to related works on cellulose derivatives and other natural polysaccharides.

3.1
Cellulose-Incorporated Systems

3.1.1
Blends with Synthetic Polymers

(a) Fundamental Aspects Earlier studies [42] demonstrated that highly compatible blends were obtainable for some pairs of cellulose/synthetic polymers by using a suitable, nonaqueous solvent and a procedure for each case, despite the generally believed intractability of the natural fibrous polysaccharide, which shows poor solubility in most organic solvents. It was also stressed that strong intermolecular interaction, superior or comparable to the self-associating nature of the respective components, was indispensable for intimate mixing. Table 1 summarizes the results of the estimation of the scale of mixing for film samples of representative blends of cellulose with synthetic polymers [43–48], quoted in tabular from the author's previous review (Sect. 5.3.2 in [42]). The binary compositions situated in an enclosed area in the table, denoted by the dotted lines, are assumed to have reached a compatible state of mixing.

Later reports [49–51] provided further insight into the blend morphology for several of the cellulose/synthetic polymer blend systems listed in Table 1. In particular, refining efforts were devoted to precise quantification of the scale of homogeneity in the blends. A dielectric relaxation

Table 1 Scale of homogeneous mixing estimated for representative blends of cellulose/synthetic polymer pairs, quoted from a previous review [42]

Composition	Scale of homogenous mixing */nm						
	CELL/P4VPy[a]	MC/P4VPy[b]	MC/PVP[b]	CELL/PVA[c]	CELL/PAN[c]	CELL/PCL[c]	CELL/Ny6[c]
CELL < 30 wt %	<2.5	2.5–15	ca. 2.7	15–39	15–35	ca. 27	>31
CELL ≈ 50 wt %				<15–36	<15–31	ca. 27	
CELL> 70 wt %		ca. 2.5	2.7–15	1–15	4–15	>28	

* The estimation was made by putting different data together, which included thermal transition data obtained by dynamic mechanical analysis and spectroscopic data quantified mainly by solid-state NMR techniques [43–48]. The binary compositions situated in an enclosed area, denoted by connection of *dotted lines* in the scale list, can be assumed to be a highly compatible state of mixing.

[a] Obtained by hydrolysis treatment of MC/P4VPy

[b] Films cast from solutions in dimethyl sulfoxide-paraformaldehyde (DMSO-PF)

[c] Films coagulated from solutions in *N,N*-dimethylacetamide-lithium chloride (DMAc-LiCl)

Notations: CELL, cellulose; P4VPy, poly(4-vinyl pyridine); MC, methylolated cellulose; PVP, poly(*N*-vinyl pyrrolidone); PVA, poly(vinyl alcohol); PAN, poly(acrylonitrile); PCL, poly(ε-caprolactone); Ny6, nylon 6.

study combined with small-angle X-ray scattering measurements was carried out for cellulose/poly(vinyl alcohol) (CELL/PVA) blends prepared from solutions in N-methyl-2-pyrrolidone-lithium chloride (NMP-LiCl) by a coagulation method [49]. As a result of the structure and relaxation analyses, it was concluded that the blends (CELL $\geq$ 60 wt %) displayed no residual crystallinity and were almost characterized by a molecular mixing of the components, but with microheterogeneities on a nanometer scale (1–2.3 nm). Heterogeneities in a similar range ($\sim$ 3 nm) were estimated for the CELL/PVA blends by 2D ^{1}H – ^{13}C heteronuclear wide line separation (WISE) NMR spectroscopy through observation of the organization of water adsorbed into the polymer matrix [50]. The study revealed that the water molecules were predominantly associated with the cellulose backbone and no water was detectable in the immediate vicinity of PVA segments. Another rigorous work [51] was concerned with spin-diffusion measurements in solid-state proton NMR spectroscopy, for evaluation of the mixing state in two representative cellulosic blends: cellulose/poly(4-vinylpyridine) (CELL/P4VPy) and cellulose/poly(acrylonitrile) (CELL/PAN), prepared from dimethyl sulfoxide-paraformaldehyde (DMSO-PF) and N,N-dimethylacetamide-lithium chloride (DMAc-LiCl) solutions, respectively. It was pointed out that the polymer pairs were essentially immiscible thermodynamically, in spite of the attainability of a very intimate level of mixing in those blends. The kinetics of solidification during processing into the binary blends are likely to be rather important in defining the resulting phase morphology, which is therefore influenced subtly by the external preparation conditions. Actually, the diffusion study provided an average domain size of $\sim$ 10 nm for CELL/P4VPy and $\sim$ 20 nm for CELL/PAN, the former data being noticeably greater than the previous assessment (Table 1).

The importance of attractive intermolecular interactions was emphasized again for two solution-coagulated blend systems, cellulose/poly(ethylene glycol) (CELL/PEG) [52] and CELL/PVA [53], in interpretations of the phase-transition behavior and of the gas sorption and transport phenomena, respectively. An interesting observation for the former system was that, when a blend containing 95 wt % PEG was heated to temperatures above the melting point of PEG ($\sim$ 60 °C), the crystalline PEG component did not melt into the usual flowing liquid but changed to an amorphous solid state. It may be assumed that a kind of organogel of cellulose was realized at these high temperatures, owing to the motional restraint of the PEG molecules, which were probably tied to the semi-rigid cellulose chains via hydrogen bonds between the constituents. In the CELL/PVA study on the membrane properties, the solubility, diffusion, and permeability coefficients for carbon dioxide in the blends were all lower than expected from the simple semilogarithmic additivity rule. This is analogous to the gas sorption and transport behavior of other miscible polymer systems in which strong intermolecular interactions are present. Prior to these two works, there was an infrared spectroscopic study by Kondo et al. [54] for similar blend systems with PEG and PVA. This characterized the formation of hydrogen bonds in the cellulose/synthetic polymer blends in terms of the specific regiochemistry of hydroxyl groups and glucose-ring oxygen in the cellulose component.

(b) Practical Aspects There were also some blend works that were useful in practice. Lyocell fibers, produced industrially by dry-jet wet spinning of cellulose solution in N-methylmorpholine N-oxide (NMMO) hydrate, show better drape and mechanical properties in the wet state than other cellulose fibers, but they are prone to undergo unfavorable fibrillation under wet abrasion. For the improvement, a blend effect of an aromatic polyamide, poly(2,6-dichloro-1,4-phenylene terephthalamide) (P26ClPTA), as additive to cellulose filaments was investigated [55]. Homogeneous solutions containing cellulose and P26ClPTA could be prepared in the common solvent NMMO monohydrate and spun to blend filaments. The wet abrasion resistance was found to assume a maximum at a P26ClPTA content of 10 wt %. A different modification of Lyocell fibers was made by spinning a solution of the mixture of cellulose with a copolymer, cellulose-*graft*-PVA, in NMMO monohydrate [56]. The copolymers of various vinyl alcohol (VA) contents were obtained by saponification of the corresponding precursor copolymers in which vinyl acetate was grafted onto the cellulose backbone with cerium ammonium nitrate. It was reported that addition of the copolymer was decisively effective in reducing the fibrillation of Lyocell fibers. Prior to these works relating to Lyocells, blend filaments of cellulose/aliphatic polyamide 66 (CELL/PA66) precipitated from solutions in a mixed solvent NMMO–phenol (80 : 20 w/w) were examined for estimations of the phase construction [57] and mechanical characteristics [58]. The two components were mainly immiscible, but PA66 reinforced the CELL-rich fibers against breaking at temperatures higher than T_g of PA66 due to a strong adhesion at the interface between the constituent domains in their respective amorphous state. This can also offer a practical applicability, since cellulose becomes brittle at high temperatures as its water content decreases.

A limited number of papers has dealt with the degradation of cellulose/synthetic polymer blends. CELL/PVA gelatinous fibers obtained from DMAc-LiCl solutions were treated with cellulase (*Trichoderma viride*) [59]. Interestingly, a blend fiber (CELL = 40 wt %) swollen in an aqueous buffer remained almost unchanged in diameter size during the enzymatic elimination of the cellulose component, while the degree of optical anisotropy diminished. Enzymatic degradation of cellulose/poly(L-lactic acid) (CELL/PLLA) blends was investigated with either proteinase K or cellulase for films cast from solutions in trifluoroacetic acid [60]. In this series, the cellulose component should be trifluoroacetylated. Although the blend system was incompatible, the weight loss of PLLA-rich samples when treated with proteinase K was more prominent compared with a control PLLA due to depression of the crystallinity of the polyester component. In contrast, cellulase did not cause degradation of the cellulose backbone in the blends, but the trifluoroacetyl side-group was completely hydrolyzed during the treatment. Regarding the thermal degradation of cellulose blends, Nishioka et al. produced a series of works using thermogravimetry [61, 62]. For several cellulose/synthetic polymer systems, to which a copolymer of cellulose grafted with the blending partner was added on demand as a compatibilizer, the composition dependence of the thermal decomposition profiles was discussed in relation to the extent of blend compatibility.

In contrast to the usual blending of cellulose via a solution medium, a fine-powdered composition was obtained by ball-milling a dry mixture of fibrous cellulose and granulated PEG without any solvent [63]. A certain extent of compatibilization of PEG with cellulose appears to occur even in their solid state, probably by insertion of parts of PEG molecules into the cellulose domains, followed by the formation of hydrogen bonds between cellulose hydroxyls and polyether oxygens. It was shown that such a mechano-chemically prepared powder was moldable three-dimensionally by facile hot-press treatment, e.g., at 120 °C for a composition of CELL/PEG = 80/20 (wt/wt) [64].

3.1.2
Blends with Natural Polymers

Zhang et al. [65–67] carried out a series of works on cellulose blends with other polysaccharides such as alginic acid, konjac glucomannan, and chitin, and with natural proteins such as silk fibroin and casein. Their studies were mainly concerned with the preparation and morphological characterization of microporous membranes applicable for selective separation of low-molecular-weight substances. In the preparation of the blends, a classical metal–complex solvent, aqueous cuprammonium hydroxide (cuoxam), or an inorganic/organic solvent mixture, NaOH aqueous solution with addition of urea or thiourea [68], was utilized for dissolution and mixing of cellulose and other natural polymer components.

As an example using NMMO solution, the spinning of fibers consisting of blends of cellulose with cationic starch was investigated for the purpose of improving moisture absorbability, dyeability, and enzymatic degradability compared with standard Lyocell fibers [69].

3.1.3
IPNs and Microfibril-Reinforced Composites

In the preparation of the cellulose blends listed above, two methods were mainly adopted: casting from solutions by evaporation of solvent, and coagulation of solutions in a non-solvent followed by drying. Both processes involve dissolution and mixing of two polymer components in a common solvent. As an alternative route leading to unique microcomposites of cellulose, Nishio and Miyashita et al. developed the chemical blending technique of *solution coagulation/bulk polymerization* [42, 70, 71]. In situ polymerization of a vinyl monomer as coagulant and/or impregnant used to form cellulose gels is an essential part of the method, enabling us to obtain an interpenetrating network (IPN) consisting of cellulose/synthetic polymer. This method can also be a means of reinforcing thermoplastic polymers by cellulose. In studies published since 1994, the effectiveness may be exemplified well by a result of dynamic mechanical measurements carried out for cellulose/poly(acryloyl morpholine) (CELL/PACMO) composites, optically clear, hard films, which were synthesized by photoinitiated polymerization in a gel state of regenerated cellulose swollen with reactive ACMO monomer [72]. As demonstrated in Fig. 4 for two CELL/PACMO compositions, the extent of lowering of their modulus E' in the glass-transition region became extremely suppressed in contrast to the sharp drop of E' observed for a reference sample of PACMO homopolymer. This suppression reflects a consequence of successful perpetuation of an original network structure of cellulose gels into the polymerized bulks.

A novel attempt was made to realize different IPN architectures in a composite system of cellulose/poly(*N*-vinyl pyrrolidone-co-glycidyl methacrylate) (CELL/P(VP-co-GMA)) [73]. The copolymer-rich compositions (4–19 wt % CELL) were synthesized successfully via photopolymerization in the gel state of cellulose impregnated with a monomer mixture of VP and GMA, resulting in the development of a semi-IPN-like organization (Fig. 5a) within the copolymerized bulks. An additional treatment of the as-polymerized composites ([O]-series) with ∼ 50% formic acid or ∼ 3% sodium hydrox-

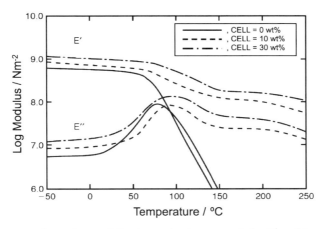

Fig. 4 Temperature dependence of the dynamic storage modulus E' and loss modulus E'' for CELL/PACMO composites. (Reproduced from [72])

ide aqueous solution gave rise to a drastic elevation of the T_g of the sample used, as demonstrated by dynamic mechanical data in Fig. 6. The T_g shift can be attributed to the formation of a tighter network structure (full IPN, Fig. 5b), occasioned by a certain crosslinking reaction in the copolymer constituent: viz., in the formic acid treatment, a ring-opening reaction of the epoxide group of GMA with the carboxylic acid was involved in the bridge oc-

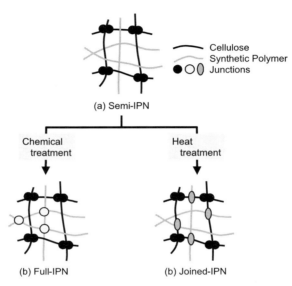

Fig. 5 Schematic illustrations of different IPN architectures attainable for CELL/P(VP-co-GMA) composites. **a** Semi-IPN; **b** Full-IPN; **c** Joined-IPN

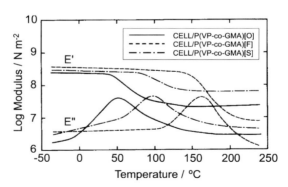

Fig. 6 Dynamic mechanical behavior for different film species of a cellulose composite synthesized with a monomer mixture of VP/GMA = 3/7, representing an effect of the chemical treatment of an original sample (CELL/P(VP-co-GMA)[O]) with aqueous formic acid [F] or sodium hydroxide [S] solution. CELL content = 4.5 wt %. (Reproduced from [73])

currence between copolymer chains, while the sodium hydroxide treatment transformed the copolymer component into a kind of ionomer as a result of hydrolysis of the amide linkage in the pyrrolidone rings. A similar T_g elevation was noted when cellulose composites with a GMA-rich copolymer were subjected to a prolonged heat treatment at temperatures above $\sim 170\,^{\circ}$C. This may be ascribed to the development of a joined IPN organization (Fig. 5c) containing some amount of intercomponent crosslinks, in consequence of the etherification possible between GMA epoxides and cellulose hydroxyls.

In a way partly analogous to the above method, if a fibrillar network of bacterial cellulose is interpenetrated with various monomers or reactive resins that are subsequently solidified, the resulting composites may be expected to exhibit a significant reinforcement of the substrate polymers in mechanical and/or thermal performance. This has been suggested recently by Yano et al. [74].

An additional example is the preparation of gelatinous IPNs of poly(N,N-dimethylacrylamide) (PDMAm) containing 6–25 wt % cellulose, which were obtained in DMAc-LiCl solution as the homogeneous reaction medium for polymerizing DMAm monomer [75]. It was reported that an IPN of CELL = 25 wt % possessed a sixfold higher modulus than a PDMAm control, when compared to their water-swollen state.

Another approach to the design and fabrication of advanced microcomposites based on cellulose is that conducted by the group of Dufresne, Cavaillé, and co-workers. They accomplished a sequence of studies on the processing and mechanical characterization of polymer nanocomposites reinforced with cellulose microcrystallites, or whiskers, which were extracted from a mantle of tunicate (a sea animal) [76–81]. The mantle (tunicine) is made up of cellulose microfibrils that are particularly well organized and crystallized, which can lead, after acid hydrolysis, to discrete microcrystallites with a high aspect ratio of around 70 and an average diameter of 10–20 nm. The preparation of composite materials was usually done by casting into a film form from aqueous suspension mixtures of the cellulose microcrystallites (as a filler) with a polymer latex (as a matrix). As illustrated in Fig. 7 for a thermoplastic styrene-butyl acrylate (35 : 65) copolymer sample ($T_g =$ 0 $^{\circ}$C) [77], the mechanical property (shear modulus G) was enhanced by more than two orders of magnitudes in the rubbery state at $T > T_g$ of the copolymer matrix, even when the cellulosic filler was loaded at quite a low concentration of 6 wt %. This striking reinforcing effect was ascribed to the formation of a percolating cellulose network, possibly resulting from strong interactions between adjacent whiskers such as hydrogen bonds. Subsequent reports described similar reinforcement phenomena of cellulose microfibrils for several nanocomposites obtained using different polymer matrices which included bacterial polyesters [78], plasticized starch [79], poly(ethylene oxide) (PEO) [80], and lithium imide-containing PEO electrolytes [81]. The observations of high mechanical performance were commonly interpreted on the

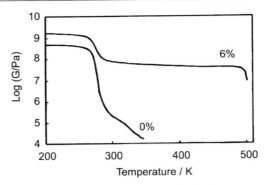

Fig. 7 Logarithmic plot of the shear modulus G (Pa) for a composite specimen of poly(styrene-co-butyl acrylate) reinforced by 6 wt % cellulose whiskers, as a function of temperature. This specimen is compared with a control sample that does not contain any whisker. (Reproduced from [77])

basis of the percolation mechanism governing the evolution of a rigid network of cellulose whiskers as the nanosized filler.

3.2
Other Systems Based on Natural Polysaccharides

Coupled with the extensive studies of cellulose-based polymer blends described in Subsects. 3.1.1 and 3.1.2, similar interest and attention have been directed to other carbohydrate polymer blends, for evaluations of compatibility, synergistic properties, and possible functionality. In particular, blend works on a cellulose-relative poly(β-1,4-glucan), chitin, have been facilitated [82–91] and the number of examples is still increasing. Among them, chitinous blends with water-soluble or hydrophilic synthetic polymers [82–88], and those with biodegradable aliphatic polyesters [89] may be of importance from the standpoint of application as biomedical materials. In some studies [82, 87, 90, 91], the blending component was a polymer having an amide linkage in the side or main chain. This selection was made based on the use of formic acid as a common solvent, and/or the possibility of observing specific intermolecular interactions in which the amide group rather than the hydroxyl participates closely. A particular interest was often focused on the effect of the degree of deacetylation of chitin (or acetylation of chitosan) on the blend compatibility [83, 85, 90], taking into account alternative structures in the C-2 side group of the amino polysaccharide.

There have also been some prominent studies of IPNs based on chitin [92, 93], alginate [94], etc. For example, the microscopic incorporation of chitin with methacrylate polymers such as poly(glycidyl methacrylate) [92] and poly(2-hydroxyethyl methacrylate) [93] was achieved by the combination of solution coagulation and bulk polymerization processes (Sect. 3.1.3). From an estimate by solid-state ^{13}C NMR spectroscopy, the preparation method proved to be a potential route leading to intimate mixtures of chitin/synthetic polymers on the scale of a few nanometers, even though the relevant polymer pair is thermodynamically immiscible.

By analogy with the works which dealt with cellulose microcrystal-reinforced nanocomposite materials, microcrystals of starch [95] or chitin [96, 97] were used as a reinforcing phase in a polymer matrix. Poly(styrene-co-butyl acrylate) [95, 96], poly(ε-caprolactone) [96], and natural rubber [97] were reinforced, and again the formation of aggregates or clustering of the fillers within the matrices was considered to account for the improvement in the mechanical properties and thermal stability of the respective composites processed from suspensions in water or suitable organic solvents.

3.3
Cellulose Ester Blends

Among industrially established cellulosic products, organic ester derivatives of cellulose form a valuable family, finding general acceptance in various application fields such as packaging, coating, release-controllable excipients, molded plastics, optical films, fibers, membranes, and other separation media [25, 98]. Properties of such cellulose esters (CEs) can be varied over a range by controlling the type and number of ester substituents in the chemical manufacturing process. Since Buchanan et al. reported that many CEs, e.g., acetate, propionate, and their mixed ester, are in fact biodegradable under a variety of conditions [26, 98–100], the status of the ester family appears to have been heightened from both commercial and academic aspects.

It is further attractive that polymer blending may offer opportunities not only to improve the processability and modify the physical properties of CEs, but also to alter the thermal instability and/or mechanical brittleness of the second component polymers, e.g., many aliphatic polyesters including bacterial poly(hydroxyalkanoate)s. Recent developments in the area of cellulose ester/polymer blends are reviewed below.

3.3.1
With Vinyl and Other Conventional Polymers

Many of the applications in which CEs are involved require clarity or uniform properties on a microscopic structural level, so miscibility or molecular-scale mixing is important when CE-based blends are formed. There are some examples of miscible or homogeneous CE blends with vinyl polymers. One interesting class of these is a mixture of CE and poly(vinyl phenol) (PVPh) [101, 102]. This phenolic polymer is known to form a miscible blend with a wide variety of polymers with potential hydrogen-bond-accepting groups [101], the latter set including commercially available CEs, i.e., cellulose acetate (CA), cellulose propionate (CP), cellulose acetate propionate (CAP), and cellulose acetate butyrate (CAB). For miscible CA/PVPh and CAB/PVPh blends, Kelley et al. have clarified the hydrogen-bonding interactions occurring between the ester linkage on the CEs and the phenolic hydroxyl on the PVPh component by means of ^{13}C cross-polarization magic-angle-spinning (CP/MAS) and ^1H combined rotation and multiple pulse spectroscopy (CRAMPS) NMR spectroscopy [102]. His group has also prepared polymer networks from a series of miscible CE/PVPh blends containing a latent formaldehyde source [103]; PVPh has potential for secondary chemical reactions with the crosslinking agent, leading to chain extension. This attempt was made to improve the strength and thermal stability of the whole system and to expand the useful properties of the renewable cellulosic polymers.

A particularly important ester derivative CA is utilized for many industrial applications due to its desirable physical properties such as good optical clarity in film form, and a comparatively high modulus and adequate flexural and tensile strengths in fiber form. Additionally, it is practically biodegradable, to an extent that depends on the degree of acetyl substitution. For this cellulose derivative, therefore, new functionalization may be expected in medical, pharmaceutical, and other bio-related fields, coupled with further applicable developments in a huge consumption type of commodities. However, the trouble is that CA alone cannot be molded easily by thermal processing, because the T_g and flow temperature are fairly high, where significant thermal decomposition can take place. Usually, a large amount of plasticizers such as phthalate compounds are mixed with CA for thermal molding. The use of low-molecular-weight plasticizers, however, can cause fume generation in the molding process due to their volatility or decomposition. In addition, bleeding out of plasticizers from molded CA products can be pronounced in long-term use. One proposed remedy for these drawbacks is the synthesis of thermoplastic derivatives of CA by grafting suitable substituents, as described in Sect. 2.2. Another effective method is the mixing of flexible polymers as a plasticizer that can alter the thermal properties of CA by molecular-level interactions.

Miyashita et al. carried out miscibility characterization of CA blends with poly(N-vinyl pyrrolidone) (PVP), poly(vinyl acetate) (PVAc), and poly(N-vinyl pyrrolidone-co-vinyl acetate) random copolymers [P(VP-co-VAc)s] [104]. On the basis of thermal transition data obtained by differential scanning calorimetry (DSC), a miscibility map (Fig. 8) was completed as a function of the degree of substitution (DS) of CA and the VP fraction in P(VP-co-VAc). Figure 9 compares results of the DSC measurements between two blending pairs of CA/P(VP-co-VAc) corresponding to the polymer combinations marked as **A** and **B** in Fig. 8. In the data (Fig. 9b) for the blends of CA (DS = 2.95) with P(VP-co-VAc) of VP = 51 mol %, we can readily see a sign of poor miscibility, as is evidenced from the lack of an appreciable shift in the

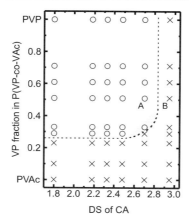

Fig. 8 Miscibility map for CA/P(VP-co-VAc) blends, as a function of DS of CA and VP fraction in P(VP-co-VAc). Symbols indicate whether a given pair of CA/P(VP-co-VAc) is miscible (○) or immiscible (×). (Quoted from [104] with additional data)

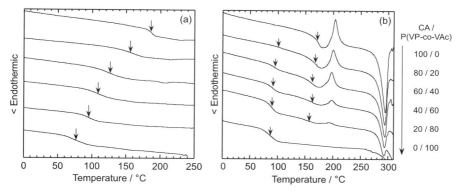

Fig. 9 DSC thermograms of CA/P(VP-co-VAc) blends: **a** data for combination A of acetyl DS = 2.70 and VP/VAc = 51/49; **b** data for combination B of acetyl DS = 2.95 and VP/VAc = 51/49 (Fig. 8)

T_g of either polymer components. In addition to the two distinguishable T_g signals, the melting endotherm and cold-crystallization exotherm of CA are both detectable for every blend at almost the same temperature positions as those for the CA (DS = 2.95) alone, with a proportional reduction of the respective peak areas. In contrast to the result, the thermograms compiled in Fig. 9a for the pair of CA (DS = 2.70) and P(VP-co-VAc) (VP/VAc = 0.51/0.49) (combination A) indicate a definitely single T_g that shifts to the higher temperature side with increasing CA content. As summarized in Fig. 8, CA/PVAc blends are immiscible irrespective of the DS of the CA component, while PVP forms a miscible monophase with CAs unless the acetyl DS exceeds a value of about 2.8. The vinyl copolymers of VP and VAc are immiscible with CA when the VP fraction in P(VP-co-VAc) is less than 25 mol% or the acetyl DS used is higher than 2.75, whereas the other combinations of DS and VP values lead to a miscible series of blends, with a possible exception such as the mixing pair of DS = 2.70 and VP = 29 mol%. The miscibility map is useful for designing improvements in properties and functionalities of cellulose acetates. For instance, the water-absorption property of the blend system would be widely changeable by altering the binary polymer composition and the ratio of hydrophilic/hydrophobic (VP/VAc) residues in the copolymer. This may be of significance, especially in applications in medical, pharmaceutical, and sanitary areas.

In relation to the intermolecular interaction possible in the above system, it was found that when CA and P(VP-co-VAc) solutions in THF were mixed with each other, the two polymer components precipitated spontaneously to form a complex-like agglomerate; the yield of the precipitate diminished with increasing DS of CA and with decreasing VP fraction in the vinyl copolymer [105]. From this observation and spectral data in Fourier-transform infrared (FT-IR) spectroscopy and ^{13}C CP/MAS NMR measurements it was proved that blend miscibility and complex formability of a CA/P(VP-co-VAc) pair depend on whether or not an effective number of hydrogen-bonding interactions occur between the residual hydroxyl groups of CA and the carbonyl groups of VP residues in the mixing system. Also, quantification of the proton spin-lattice relaxation times in the rotating frame ($T_{1\rho}^{H}$) showed that the miscible CA blends and complexes with PVP and P(VP-co-VAc) are homogeneous within limits of approximately 3 nm.

In a similar manner to that for the CA/P(VP-co-VAc) system, a miscibility map was constructed for cellulose butyrate (CB)/P(VP-co-VAc) blends by Ohno et al. [106]. Unlike the situation for the CA blends, there appeared a specific range of VP/VAc composition (VP ≈ 30–60 mol%) where CBs of butyryl DS ≈ 2.5–2.95 were miscible with the vinyl copolymers in spite of the immiscibility with both PVP and PVAc homopolymers. In view of the finding of such a *miscibility window*, the mixing state of CB/P(VP-co-VAc) blends seems to be determined by the balance of three factors: (1) the hydrogen-bonding interaction between the residual hydroxyls of the cellulose ester and the carbonyls of VP units in the copolymer; (2) steric hindrance of bulky butyryl groups, leading to suppression of the above interaction; and (3) another attractive force derived from the repulsion between monomer residues constituting the vinyl copolymer (i.e., an intramolecular copolymer effect).

From a functional point of view, Nguyen et al. examined the pervaporation characteristics of CA/P(VP-co-VAc) blends for application as alcohol-selective membrane materials [107]. The blend membranes were shown to be very efficient in the removal of ethanol from its mixture with ethyl *tert*-butyl ether (ETBE). *tert*-butyl ethers are octane-value enhancers for gasoline, and the synthesis requires an excess of alcohol in the reaction to reach high conversion degrees of iso-olefin to the corresponding ether. The excess of alcohol has to be separated from the reaction system and recycled. A problem in the conventional distillation is occurrence of an azeotrope, as in the case of ethanol–ETBE mixtures.

For gas separation membranes, for example, He, O_2, and N_2 gas transport properties of CA/poly(methyl methacrylate) (PMMA) blends have been measured [108]. This article reported that CA/PMMA blends exhibited phase separation with limited intermiscibility between the components, but they were possibly useful as membrane materials to produce high-purity helium gas streams combined with high helium recovery.

It has been shown that cellulose acetate hydrogen phthalate (CAPh) of 19–24% acetyl and 30–40% phthalate contents forms a miscible blend with PMMA as well as with PVP in the entire composition range [109–111]. The miscibility of CAPh with PVP may be due to the hydrogen-bonding attraction between the free hydroxyl group of CAPh and the carbonyl group of PVP; however, the major interaction in the CAPh/PMMA blends is possibly formed between the OH in the phthalic acid group of CAPh and the $C = O$ of PMMA, taking account of the immiscibility of the CA/PMMA blends.

As an impressive phase study, it was observed that a blend system composed of CAB having 54 mol % butyryl side groups and poly(ethylene oxide) (PEO; $M_n = 35\,000$) yielded a phase diagram with a lower critical solution temperature (LCST) boundary, and a cloud point of 168 °C at the critical composition of CAB/PEO = 40/60 (w/w) [112].

Although it is not a polymer blend but a nanocomposite, a series of biodegradable nonwoven mats comprising CA and poly(vinyl alcohol) (PVA) nanofibers have been prepared via multi-jet *electrospinning* [113]. Such electrospinning has recently attracted great attention, because polymer nanofibrils with a high surface-to-volume ratio can be fabricated by using the technique [114, 115]. Attempts to use homogeneous blend solutions as the spinning dope should be interesting in further developments of this method.

3.3.2
With Biodegradable Aliphatic Polyesters

Currently there is a fast-growing demand for biodegradable polymer materials for limited-term applications mainly in the medical, marine, and agricultural areas. Aliphatic polyesters, including bacterially produced poly(hydroxyalkanoate)s, have attracted considerable interest as materials of great promise possessing good biocompatibility and environmental degradability. Unfortunately, solids of aliphatic polyesters generally lack mechanical toughness and ductility, and they are often badly affected by heat, due to their rather low melting points (e.g. polylactones) or decomposition temperature (e.g. poly(α-hydroxy acid)s). These undesirable physical properties prevent their wider-scale use in various fields, in spite of their susceptibility to biodegradation. A means of overcoming such drawbacks could be the modification of the biodegradable polyesters by blending with other suitable polymers. A blending partner could be cellulose ester derivatives, as suggested by Scandola et al. [116–119] and Buchanan et al. [120–122] in their publications in the early 1990s.

Scandola's group reported miscibility in binary blends of poly(3-hydroxy-butyrate) (PHB) with CAB [116–118] and CAP [116]. Similarly, blends of poly(3-hydroxybutyrate-co-3-hydroxyvalerate) (PHBV) with CAB and CAP were shown to be miscible in the amorphous mixing state [119]. Following these reports, they studied biodegradation of blend films of the bacterial polyester/cellulose ester pairs in activated sludge [123–125]. Differences in the degradability observed between the test specimens were discussed mainly in connection with the phase morphology (involving intra-spherulitic structure), which varied depending on the composition and thermal history in the blend preparation.

On the other hand, Buchanan and coworkers also reported the miscibility, mechanical property, and spectroscopic characterization of the CAB/PHBV system [120]. Similar investigations were extended to CAP blends with a series of condensation polyesters consisting of a C5 dicarboxylic acid (glutaric acid) and C2 to C8 straight-chain diols [121, 122]. They showed that the polyesters prepared from C2 to C6 diols formed transparent amorphous glasses which exhibited a single composition-dependent T_g when blended with CAP (acetyl DS = 0.10, propionyl DS = 2.65), and indicated further that the polyester prepared from C4 diol, i.e., poly(tetramethylene glutarate), had the highest level of miscibility with the cellulose mixed ester. In further works by the same group [126–128], biodegradation behavior of many blends of cellulose ester (CAP or CA) with different aliphatic polyesters including the aforementioned was explored mainly in a bench-scale simulated municipal compost environment. It was observed for CAP/polyester blends that, at a fixed DS of CAP, as the content of the polyester component was increased, the rate of composting and the weight loss due to composting increased as a whole. When the CAP DS was below approximately 2.0, both components degraded; however, when the CAP was highly substituted, almost all the weight loss (albeit relatively small) was ascribed to loss of the polyester. It was also suggested that initial degradation of the polyester was due to chemical hydrolysis, followed by a biological assimilation process. General views of the biodegradability or destructibility of cellulose esters and their blends with low-molecular-weight diluents, polyesters, and polysaccharides have been summarized in a review [98] by Edgar, Buchanan, and other members of Eastman Chemical Co.

There have been many related works by other research groups that are concerned with the phase behavior, isothermal crystallization kinetics, and tensile properties of cellulose propionate (DS = 2.75)/PHB blends [129], the miscibility and crystallization behavior of CAB/PHB blends [130], or the melt processability and mechanical properties of butyryl-rich CAB/poly(butylene succinate-*co*-butylene carbonate) blends [131], to mention a few. These blends all form a miscible amorphous phase. As an example dealing with a blend compatibilizer, Ogata et al. found a transition-metal alkoxide, tetraisopropyl titanate, acting as an additive to improve the optical and thermal properties of immiscible blends of CA with poly(butylene succinate) (Bionolle, Showa Highpolymer Co., Ltd, Japan) [132].

Meanwhile, Nishio et al. provided an insight into the relationship between the ability of cellulose alkyl esters, CELL-OCOC$_{n-1}$H$_{2n-1}$ ($n = 2-7$), to form a miscible blend with poly(ε-caprolactone) (PCL) and the ester derivative structure in terms of the side-chain length (n) and acyl DS [133] (R. Kusumi et al., personal communication). It was deduced that cellulose butyrate (CB) with $n = 4$ and cellulose valerate (CV) with $n = 5$ exhibit the highest miscibility on blending with PCL of all the cellulose aliphatic esters examined. This estimation is, however, based on the comparison of thermal transition data obtained by using relatively highly substituted cellulosic samples (DS > ~ 2). When the acyl DS was less than 1.5, even the CB/PCL blends became immiscible, as exemplified in Fig. 10. In this figure, the dependence of T_g and T_m (melting temperature of PCL) on the blend composition is compared between two CB/PCL series, differing in the DS of the CB component; one series has DS = 1.2 and the other has DS = 2.4, showing double- and single-T_g behavior, respectively. Since a CB/PCL series using butyryl DS = 2.94 also gave a single T_g–composition curve, a hydrogen-bonding specific interaction should be ruled out from the possible factors for achieving blend miscibil-

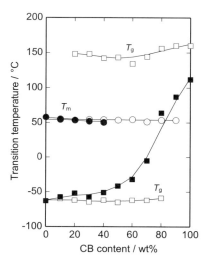

Fig. 10 Composition dependence of T_g and T_m, compared for two series of CB/PCL blends. *Open and filled symbols* represent the data for CB(DS = 1.2)/PCL blends and CB(DS = 2.4)/PCL blends, respectively. (Reproduced from [133])

ity, in contrast to the case of cellulose ester/VP-containing polymer blends described in Sect. 3.3.1. At the present stage, there is no definitive explanation for why butyryl and valeryl substitutions are more effective for the enhancement of miscibility in the binary system of alkyl-esterified cellulose/PCL mixture, but the structural similarity appears to be an important factor: CB and CV have, most abundantly, the same structural unit as that of PCL $(C - C - O - (CO) - C - C - C$ or $C - O - (CO) - C - C - C - C)$, if the carbon atoms C-1 to C-6 in a glucopyranose unit are taken into account.

4
Nanocomposites by Polysaccharide/Inorganic Hybridization

Polymer/inorganic nanocomposites, usually defined as polymer materials containing a smaller amount of inorganic particles structured at an average scale (in diameter or thickness) of less than several tens of nanometers, are expected to be indispensable in future engineering technologies including biomaterial designing. Well-designed nanocomposites may have properties superior to those of the parent polymer and/or inorganic species, which has been partly demonstrated in the field of engineering plastics [134, 135]. A variety of polysaccharide/inorganic nanocomposites has also come out in recent literatures.

4.1
Layered Clay Hybrids

The concept of reinforcing polymer matrices by exfoliation and microdispersion of layered clays as fillers was first successfully applied to a polyamide 6/organophillic silicate system by researchers from Toyota (Japan) [134, 135]. Nanocomposites based on this concept offer the potential for diversification and application of polymers, due to their excellent properties such as high heat-distortion temperature, dimensional stability, improved barrier properties, flame retardance, and enhanced thermo-mechanical properties. Nanoreinforcement of cellulosics with organoclay would also create new value-added applications as high-performance green polymer materials. Park et al. have recently demonstrated the fabrication of injection-moldable green nanocomposites from cellulose acetate (CA) powder, ecofriendly triethyl citrate (TEC) plasticizer, and organically modified montmorillonite (Cloisite 30B) [136]. As an optimized condition of the processing, the addition of TEC plasticizer at 20 wt % was shown to lead to better intercalation and exfoliation of clay (5 wt %) as well as better thermal and mechanical properties of the resulting nanocomposites. According to their subsequent paper [137], the loading of a compatibilizer, maleic-anhydride-grafted cellulose acetate butyrate (CAB-g-MA) at 5 wt % resulted in further improvement in the mechanical properties of the nanocomposites due to enhancement of the adhesion between the organoclay and the CA molecular chains. There is significant potential for a steady increase of this kind of work on renewable-resource-based biodegradable nanocomposites. Additionally, various layered double hydroxides (LDHs) [138, 139] may also be candidates for inorganic fillers as well as montmorillonite-type silicates.

4.2
Silica and Metal Oxides Hybrids Using Alkoxide Precursors

The hybridization of polysaccharides with silica has been investigated in progression by applying a so-called sol-gel process in which the inorganic phase is usually made via hydrolysis and polycondensation of alkoxysilane (typically tetraethylorthosilicate, TEOS) in the sol state. For instance, cellulose acetate (CA)/SiO_2 hybrid sheets and membranes can be easily obtained by casting of CA/TEOS mixtures, followed by phase inversion of TEOS. Cellulosic graft copolymer/SiO_2 hybrids are also possible, e.g., through a copolymerization reaction between acrylic acid monomer and hydroxyethyl cellulose in the presence of a silica sol [140]. The control of the polymer/inorganic phase separation is a subject that should be pursued in the future, however. In this regard, there is great interest in the recent successful synthesis of mesoporous silica monoliths by the sol-gel mineralization method utilizing some specific assemblies of cellulose or its derivatives as templates [141–144]. For example, the porous structure of solution-coagulated CA membranes and even a liquid-crystalline ordered phase of cellulosics can be replicated on an inorganic frameworks obtained after the organic templates are burned out by calcination at 400–500 °C (Sect. 5.2).

Similar results with metal alkoxide precursors, including the preparation of highly dispersed alumina (Al_2O_3) and other metal oxides such as TiO_2, ZrO_2, Fe_2O_3, Nb_2O_5, and Sb_2O_3 on cellulose and its derivative surfaces, which offer numerous practical applications, have been described in recent years [141, 145, 146]. Further elaborate modifications of such cellulose/metal oxide composites, undertaken by Gushikem et al., are of particular interest. They prepared Al_2O_3 (hydrated) on the surface of cellulose (CELL) fiber or on a membrane of CA and further modified the composites with a series of silane couplers ((C_2H_5O)$_3$SiL, where L represents – (CH_2)$_3NH_2$, – (CH_2)$_3NH(CH_2$)$_2NH_2$, etc. The Al – OH groups of the alumina, distributed uniformly on or in the cellulosic substances, can easily react with the coupling reagents to form stable Al – O – Si bonds (Scheme 5). The resulting organofunctionalized cellulosic materials, CELL/Al_2O_3/SiL and CA/Al_2O_3/SiL, can adsorb metal halides ($FeCl_3$, $CuCl_2$, $ZnCl_2$, etc.) due to complexation between the attached SiL ligand and transition-metal ions [146, 147]. They also suggested that the CA/Al_2O_3/Si-aminopropyl Cu(II) complex, a form of the hybrid CA membrane after capturing $CuCl_2$

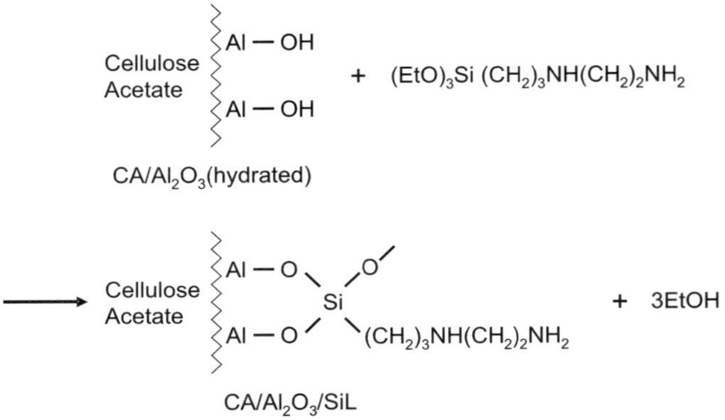

Scheme 5 Preparation of a hybrid membrane CA/Al$_2$O$_3$/SiL [L = $-$(CH$_2$)$_3$NH(CH$_2$)$_2$NH$_2$] [146]

in ethanol solution, would be potentially useful for developing a new oxygen sensor utilizing electrochemical properties of copper [146]. Another CA/Al$_2$O$_3$ composite phosphorylated with H$_3$PO$_4$ was suggested as a separator for mixed alkali metal ions, based on the observation of differences in capacity between the respective metallic cation–proton exchanges on a aluminum hydrogen phosphate phase [148].

Meanwhile, Kurokawa and coworkers developed a method of entrap-immobilization of biocatalysts on fibers, using a gel formation of CA and metal (Ti, Zr) alkoxide [149, 150]. In their standard procedure, a concentrated CA solution in acetone, in which an appropriate amount of biocatalyst (e.g. β-galactosidase, urease) was dispersed, was extruded in gelatinous fiber form into a stirred acetone solution containing 5–10 wt % titanium iso-propoxide or zirconium tetra-n-butoxide, through a glass nozzle with N$_2$ compressed gas. They mentioned that the gel fiber was stable in both organic and aqueous solutions and easy to handled, and that the resultant immobilized biocatalyst showed a high reusability while maintaining the sufficient activity. The gelation was ascribed to a coordination of the polyvalent metal species with carbonyl and hydroxyl groups of CA.

4.3
Zeolite Hybrids

Natural and synthetic zeolites, a family of aluminosilicates with pores and cavities in the range 0.4–1.5 nm, are well-known heterogeneous catalysts and sorbents. Zeolite-incorporated cellulosic fibers and membranes could be suitable for medical antibacterial materials, deodorizers, absorbent pads, sanitary napkins, gas separators, ion exchangers, and so forth; however, the complete and continual use of the whole zeolite surface is not easy in the grossly combined form with such a fibrous organic matrix. One area of interest for research in this field is how to load a zeolite phase into the cellulosic matrices and how to stabilize this phase.

Mintova et al. studied deposition of zeolite A on various cellulose fibers pretreated chemically and/or mechanically [151]. It was shown that the amount of zeolite deposited was controllable by suitable fiber pretreatment with ball-milling or with diethyl ether under ultrasonic action. The reactive high-concentration hydroxyl groups on the structurally loosened celluloses seem to interact with aluminosilicate species and thus promote the formation of nuclei for zeolite crystallization. Pretreatment of natural cellulose fibers with alkali provides another simple route for anchoring preformed zeolite crystallites onto the cellulose surface.

Larsen et al. synthesized zeolite (Y and L)/cellulose composites from Kraft pulp treated with aqueous NaOH and preformed zeolite powders, the zeolite content becoming stable (18 and 8 wt % for Y and L, respectively) on washing with boiling water for several hours [152]. If sodium silicate and sodium aluminate are previously fed into the NaOH solution to swell the cellulose fibers; both seeding and crystal growth of zeolites should be simultaneously realized in the inner area of the fibers under appropriate reaction conditions. By application of this idea, novel high-performance cellulose composites with zeolites were produced, the zeolite loading rate lying in a range 32–58 wt % that changed according to the cellulose source [153].

As an example of hybridization of zeolites with cellulose derivatives, self-supporting zeolite membranes with a sponge-like architecture and zeolite microtubes were prepared by using CA filter membranes as a template [154]. The hierarchical structure with sub-nanometer- to micrometer-sized pores is a characteristic of great promise for a wide range of applications such as catalysis, adsorption, and separation. There was also an attempt to prepare alginate membranes incorporated with zeolites, e.g., for pervaporation separation of water/acetic acid mixtures [155].

Kitaoka et al. recently developed a new route to fabricate a paper-like composite containing titanium dioxide (TiO_2) and zeolite by a papermaking technology using cellulose and ceramic (Al_2O_3/SiO_2) fibers as a sheet matrix [156, 157]. The composite sheets were suggested as promising materials showing a high efficiency for the removal of pollutants or toxic substances, such as NO_x, acetaldehyde, and endocrine-disrupting chemicals from the atmospheric or water environments, by a synergistic effect obtained through the combined use of the TiO_2 photocatalyst and the zeolite adsorbent.

4.4
Magnetic Iron Oxides Hybrids

As a functional development towards advanced materials, the design of magnetic nanocomposites based on cellulose and related polysaccharides has attracted considerable attention in the past decade, with potential applications including information transfer or storage media, fabrics for electromagnetic shielding, new filtration and separation systems, magnetic drug-delivery sys-

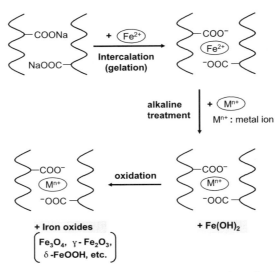

Scheme 6 A typical route for preparing magnetic nanocomposites of gel-forming polysaccharide/iron oxide by the in situ chemical loading technique. The scheme is also applicable for the replacement of the carboxyl anion by a sulfo group ($-SO_3^-$)

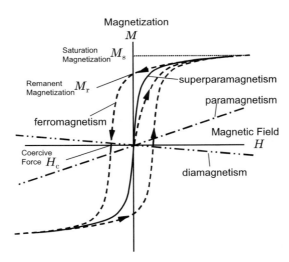

Fig. 11 Different types of curves of magnetization (M) versus applied field (H). In superparamagnetism, the relevant material displays no remanence and coercivity phenomena, but with a saturation magnetization

tems. A major technique for the compositions is in situ synthesis [158–165] of iron oxide nanoparticles in fibrillar suspensions or gels of polysaccharides, rather than physical loading of magnetic pigments into the polymer matrices. The standard of the chemical loading method consists of three steps (Scheme 6): (1) ferrous ion-absorption of the original polymer materials following their swelling or gelation in a ferrous salt solution (typically aqueous $FeCl_2$ or $FeSO_4$); (2) in situ precipitation of ferrous hydroxide by an alkaline treatment of the swollen polymers usually with aqueous hydroxide of alkali or alkali-earth metal; (3) oxidation of the ferrous hydroxide with an oxidizing agent (e.g. H_2O_2) or O_2-bubbling. A significant feature in the adoption of this chemical method is that the composite materials prepared can be superparamagnetic at ambient temperature; that is, there appears no remanent magnetization (M_r) and coercive force (H_c) in magnetic hysteresis-loop measurements, in contrast to the common ferro(i)magnetism (see Fig. 11). The unique magnetic character can be observed when the magnetic particles are dispersed in the matrix on a scale of less than a few tens of nanometers.

(a) With Cellulose and its Derivatives In the beginning of a sequence of studies by Marchessault et al., carboxylated or sulfonated celluloses were used as the polymer matrix to absorb ferrous ions [159, 160]. For example, magnetic cellulosic fibers containing ferrite particles of ~ 10 nm in size were successfully obtained by the in situ synthesis method with carboxymethyl cellulose (CMC), sulfonated thermomechanical wood pulp (TMP), and similarly functionalized rayon filaments. All the hybrid materials were shown to be superparamagnetically responsive to an external magnetic field, i.e., they are only magnetic in the presence of the field.

The reaction process depicted in Scheme 6 may be repeated several times for the purpose of elevating the iron content and enhancing the oxidization or modification of the iron compounds precipitated. A superparamagnetic (SPM) cellulose fiber exhibiting a very high saturation magnetization (M_s), ~ 25 J/T/kg (= emu/g) of sample, was prepared by synthesizing ferrites in the sulfonated TMP via five cycles of the reaction [160], as demonstrated in Fig. 12. Such an SPM character with high M_s implies that while the material is easily magnetized it never retains the magnetization once the field is removed. Major components of the iron oxides were assigned as maghemite (γ-Fe_2O_3) and magnetite (Fe_3O_4) by diffractometry.

Later the same research group claimed that, in the first step of the in situ synthesis scheme, ionic functional groups as such are not necessary for the introduction of ferrous ions into a cellulose matrix [161, 162]. This suggestion was made based on a comparative study of ferrite synthesis between a case with anionically modified cellulose materials and the other case with non-ionic cellulose gels, which included a never-dried bacterial cellulose (BC) membrane and a never-dried cellulose wet-spun filament or cast film (Lyocell) using N-methylmorpholin-N-oxide as the solvent. SPM proper-

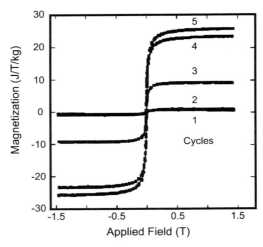

Fig. 12 Magnetization curves observed for sulfonated TMP after one to five cycles of the in situ reaction for iron oxide loading. (Reproduced from [160])

ties were again demonstrated for all the cellulosic composites by magnetic-susceptibility curve measurements. The neutral BC and Lyocell gels turned out to be rather more effective substrates, and BC in particular gave the highest yield and most uniform loading of nanometer-sized ferrites [161]. The dimensions of the fibrillar texture and inner spaces (microvoids) of the cellulose substrate used would seriously influence the diffusion and accessibility of ferrous ions and oxidizing agents inside the swollen matrix. For the magnetic Lyocell and BC membranes obtained by repeating the reaction cycle, the presence of at least two types of ferrites, acicular FeOOH and round Fe_3O_4 or γ-Fe_2O_3, were confirmed by the combined use of transmission electron microscopy and Mössbauer spectroscopy [162]. It was also pointed out that cellulose microfibrils, developed well in BC, could provide a site for nucleation of crystal growth of the magnetic particles [163].

There were a few extensions by other research groups that dealt with the structure and magnetometric characterization of cellulosic-based magnetic nanocomposites prepared by in situ loading of iron oxides [166, 167]. The role of the polymer matrix was discussed in connection with the control of the size and distribution of magnetic nanoparticles. Concerning the synthesis of the magnetic fillers, a reaction scheme relying on coprecipitation in alkaline medium from a mixed solution of iron(II) and iron(III) salts ($Fe^{2+} : Fe^{3+} = 1 : 2$) was also considered. Except for the use of the in situ precipitation method, a successful preparation of magnetic gels composed of hydroxypropyl cellulose particles (< 100 nm) incorporated with nanosized maghemite ($\sim$ 7 nm) may call our attention to the importance of colloid techniques [168].

(b) With Electrolytic Polysaccharides from Natural Algae, etc.

According to the synthetic route outlined in Scheme 6, a few attempts were made to prepare alginate-based magnetic nanocomposites in the form of spherical beads [164, 165], for possible use in the biomedical and pharmaceutical fields, e.g., as a novel drug-delivery system.

Alginate is an electrolytic polysaccharide available in bulk from marine algae. The molecular structure is composed of β-D-mannuronate (M) and α-L-guluronate (G) residues; however, they are arranged in a block-wise fashion, constructed not only of homo-polyuronate blocks (MM or GG) but also of alternating blocks (MG). An aqueous solution of alginate (usually sodium salt) is readily transformed into a hydrogel on addition of metallic divalent cations such as Ca^{2+}, Ba^{2+}, and Fe^{2+}. It has been generally accepted that crosslinks are formed by coordination of the divalent cations to the interchain cavities made up of GG blocks, resulting in the development of a so-called egg-box junction zone [169]. Therefore, this crosslinking zone can be, in the same way as the reaction space, suitable for the in situ synthesis of iron oxides. Actually, the above efforts [164, 165] demonstrated nanostructured iron oxide/alginate beads with SPM properties. Llanes et al. also showed that guluronic-rich alginate (M/G = 0.6) adsorbed a larger amount of iron and thus retained many more oxide particles inside the gel matrix compared with mannuronic-rich alginate (M/G = 1.6), but this was the case using a single cycle of the in situ alkaline oxidation of ferrous ions [165].

As an embodied application of such magnetic beads, Winnik et al. reported the preparation of an iron oxide/alginate-based *ferrofluid* solution (a stable suspension of magnetic nanosized particles) that is suitable for the fabrication of microcapsules visible with magnetic resonance imaging (MRI) [170]. Temperature- and field-dependent magnetization measurements for the ferrofluid solution revealed that the SPM character was maintained at temperatures down to 25 K, but at 5 K the fluid was ferromagnetic to give a hysteresis on the M–H curve.

It should be added here that there was a report of preparing magnetic alginate beads by another in situ technique [171] prior to the above studies. A hydrolysis of urea by urease, producing ammonia, was utilized to make an alkaline atmosphere inside the alginate gel matrix, instead of the treatment with caustic alkali in Scheme 6.

Alginate gels are relatively rigid, but usually fragile; this may be a disadvantage in processing into non-spherical forms such as films and filaments via the gelatinous state. Fortunately, as was shown formerly [94], alginate is compatible with flexible poly(vinyl alcohol) (PVA), and their blend can be rendered as an interpenetrating polymer network (IPN, Fig. 5) by treatment with calcium tetraborate solution. This is due to the occurrence of chelate complexing of alginate with calcium cation and borate-ion-aided crosslinking between PVA chains. Nishio et al. have recently succeeded in constructing such a full-IPN type of alginate/PVA gels containing iron oxides from mixed polymer solutions by a modified in situ ferrite synthesis procedure; the gelation and alkali treatment were carried out with the aid of a metallic borate [172]. The viscoelasticity of the magnetic IPN composites in the gelatinous state was controllable by changing the mixing ratio of the alginate/PVA pair. Their magnetic character was also variable according to the polymer composition, alkaline reagent adopted, and measurement temperature. The observations of superparamagnetism at room temperature for the majority of the IPNs may be of significance for designs and diverse applications of new magnetic materials that can be mechanically processed, in addition to the perceptibility to an external magnetic stimulus only on demand and with restraint of energy loss.

Inorganic nanoparticles are found in relative abundance in natural organisms, which possess an ability to mineralize the particles under relatively mild, aqueous conditions, so that the inorganic phases are organized in association with proteins or polysaccharides [173, 174]. With a view to understanding such biomineralization processes, there have also been studies [175–178] dealing with the formation and stabilization of mainly iron(III) oxyhydroxy nanoaggregates in the presence of alginic acid, κ-carrageenan, chitosan, or other functional water-soluble polymers.

4.5
Apatite Deposition

One of the important fields where carbohydrate polymer/inorganic hybrids may be successfully applied is bioactive materials, e.g., artificial bones expedient for surgery to accelerate the recovery of living bones. There has been increasing interest in hydroxyapatite (HAp) deposition onto the matrix surface of cellulose or related polysaccharide. HAp is a form of calcium phosphate, a main constituent of the inorganic phase of human bone. This kind of study is also a step on the way to exploitation of new biomimetic mineralization methods.

For instance, microcrystalline cellulose, previously modified with titanium iso-propoxide, was shown to be effective for HAp crystal growth [179]. It is likely that the alkoxide can react with the primary hydroxyl of cellulose to produce three new labile hydroxyl groups as true active sites to react with the calcium and phosphate ions to start the growth of HAp. There are also some examples of HAp preparation using chitin, chitosan, and their modified polymers as scaffolds [180–182]. As a suggestive guidance, Kawashita et al. draw an inference that carboxyl-group-containing organic polymers form apatite on the surfaces of their gels in a simulated body fluid, if their carboxyl groups have been previously fully combined with calcium ions, e.g., in a saturated $Ca(OH)_2$ solution [182]. This conclusion was based on a comparative soaking experiment with and without the $Ca(OH)_2$ pretreatment between gel samples of carboxymethylated chitin and gellan gum, both containing carboxyl groups, and of curdlan which does not contain the group.

5
Mesomorphic Ordered Systems

Cellulosic polymers are capable of forming an optically anisotropic mesophase in a condensed fluid state due to the inherent semi-rigidity in the carbohydrate backbone. The supermolecular structure in the mesophase is usually cholesteric (or chiral nematic) (Fig. 15), with possible exceptions such as a columnar type found for thermotropic cellulose trialkanoates. Although there remain unresolved problems concerning the formation of lyotropic and thermotropic liquid-crystalline phases based on cellulose, the general features of their mesomorphic state and optical or chiro-optical behavior are firmly established [183–188]. Further progress in this field may be motivated by efforts to obtain new functional or high-performance materials from liquid-crystalline cellulosics. In the past decade, in fact, there have been many noteworthy works along this line. The following review aims to give an outline of recent advances in the functionalization and application of cholesteric liquid crystals of cellulosics, mainly those involving other polymeric or inorganic components. Some related fundamental studies and preliminary observations are also quoted from necessity.

5.1
Ionic Additive Effects on Mesomorphic Properties

Liquid-crystalline solutions and melts of cellulosic polymers are often colored due to the selective reflection of visible light, originating from the cholesteric helical periodicity. As a typical example, hydroxypropyl cellulose (HPC) is known to exhibit this optical property in aqueous solutions at polymer concentrations of 50–70 wt %. The aqueous solution system is also known to show an LCST-type of phase diagram and therefore becomes turbid at an elevated temperature [184].

The optical characteristics of aqueous HPC solutions are significantly affected by the addition of neutral salts as the third component [189–192]; in brief, the wavelength (λ_M) of maximum light reflectance and the cloud point (T_c), which are correlated with the cholesteric pitch (P) and phase-separation temperature, respectively, vary systematically with a change in the strength of a so-called chaotropic effect of the ions constituting the additive salts. For instance, the effectiveness of increasing P and T_c becomes more pronounced in the order $Cl^- < Br^- < NO_3^- < I^- < SCN^-$ and the order of $Cs^+ < K^+ < Na^+ < Li^+ < Gu^+$ (guanidinium ion), when compared in a series of salts with a common cation or anion, where Cl^-, Br^-, and most alkali-metal cations are antichaotropic (enhancing the hydrophobicity of the solute substance) for the aqueous cellulosic solutions. In a deuteron NMR study [189], it was suggested that the added salts would be able to regulate subtly a conformational state of liquid-crystalline HPC, to induce a marked change of the cholesteric twisting power (P^{-1}) without disturbing the molecular orientation order in each of the nematic layers. (However, the viscoelasticity of the liquid crystals may change depending on the ion species of the coexistent salt [191].) The degree and directional sense of the effectiveness of a given salt at altering the optical parameters λ_M and T_c are determined by the algebraic sum of the respective effects of its constituent ions, and thus the coloration and turbidity of the aqueous HPC lyotropic system can be controlled as desired by selecting the combination of cation and anion species, as illustrated in Fig. 13a.

Quite recently, a series of N-alkyl substituted imidazolium salts has been evaluated for additive effects on the mesomorphic behavior and ensuing optical properties of HPC aqueous solutions, followed by characterization of the thermotropicity of novel cellulose derivatives with such an ionic liquid structure in the side-chains [193].

Aqueous suspensions of cellulose microcrystallites obtained by acid hydrolysis of native cellulose fibers can also produce a cholesteric mesophase [194]. Sulfuric acid, usually employed for the hydrolysis, sulfates the surface of the microcrystallites and therefore they are actually negatively charged. Dong et al. performed some basic studies on the ordered-phase formation in colloidal suspensions of such charged rod-like cellulose crystallites (from cotton filter paper) to evaluate the effects of addition of electrolytes [195, 196]. One of their findings was a decrease in the chiral nematic pitch P of the anisotropic phase, with an increase in concentration of the trace electrolyte (KCl, NaCl, or HCl of < 2.5 mM) added. They assumed that the electric double layer on

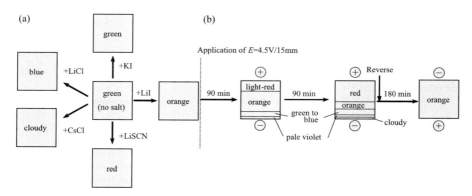

Fig. 13 a Salt addition effect on the selective light-reflection phenomenon of aqueous HPC liquid crystals; **b** variation in the cholesteric reflection color with time, observed for an HPC/LiI/water system under the action of an electric field $E = 4.5$ V/15 mm. HPC conc., 62.5 wt %; salt concentration, 0.5 M; temp., 20 °C

the crystallite rods was suppressed at higher ionic strengths, resulting in the stronger chiral interaction between the crystallites leading to a lower P.

Araki and Kuga investigated similar microcrystal suspensions prepared by sulfuric acid hydrolysis of bacterial cellulose (BC) [197]. The system also underwent spontaneous mesophase separation after complete desalination, but the anisotropic phase was of a *nonchiral* nematic order, unlike the case of other cellulose microcrystallites from higher plants which form a cholesteric phase (the supermolecular helicoidal arrangement is usually left-handed [198, 199]). Addition of < 1 mM NaCl induced cholesteric chirality in the originally nematic BC phase, however. This phenomenon was explained as being due to the change in the effective particle shape, from a cylindrical to a clearly twisted rod as a result of screening of the surface charge, which agrees with the concepts stated by Dong et al.

A regioselectively substituted cellulose ether, 6-O-trityl-2,3-O-diPEOcellulose (TPEOCELL), as formulated in Scheme 7, was synthesized due to interest in its ion-conducting function (Sect. 5.4) [200]. The poly(ethylene oxide) (PEO) side-chains are about seven units long. This cellulose derivative forms, by itself, a right-handed cholesteric mesophase at room temperature with a selective reflection band in the visible region. The T_g and a clearing temperature (isotropic point) as a thermotropic material were estimated to be about $-50\,°C$ and $140\,°C$, respectively. Due to the complexing ability of the PEO side-chains, TPEOCELL can be a good solid solvent for salts such as $LiCF_3SO_3$, $NaCF_3SO_3$, $(CF_3SO_3)_2Cu$, and $LiClO_4$. Generally, the doping of the salts caused a significant untwisting of the cholesteric structure, causing a red-shift in the selective light reflection.

The observations introduced above are all suggestive of subsequent developments of high-added-value materials or modern functional systems

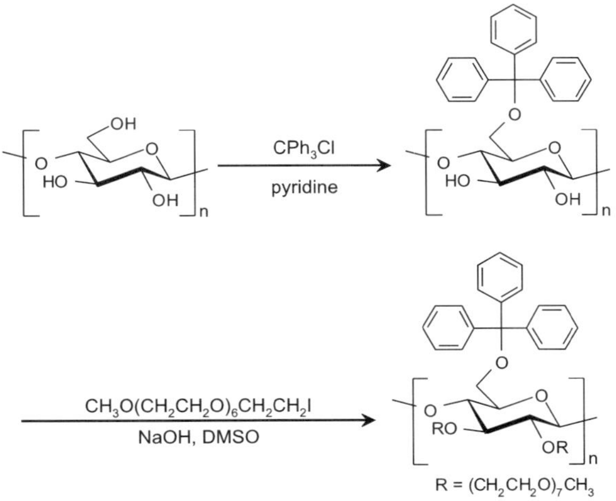

Scheme 7 Preparation of 6-O-trityl-2,3-O-diPEOcellulose (TPEOCELL) [200]

designed by using the cellulosic liquid-crystallinity, as described in the following sections. Among other works, there was an examination of the effect of addition of inorganic salts on the mesophase formation of hydroxyethyl cellulose acetate (HECA) in dimethylformamide [201]. In the presence of the transition-metal salts $CuCl_2$ or $CoCl_2$ the ability of HECA to form liquid crystals seems to be promoted by complexation with the transition-metal ions, unlike the case using LiCl or $CaCl_2$, although the details are not clear. Kryszewski et al. reported on the preparation of cellulose microcrystallites complexed with Fe(III), the products being brownish, stable in air and redispersible in water [202]. Such iron-attached cellulose nanorods may be useful as an anisotropic component to develop novel functional materials, although this possibility was not explicitly embodied in that work.

5.2
Locking-in Mesomorphic Order in Solids and Gels

There have been a lot of studies of cholesteric films and gels in order to exploit their potential as specific optical media and as other functional materials. Most of the preparations were achieved by modification or improvement of previous attempts to immobilize the cholesteric structure of cellulose derivatives into the bulky networks either by crosslinking of cellulosic molecules with functional side-chains in the liquid-crystalline state [203], or by polymerization of monomers as lyotropic solvents for cellulose derivatives [204–206].

Huang et al. [207–209] studied the optical, mechanical, and swelling properties of ethyl-cyanoethyl cellulose [(E-CE)C]/poly(acrylic acid) (PAA) composites endowed with cholesteric architectures, the samples being prepared by photopolymerizing acrylic acid and, if necessary, with simultaneous crosslinking of PAA. In Cowie's laboratory, concentrated solutions of cellulose tricarbanilate (CTC) in methyl acrylate (MA) and in methyl methacrylate (MMA) were solidified by photopolymerizing the monomers, and the resulting opalescent composites were characterized by dynamic mechanical analysis, tensile testing, and electron microscopy [210]. In a similar manner, cholesteric films of 4-methylphenylurethane of cellulose with poly(N-vinyl pyrrolidone) [211], and ternary blend films of ethyl cellulose (EC)/acetylated EC (AEC)/PAA [212] were prepared to allow the evaluation of their chirooptical properties and morphological characteristics.

There have been continuing works by Suto et al. [213–217] on the preparation of cholesteric solid films of hydroxypropyl cellulose (HPC) crosslinked preferably with glutaraldehyde. Coloring conditions [213, 217], swelling [215, 216], and tensile-creep behavior [214] of the crosslinked HPC films were examined. Also, the permselectivity of O_2/N_2 gasses for such liquid-crystalline cellulosic films was reported to become greater than that for the corresponding amorphous ones [218].

Before and after the works described above, contributions to the design and fabrication of similar multicomponent films or gels of cholesteric character, mainly based on HPC, EC, or their derivatives were also made [202, 219–224]. Some of these [219, 220, 224] dealt with shear-deformed network systems preserving a unique banded structure, so that the disappearance and recovery of the optical anisotropy could be controlled thermo-reversibly. Special mention should be made of the successful preparation of two novel classes of solid materials maintaining cholesteric liquid-crystalline order. One consists of essentially pure cellulose only, and the other is a ceramic silica with an imprint of cellulosic chiral mesomorphy.

According to Revol et al. [198], iridescent films can be produced from cholesteric liquid-crystalline suspensions of acid-treated cellulose microcrystallites in water by simple evaporation. They also described that the cholesteric pitch of the resulting films could be controlled by changing the ionic strength of the casting suspensions, e.g., by addition of NaCl, KCl, etc. (Sect. 5.1). Such a cellulose assembly reflecting colored light is certainly attractive for applications as components of security papers, and as decorative films and pigments.

On the other hand, Mann et al. [143] and Antonietti et al. [144] used the lyotropic liquid-crystalline phase of cellulose microcrystallites and that of HPC, respectively, as mesoscopic templates to generate porous silicas by the sol–gel conversion of tetramethoxyorthosilicate (TMOS). In the latter example, HPC was mixed with aqueous hydrochloric acid and TMOS at a polymer concentration of ≥ 58 wt %, and the mixture was sealed in a rubber tube and matured to organize a cholesteric mesophase in the solution. The condensation of TMOS promoted in an open system using low-vacuum application resulted in the production of an iridescent organic/inorganic hybrid material, but with the average reflection wavelength shorter than the parent solution. The HPC/silica hybrids thus obtained were transformed into monolithic silicas with a cholesteric pore structure by removing the organic template using calcinations at 550 °C. Such mesoporous materials may offer potential applications in enantioselective separation techniques, and color-displaying and other modern technologies requiring chiral photonic elements.

5.3
Control of Mesomorphic Structure and Properties by External Stimulation

(a) Magnetic Field Application Liquid-crystalline aqueous suspensions of cellulose microcrystallites can be oriented by applying a magnetic field (ca. 2–7 T), where the cholesteric axis aligns parallel (i.e., individual nematic planes being perpendicular) to the field direction, because of the negative diamagnetic susceptibility of the cellulose crystallites [225, 226]. This magnetic alignment was made available in fabrication of uniformly colored cholesteric films of cellulose with a homogeneous planar orientation [198], as well as in basic structural characterizations of the cellulose microcrystal per se and its assembly [226, 227].

Similar effects in an external magnetic field were investigated for cholesteric mesophases of ethyl-cyanoethyl cellulose [(E-CE)C] in dichloroacetic acid (DCA) [228, 229] and in acrylic acid [228]. The cholesteric axis was also found to align essentially parallel to the direction of the magnetic field (9.4 T) after the mesomorphic solutions were treated in the field, indicating a negative diamagnetic anisotropy of the cellulose derivative molecule. It was emphasized, however, that the effect of the magnetic field on the orientation and helical pitch of the cholesteric phase of (E-CE)C/DCA depended on the concentration of the solution used, and, particularly in the anisotropic/isotropic biphasic samples of lower concentrations, the magnetic effect was largely restrained by the surface tension of the discrete cholesteric domains [229].

(b) Electric Field Application Chiba and Nishio et al. [230] made an attempt to control the optical properties of the aqueous HPC/salt system electrically (Sect. 5.1). In general, it is not easy to manipulate the mesomorphic structure and properties of main-chain-type liquid-crystalline polymers in such an external electrical way due to their low sensitivity to an electric field at practical magnitudes. By virtue of the electrophoretic motility of the coexistent salt ions as P and/or T_c shifts, however, dynamic variation in the cholesteric coloration and/or optical turbidity of the HPC solutions can be achieved under the action of a relatively weak electromotive force (e.g. 4.5 V) to appropriate electrodes in contact with a layered sample. Figure 13b represents diagrammatically a time-evolving gradation in reflection color, observed for a LiI-containing HPC liquid crystal that was uniformly colored orange in the initial stage. It was reasonably assumed that the electrical stimulation allowed the ion particles to migrate and localize in the lyotropic system, resulting in a certain gradient of P and/or T_c in response to the nonuniform distribution.

As an extension of the above study, cholesteric colored films were prepared from HPC anisotropic solutions in a mixed solvent containing methacrylate monomer and crosslinking agents via photopolymerization and crosslinking steps. When the films were immersed in aqueous salt solutions, their coloration and optical turbidity varied depending on the strength of the chaotropicity of the impregnant ions [231]. The visual appearance of the swollen networks was also changed by the application of an electric potential between two edges of the rectangular samples (R. Chiba et al., personal communication) in a manner similar to that observed for the electrified liquid-crystalline HPC/salt aqueous solutions.

(c) Mechanical Compression An impressive work on the functional design of liquid-crystalline cellulose derivatives was accomplished by Müller and Zentel [232, 233], mainly concerned with the manipulation of the wavelength

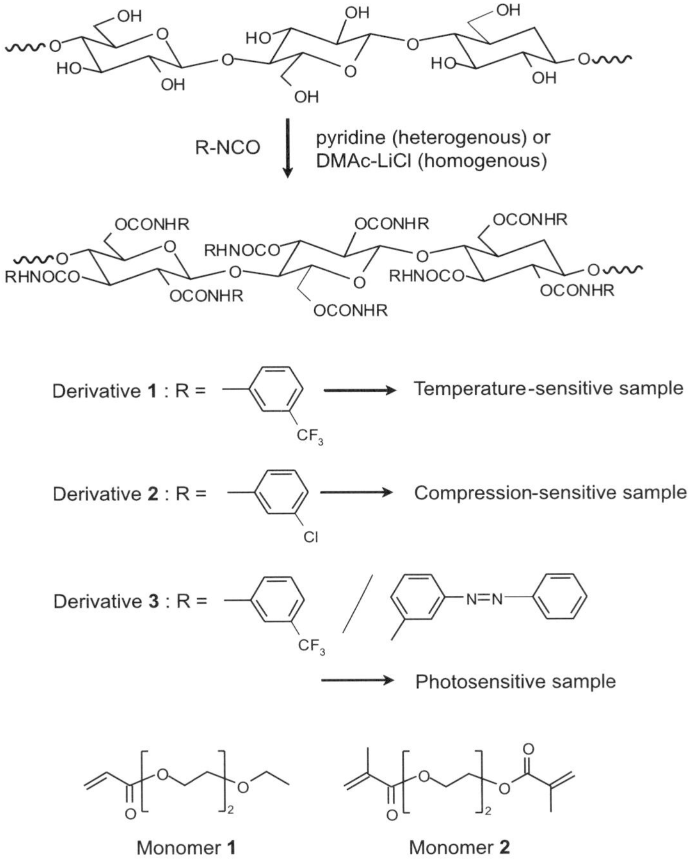

Scheme 8 Cellulose urethane derivatives, and their mixing partners to make cholesteric liquid crystals [233]

of selective light reflection by various external stimuli. As illustrated in Scheme 8 [233], they synthesized different urethane derivatives (carbanilates) of cellulose by full substitution reactions with phenyl isocyanate having a – CF$_3$ or – Cl group in the *meta*-position and, if necessary, also with iso-cyanoazobenzene (see item **d**) below). The derivatives were used to make a lyotropic cholesteric mesophase using mono- and bisacrylates as solvents.

For example, the selective reflection of a cholesteric mixture of deriva-tive 1 ($\sim$ 45 wt %)/monomer 2 (diethylene glycol dimethacrylate) covered the whole visible spectrum in a narrow temperature range; i.e., the mixture was bluish at 20 °C, greenish at 30 °C, and reddish at 50 °C. Therefore, by succes-sive photo-crosslinking of the liquid-crystalline sample at different tempera-tures through a mask, a colored picture or pattern could be fixed onto the resulting composite film.

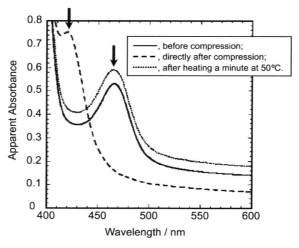

Fig. 14 UV-VIS spectra of a cholesteric elastomer, based on a mixture of derivative 2 (44 wt %), monomer 1 (52 wt %), and monomer 2 (3 wt %), crosslinked at 32 °C. *Arrows* indicate the maximum position of the selective light-reflection band. (Reproduced from [233])

When a lyotopic system of derivative 2 (44 wt %)/monomer 1 [2-(2-ethoxyethoxy) ethyl acrylate] (52 wt %)/monomer 2 (3 wt %) containing a photo-initiator at 1 wt % was photo-crosslinked at 32 °C, a flexible film was produced, giving a selective reflection band centered at 465 nm (colored bluish), as shown in Fig. 14. If the film was compressed between glass slides so that the thickness decreased to 90% of its original size, then the wavelength of selective reflection dropped to 420 nm (dark violet hue). After the film was relaxed, accelerated by heating at 50 °C, the original reflection color was recovered. This suggests applicability as a rewritable recording medium.

(d) Photoisomerization of Pendant Groups According to the continued efforts by Müller et al. [233], an azobenzene moiety containing co-urethane of cellulose, derivative 3 in Scheme 8, formed a right-handed cholesteric phase in monomer 2; the maximum reflection peak appeared at a wavelength longer than 550 nm at ~ 45 wt % of the derivative, which contained 3.6 wt % of 3-phenylazopheny carbamoyl side groups. As is well known, such azo chromophores undergo a *trans–cis* isomerization photochemically, e.g., by 365 nm irradiation for the *trans* to *cis* and > 400 nm irradiation for the *cis* to *trans* transitions. It was demonstrated for the lyotropic system of derivative 3 (44.3 wt %)/monomer 2 (55.7 wt %) that the wavelength of the selective reflection can be changed reversibly by photoisomerization of the pendant azobenzene groups. This may be ascribed to the photochemical variation of the cholesteric pitch in a reversible manner, as represented diagrammatically

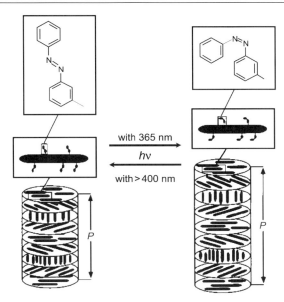

Fig. 15 Schematic representation of the photoisomerization effect on a cholesteric meso-phase based on azo-chromophore-containing cellulose derivative molecules. (Quoted from [233] with an adequate modification)

in Fig. 15. However, the observation of a comparatively larger pitch in the *cis* form of the azo side-groups was not explained satisfactorily.

As an additional remark, the same research group designed a cholesteric network that acted as a humidity sensor, using a derivative of HPC esterified with acryloyl chloride to a low degree of substitution (0.22). The HPC acrylate hydrogel can change the cholesteric pitch, and thereby the reflective color changes with water uptake.

5.4
Other Novel Functionalities

An effort to combine the thermotropic liquid-crystalline properties of selected cellulose derivatives with the ion-conductive characteristics of poly(ethylene oxide) (PEO) has currently been pursued [234, 235]. TPEO-CELL shown in Scheme 7 is a representative sample.

The ion conductivities of a range of TPEOCELL complexes with $LiCF_3SO_3$ were examined by AC impedance spectroscopy from 293–373 K, so that all the complexes remained in the liquid-crystalline state during the measurements [235]. The inorganic salt concentration, indicated by the molar ratio of lithium ions to oxygens in the polymer chain, [Li]/[O], ranged from 0.038 to 0.125. The bulk conductivities were $\sim 10^{-6}$ Scm^{-1} at ambient temperatures and $\sim 10^{-4}\,Scm^{-1}$ at 373 K, comparable in magnitude to other PEO-based systems, and assuming a maximum at a certain salt concentration that became higher as the temperature was raised. (However, the conductivities may be taken as rather medial values, considering a recent report [236] of $\sim 10^{-4}\,Scm^{-1}$ at 300 K for a fully cyanoethylated HPC derivative dissolving $LiClO_4$). Interestingly, the salt-incorporated TPEOCELL liquid crystals exhibited a linear Arrhenius dependence of log (conductivity) on reciprocal temperature, in contrast to the majority of amorphous solid polymer electrolytes for which a nonlinear dependence is generally observed. Nevertheless, the activation energies for ion transport in the $TPEOCELL/LiCF_3SO_3$ complexes were in the range 53–73 kJ mol^{-1}, being the same order of magnitude as that for many amorphous systems.

Wendorff et al. synthesized various types of liquid-crystalline HPC esters with aliphatic or aromatic acyl substituents, acting on the potential of the products as a nonlinear optical medium for applications in optoelectronics, e.g., in organic light-emitting devices or sensors [237, 238]. Most series of derivatives prepared at different degrees of substitution displayed cholesteric reflection colors without solvents over a range of temperatures covering room temperature. They studied the light absorption and emission properties for guest–host systems composed of a pair of the HPC esters and chromophores such as coumarins [237], and further for some ester derivatives with chromophores linked covalently onto the original HPC molecule [238]. The characterization of the photo- and electroluminescence was detailed, but a limiting factor was pointed out with respect to optoelectronic applications for liquid-crystalline cellulose derivatives: it seems difficult to obtain relevant films with uniform orientation of the cholesteric axes as well as better helical periodicity.

Further challenges need to be addressed for the functional development of liquid-crystalline cellulosics, coupled with the elucidation of fundamental aspects of their chiral nature. In addition, the potential thermotropicity of cellulose microcrystallites with the surfaces modified adequately and manipulation of the cholesteric sense (handedness) in the helically arranged molecular assembly are embryonic but fascinating subjects in this research field.

6
Concluding Remarks

The present review is mainly concerned with the preparation and functionalization of microcompositional materials with cellulosic polysaccharides as the principal component, including four major categories: graft copolymers, miscible or compatible polymer blends and networks, polysaccharide/inorganic nanohybrids, and mesomorphic ordered systems. Ultrathin layers of cellulosic polymers were not within the scope of this review, but there have been extensive studies accompanied by marked progress in that field, especially in Germany, as was demonstrated for a specific class of hairy-rod macromolecule cellulose derivatives, which can form multilayered assemblies [239].

Looking over the studies summarized in this article, we find that significant synergetic effects between physical and physicochemical properties, including optical, thermal and tensile behavior, swelling, adsorption, and other conjugative responses, can be observed in cellulosic and related polysaccharide materials. In addition, microscopic incorporation of other polymeric or inorganic ingredients can cooperatively interact with relevant polysaccharides. Essential factors that are required for these achievements are the inherent characteristics of cellulosic polymers, such as side-group reactivity allowing derivatization, the ability to form hydrogen bonds, complexing ability, conditionally changeable hydrophilic/hydrophobic nature, enzymatic degradability, semi-rigidity, chirality, etc. The observation of such desirable effects indicates the acquisition of functional diversity in cellulosics-based materials, but does not directly ensure practical application. The next decade should witness the industrial use of a wide spectrum of products derived from the efforts.

An important step towards designing multicomponent systems, the formation of a so-called *microphase separation* structure, has never been accomplished for cellulosic bulk materials. More-elaborate graft copolymerization could enable the formation of such a regulated microdomain structure. Currently, however, precisely designed block copolymers composed of cellulosic/synthetic polymer segments will be required for both morphological and functional reasons. Recent progress in the synthesis of fine cellulosics, each having a definite molecular structure [240–242], should greatly contribute to reaching this goal in the near future.

Acknowledgements The author is greatly indebted to Professor S. Kobayashi of Kyoto University and to Professor D. Klemm of Friedrich-Schiller-Universität, Jena for their encouragement and helpful discussions as well as for their kind invitation to contribute to this special volume. The author also wishes to express his thanks to Drs. Y. Miyashita and Y. Teramoto, and to the graduate students R. Chiba and T. Ohno in my laboratory for their assistance in the preparation of this chapter.

References

1. Bledzki AK, Gassan J (1999) Prog Polym Sci 24:221
2. Eichhorn SJ, Baillie CA, Zafeiropoulos N, Mwaikambo LY, Ansell MP, Dufresne A, Entwistle KM, Herrera-Franco PJ, Escamilla GC, Groom L, Hughes M, Hill C, Rials TG, Wild PM (2001) J Mater Sci 36:2107
3. Mohanty AK, Misra M, Hinrichsen G (2000) Macromol Mater Eng 276/277:1
4. Mohanty AK, Misra M, Drzal LT (eds.) (2005) Natural fibers, biopolymers, and biocomposites. CRC, Boca Raton

5. Hon DN-S (1993) In: Seymour RB, Porter RS (eds) Manmade fibers, their origin and development. Elsevier Applied Science, London, p 91
6. Cardamone JM (1993) In: Seymour RB, Porter RS (eds) Manmade fibers, their origin and development. Elsevier Applied Science, London, p 107
7. Vigo TL (1998) Polym Adv Technol 9:539
8. Sudesh K, Abe H, Doi Y (2000) Prog Polym Sci 25:1503
9. Okada M (2002) Prog Polym Sci 27:87
10. Albertsson A-C, Varma IK (2002) Adv Polym Sci 157:1
11. Hakkarainen M (2002) Adv Polym Sci 157:113
12. Kowalski A, Duda A, Penczek S (1998) Macromol Rapid Commun 19:567
13. Storey RF, Taylor AE (1998) J Macromol Sci Pure Appl Chem A35:723
14. In't Veld JA, Velner EM, von de Witte P, Harnhuis J, Dijkstra PJ, Feijen J (1997) J Polym Sci Pt A Polym Chem 35:219
15. Kricheldorf HR, Kreiser-Saunders I, Boettcher C (1995) Polymer 36:1253
16. Kricheldorf HR, Kreiser-Saunders I, Stricker A (2000) Macromolecules 33:702
17. Li Y, Northnagel J, Kissel T (1997) Polymer 38:6197
18. Donabedian DH, McCarthy SP (1998) Macromolecules 31:1032
19. Dubois P, Krishnan M, Narayan R (1999) Polymer 40:3091
20. Choi EJ, Kim CH, Park JK (1999) Macromolecules 32:7402
21. Yoshioka M, Hagiwara N, Shiraishi N (1999) Cellulose 6:193
22. Detchprohm S, Aoi K, Okada M (2001) Macromol Chem Phys 202:3560
23. Teramoto Y, Nishio Y (2003) Polymer 44:2701
24. Teramoto Y, Ama S, Higeshiro T, Nishio Y (2004) Macromol Chem Phys 205:1904
25. Rustemeyer P (ed.) (2004) Cellulose acetates: properties and applications, Macromol Symp 208. Wiley, Weinheim
26. Buchanan CM, Gardner RM, Komarek RJ (1993) J Appl Polym Sci 47:1709
27. Sakai K, Yamauchi T, Nakatsu F, Ohe T (1996) Biosci Biotech Biochem 60:1617
28. Sibata T (2005) In: Kobayashi K, Shoda S (eds) The Recent Trends of Glycochemistry, Pt 2. CMC Pub., Tokyo, p 121
29. Reimschuessel HK (1979) J Polym Sci Polym Chem Ed 17:2447
30. Teramoto Y, Nishio Y (2004) Biomacromolecules 5:407
31. Teramoto Y, Nishio Y (2004) Biomacromolecules 5:397
32. Ohya Y, Maruhashi S, Ouchi T (1998) Macromol Chem Phys 199:2017
33. Ohya Y, Maruhashi S, Ouchi T (1998) Macromolecules 31:4662
34. Aoi K, Okada M (1996) Prog Polym Sci 21:151
35. Aoi K, Takasu A, Okada M (1998) Macromol Chem Phys 199:2805
36. Takasu A, Aoi K, Tsuchiya M, Okada M (1999) J Appl Polym Sci 73:1171
37. Qu X, Wirsén A, Albertsson A-C (1999) J Appl Polym Sci 74:3193
38. Heinze T, Dicke R, Koschella A, Kull AH, Klohr EA, Koch W (2000) Macromol Chem Phys 201:627
39. Ikeda I, Washino K, Maeda Y (2003) Sen'i Gakkaishi 59:110
40. Ejaz M, Yamamoto S, Ohno K, Tsujii Y, Fukuda T (1998) Macromolecules 31:5934
41. Tsujii Y, Ohno K, Yamamoto S, Goto A, Fukuda T (2006) Adv Polym Sci 197:1
42. Nishio Y (1994) In: Gilbert RD (ed) Cellulosic polymers, blends and composites, Chap. 5. Carl Hanser, Munich
43. Nishio Y, Roy SK, Manley RSJ (1987) Polymer 28:1385
44. Nishio Y, Manley RSJ (1988) Macromolecules 21:1270
45. Nishio Y, Manley RSJ (1990) Polym Eng Sci 30:71
46. Masson J-F, Manley RSJ (1991) Macromolecules 24:5914
47. Masson J-F, Manley RSJ (1991) Macromolecules 24:6670

48. Masson J-F, Manley RSJ (1992) Macromolecules 25:589
49. Schartel B, Wendling J, Wendorff JH (1996) Macromolecules 29:1521
50. Radloff D, Boeffel C, Spiess HW (1996) Macromolecules 29:1528
51. VanderHart DL, Manley RSJ, Barnes JD (1994) Macromolecules 27:2826
52. Liang XH, Guo YQ, Gu LZ, Ding EY (1995) Macromolecules 28:6551
53. Patel K, Manley RSJ (1995) Macromolecules 28:5793
54. Kondo T, Sawatari C, Manley RSJ, Gray DG (1994) Macromolecules 27:210
55. Morgenstern B, Leillinger O, Maron R (1996) Angew Makromol Chemie 243:129
56. Lim KY, Seong YJ, Kim BC (2003) Polym J 35:691
57. Garcia-Ramirez M, Cavaillé JY, Dupeyre D, Péguy A (1994) J Polym Sci Pt B Polym Phys 32:1437
58. Garcia-Ramirez M, Cavaillé JY, Dufresne A, Tékély P (1995) J Polym Sci Pt B Polym Phys 33:2109
59. Yokoyama F, Sato A, Tsugita H, Yamashita Y, Mitsuishi K, Kawano M (1996) Sen'i Gakkaishi 52:155
60. Nagata M, Okano F, Sakai W, Tsutsumi N (1998) J Polym Sci Pt A Polym Chem 36:1861
61. Nishioka N, Hamabe S, Murakami T, Kitagawa T (1998) J Appl Polym Sci 69:2133
62. Nishioka N, Itoh T, Uno M (1999) Polym J 31:1218
63. Endo T, Kitagawa R, Zhang F, Hirotsu T, Hosokawa J (1999) Chem Lett 1155
64. Endo T (2000) Cellulose Commun 7:63
65. Yang G, Yamane C, Matsui T, Miyamoto I, Zhang L, Okajima K (1997) Polym J 29:316
66. Yang G, Xiong X, Zhang L (2002) J Membr Sci 201:161
67. Zhang L, Guo J, Du Y (2002) J Appl Polym Sci 86:2025
68. Laszkiewicz B, Cuculo JA (1993) J Appl Polym Sci 50:27
69. Nechwatal A, Michels C, Kosan B, Nicolai M (2004) Cellulose 11:265
70. Nishio Y, Hirose N (1992) Polymer 33:1519
71. Miyashita Y, Kimura N, Nishio Y, Suzuki H (1994) Kobunshi Ronbunshu 51:466
72. Miyashita Y, Kimura N, Suzuki H, Nishio Y (1998) Cellulose 5:123
73. Miyashita Y, Nishio Y, Kimura N, Suzuki H, Iwata M (1996) Polymer 37:1949
74. Yano H, Sugiyama J, Nakagaito AN, Nogi M, Matsuura T, Hikita M, Handa K (2005) Adv Mater 17:153
75. Williamson SL, Armentrout RS, Porter RS, McCormick CL (1998) Macromolecules 31:8134
76. Favier V, Canova GR, Cavaillé JY, Chanzy H, Dufresne A, Gauthier C (1995) Polym Adv Technol 6:351
77. Favier V, Chanzy H, Cavaillé JY (1995) Macromolecules 28:6365
78. Dufresne A (2000) Compos Interfaces 7:53
79. Anglès MN, Dufresne A (2001) Macromolecules 34:2921
80. Azizi Samir MAS, Alloin F, Sanchez JY, Dufresne A (2004) Polymer 45:4149
81. Azizi Samir MAS, Alloin F, Sanchez JY, Dufresne A (2004) Macromolecules 37:4839
82. Miyashita Y, Yamada Y, Kimura N, Nishio Y, Suzukii H (1995) Sen'i Gakkaishi 51:396
83. Miyashita Y, Sato M, Kimura N, Nishio Y, Suzuki H (1996) Kobunshi Ronbunshu 53:149
84. Lee YM, Kim SH, Kim SJ (1996) Polymer 37:5897
85. Kimura N, Sato M, Miyashita Y, Suzuki H, Nishio Y (1997) Sen'i Gakkaishi 53:409
86. Kubota N, Konaka G, Eguchi Y (1998) Sen'i Gakkaishi 54:212
87. Nishio Y, Koide T, Miyashita Y, Kimura N, Suzuki H (1999) J Polym Sci Pt B Polym Phys 37:1533
88. Miyashita Y, Nishio Y, Akamatsu T, Kimura N, Suzuki H (1999) Sen'i Gakkaishi 55:254

89. Honma T, Senda T, Inoue Y (2003) Polym Int 52:1839
90. Ko MJ, Jo WH, Kim HC, Lee SC (1997) Polym J 29:997
91. Dufresne A, Cavaillé JY, Dupeyre D, Garcia-Ramirez M, Romero J (1999) Polymer 40:1657
92. Miyashita Y, Yamada Y, Kimura N, Suzuki H, Iwata M, Nishio Y (1997) Polymer 38:6181
93. Miyashita Y, Kobayashi R, Kimura N, Suzuki H, Nishio Y (1997) Carbohydr Polym 34:221
94. Miura K, Kimura N, Suzuki H, Miyashita Y, Nishio Y (1999) Carbohydr Polym 39:139
95. Dufresne A, Cavaillé JV, Helbert W (1996) Macromolecules 29:7624
96. Morin A, Dufresne A (2002) Macromolecules 35:2190
97. Gopalan Nair K, Dufresne A (2003) Biomacromolecules 4:657
98. Edgar KJ, Buchanan CM, Debenham JS, Rundquist PA, Seiler BD, Shelton MC, Tindall D (2001) Prog Polym Sci 26:1605
99. Komarek RJ, Gardner RM, Buchanan CM, Gedon S (1993) J Appl Polym Sci 50:1739
100. Gardner RM, Buchanan CM, Komarek RJ, Dorschel DD, Boggs C, White AW (1994) J Appl Polym Sci 52:1477
101. Landry MR, Massa DJ, Landry CJT, Teegarden DM, Colby RH, Long TE, Henrichs PM (1994) J Appl Polym Sci 54:991
102. Davis MF, Wang XM, Myers MD, Iwamiya JH, Kelley SS (1998) In: Heinze TJ, Glasser WG (eds) Cellulose derivatives: modification, characterization and nanostructures. ACS Symp Ser 688:283
103. Gaibler DW, Rochefort WE, Wilson JB, Kelley SS (2004) Cellulose 11:225
104. Miyashita Y, Suzuki T, Nishio Y (2002) Cellulose 9:215
105. Ohno T, Yoshizawa S, Miyashita Y, Nishio Y (2005) Cellulose 12:281
106. Ohno T, Nishio Y (2006) Cellulose 13:245
107. Nguyen QT, Noezar I, Clément R, Streicher C, Brueschke H (1997) Polym Adv Technol 8:477
108. Bikson B, Nelson JK, Muruganandam N (1994) J Membr Sci 94:313
109. Rao V, Ashokan PV, Shridhar MH (1999) Polymer 40:7167
110. Rao RV, Ashokan PV, Shridhar MH (2000) Polym Degrad Stab 70:11
111. Rao V, Ashokan PV, Shridhar MH (2000) J Appl Polym Sci 76:859
112. Park MS, Kim JK (2002) J Polym Sci Pt B Polym Phys 40:1673
113. Ding B, Kimura E, Sato T, Fujita S, Shiratori S (2004) Polymer 45:1895
114. Renecker DH, Chun I (1996) Nanotechnology 7:216
115. Yarin AL, Koombhongse S, Renecker DH (2001) J Appl Phys 89:3018
116. Scandola M, Ceccorulli G, Pizzoli M (1992) Macromolecules 25:6441
117. Ceccorulli G, Pizzoli M, Scandola M (1993) Macromolecules 26:6722
118. Pizzoli M, Scandola M, Ceccorulli G (1994) Macromolecules 27:4755
119. Lotti N, Scandola M (1992) Polym Bull 29:407
120. Buchanan CM, Gedon SC, White AW, Wood MD (1992) Macromolecules 25:7373
121. Buchanan CM, Gedon SC, White AW, Wood MD (1993) Macromolecules 26:2963
122. Buchanan CM, Gedon SC, Pearcy BG, White AW, Wood MD (1993) Macromolecules 26:5704
123. Gilmore DF, Fuller RC, Schneider B, Lenz RW, Lotti N, Scandola M (1994) J Environ Polym Degrad 2:49
124. Tomasi G, Scandola M (1995) J Macromol Sci Pure Appl Chem A32:671
125. Scandola M (1995) Can J Microbiol 41 (Suppl. 1):310
126. Buchanan CM, Boggs CN, Dorschel DD, Gardner RM, Komarek RJ, Watterson TL, White AW (1995) J Environ Polym Degrad 3:1

127. Buchanan CM, Dorschel DD, Gardner RM, Komarek RJ, White AW (1995) J Macromol Sci Pure Appl Chem A32:683
128. Buchanan CM, Pearcy BG, White AW, Wood MD (1997) J Environ Polym Degrad 5:209
129. Maekawa M, Pearce R, Marchessault RH, Manley RSJ (1999) Polymer 40:1501
130. El-Shafee E, Saad GR, Fahmy SM (2001) Eur Polym J 37:2091
131. Lee SH, Yoshioka M, Shiraishi N (2000) J Appl Polym Sci 77:2908
132. Uesaka T, Ogata N, Nakane K, Shimizu K, Ogihara T (2002) J Appl Polym Sci 83:1750
133. Nishio Y, Matsuda K, Miyashita Y, Kimura N, Suzuki H (1997) Cellulose 4:131
134. Okada O, Kawasumi M, Usuki A, Kurauchi T, Kamigaito O (1990) Mater Res Soc Symp Proc 171:45
135. Yano K, Usuki A, Okada A, Kurauchi T, Kamigaito O (1997) J Polym Sci Pt A Polym Chem 35:2289
136. Park HM, Misra M, Drzal LT, Mohanty AK (2004) Biomacromolecules 5:2281
137. Park HM, Liang X, Mohanty AK, Misra M, Drzal LT (2004) Macromolecules 37:9076
138. Newman SP, Jones W (1998) New J Chem 105
139. O'Leary S, O'Hare D, Seeley G (2002) Chem Commun 1506
140. Lin S, Wu J, Yao K, Cai K, Xiao C, Jiang C (2004) Composite Interfaces 11:271
141. Caruso RA, Schattka JH (2000) Adv Mater 12:1921
142. Caruso RA, Antonietti M (2002) Adv Funct Mater 12:307
143. Dujardin E, Blaseby M, Mann S (2003) J Mater Chem 13:696
144. Thomas A, Antonietti M (2003) Adv Funct Mater 13:763
145. Gushikem Y, Toledo EA (1999) In: Esumi K (ed) Polymer interfaces and emulsions, Chap. 13. Marcel Dekker, New York
146. Lazarin AM, Borgo CA, Gushikem Y (2003) J Membr Sci 221:175
147. Lazarin AM, Gushikem Y, de Castro SC (2000) J Mater Chem 10:2526
148. Lazarin AM, Borgo CA, Gushikem Y, Kholin YV (2003) Anal Chim Acta 477:305
149. Ikeda Y, Kurokawa Y, Nakano K, Ogata N (2002) Cellulose 9:369
150. Nakane K, Ogihara T, Ogata N, Kurokawa Y (2003) J Mat Res 18:672
151. Mintova S, Valtchev V (1996) Zeolites 16:31
152. Vu D, Marquez M, Larsen G (2002) Micropor Mesopor Mater 55:93
153. Sugiyama K (2003) Cellulose Commun 10:76
154. Wang Y, Tang Y, Dong A, Wang X, Ren N, Shan W, Gao Z (2002) Adv Mater 14:994
155. Kittur AA, Tambe SM, Kulkarni SS, Kariduraganavar MY (2004) J Appl Polym Sci 94:2101
156. Fukahori S, Ichiura H, Kitaoka T, Tanaka H (2003) Environ Sci Technol 37:1048
157. Ichiura H, Kitaoka T, Tanaka H (2003) J Mater Sci 38:1611
158. Ziolo RF, Giannelis EP, Weinstein BA, O'Horo MP, Ganguly BN, Mehrotra V, Russell MW, Huffman DR (1992) Science 257:219
159. Marchessault RH, Ricard S, Rioux P (1992) Carbohydr Res 224:133
160. Raymond L, Revol J-F, Ryan DH, Marchessault RH (1994) Chem Mater 6:249
161. Raymond L, Revol J-F, Marchessault RH, Ryan DH (1995) Polymer 36:5035
162. Sourty E, Ryan DH, Marchessault RH (1998) Cellulose 5:5
163. Sourty E, Ryan DH, Marchessault RH (1998) Chem Mater 10:1755
164. Kroll E, Winnik FM, Ziolo RF (1996) Chem Mater 8:1594
165. Llanes F, Ryan DH, Marchessault RH (2000) Int J Biol Macromol 27:35
166. Suber L, Foglia S, Ingo GM, Boukos N (2001) Appl Organometal Chem 15:414
167. Si S, Kotal A, Mandal TK, Giri S, Nakamura H, Kohara T (2004) Chem Mater 16:3489
168. Chatterjee J, Haik Y, Chen CJ (2003) Colloid Polym Sci 281:892

169. Draget KI, Smidsrød O, Skjåk-Bræk G (2002) In: Vandamme EJ, De Baets S, Stein-büchel A (eds) Polysaccharides II: polysaccharides from eukaryotes, biopolymers, vol. 6, chap. 8. Wiley, Weinheim
170. Shen F, Poncet-Legrand C, Somers S, Slade A, Yip C, Duft AM, Winnik FM, Chang PL (2003) Biotechnol Bioeng 83:282
171. Hayama T, Takizawa T, Hidaka H, Horikoshi K (1993) J Chem Eng Jpn 26:223
172. Nishio Y, Yamada A, Ezaki K, Miyashita Y, Furukawa H, Horie K (2004) Polymer 45:7129
173. Mann S, Webb J, Williams RJP (eds.) (1989) Biomineralization: chemical and bio-chemical perspectives. Wiley, Weinheim
174. Mann S (2002) Biomineralization: principles and concepts in bioinorganic materials chemistry. Oxford University Press, Oxford
175. Nesterova MV, Walton SA, Webb J (2000) J Inorg Biochem 79:109
176. Jones F, Cölfen H, Antonietti M (2000) Colloid Polym Sci 278:491
177. Jones F, Cölfen H, Antonietti M (2000) Biomacromolecules 1:556
178. Sipos P, Berkesi O, Tombácz E, St Pierre TG, Webb J (2003) J Inorg Biochem 95:55
179. González M, Hernández E, Scencio JA, Pacheco F, Pacheco S, Rodríguez R (2003) J Mater Chem 13:2948
180. Wan ACA, Khor E, Hastings GW (1998) J Biomed Mater Res 41:541
181. Yamaguch I, Tokuchi K, Fukuzaki H, Koyama Y, Takakuda K, Monma H, Tanaka J (2001) J Biomed Mater Res 55:20
182. Kawashita M, Nakao M, Minoda M, Kim H-M, Beppu T, Miyamoto T, Kokubo T, Nakamura T (2003) Biomaterials 24:2477
183. Guo J-X, Gray DG (1994) In: Gilbert RD (ed.) Cellulosic polymers, blends and com-posites, chap. 2. Carl Hanser, Munich
184. Gray DG, Harkness BR (1994) In: Shibaev VP, Lam L (eds) Liquid crystalline and mesomorphic polymers. Springer, New York, p 298
185. Fukuda T, Takada A, Miyamoto T (1994) In: Gilbert RD (ed) Cellulosic polymers, blends and composites, chap. 3. Carl Hanser, Munich
186. Fukuda T, Tsujii Y, Miyamoto T (1995) Macromol Symp 99:257
187. Zugenmaier P (1994) In: Gilbert RD (ed) Cellulosic polymers, blends and compos-ites, chap 4. Carl Hanser, Munich
188. Zugenmaier P (1998) In: Demus D, Goodby J, Gray GW, Spiess H-W, Vill V (eds.) Handbook of liquid crystals, vol. 3, chap. 9. Wiley, Weinheim
189. Nishio Y, Chiba R, Miyashita Y, Oshima K, Miyajima T, Kimura N, Suzuki H (2002) Polym J 34:149
190. Nishio Y, Kai T, Kimura N, Oshima K, Suzuki H (1998) Macromolecules 31:2384
191. Prevysh VA, Spontak RJ, Khan SA (1995) Mat Res Soc Symp Proc 394:137
192. Prevysh VA, Wang B-C, Khan SA, Spontak RJ (1997) Colloid Polym Sci 275:284
193. Chiba R, Kasai T, Nishio Y (2005) Polym Prepr Jpn 54:838
194. Revol J-F, Bradford H, Giasson J, Marchessault RH, Gray DG (1992) Int J Biol Macro-mol 14:170
195. Dong XM, Kimura T, Revol J-F, Gray DG (1996) Langmuir 12:2076
196. Dong XM, Gray DG (1997) Langmuir 13:2404
197. Araki J, Kuga S (2001) Langmuir 17:4493
198. Revol J-F, Godbout L, Gray DG (1998) J Pulp Paper Sci 24:146
199. Edgar CD, Gray DG (2001) Cellulose 8:5
200. Yue Z, Cowie JMG (2002) Macromolecules 35:6572
201. Dai Q, Huang Y, Shen J (1996) Polym Adv Technol 7:693
202. Kryszewski M, Wojciechowski P (1998) Polym Adv Technol 9:654

203. Bhadani SN, Gray DG (1984) Mol Cryst Liq Cryst (Lett) 102:225
204. Nishio Y, Yamane T, Takahashi T (1985) J Polym Sci Polym Phys Ed 23:1043
205. Nishio Y, Susuki S, Takahashi T (1985) Polym J 17:753
206. Kozakiewicz JJ, Maginess JE (1987) J Appl Polym Sci 34:1109
207. Jiang SH, Huang Y, Shen JR (1995) J Appl Polym Sci 57:493
208. Wang L, Wang X, Huang Y (2004) J Appl Polym Sci 92:213
209. Wang X, Wang L, Huang Y (2005) J Appl Polym Sci 95:724
210. Cowie JMG, Arrighi V, Cameron J, McEwan I, McEwen IJ (2001) Polymer 42:9657
211. Arrighi V, Cowie JMG, Vaqueiro P, Prior KA (2002) Macromolecules 35:7354
212. Cowie JMG, Rodden GI (2002) Polymer 43:3415
213. Suto S, Suzuki K (1995) J Appl Polym Sci 55:139
214. Suto S, Watanabe J, Suzuki K (1999) J Mater Sci 34:6151
215. Suto S, Inoue M (1999) Polymer 40:2455
216. Suto S, Umeda T (1999) Angew Makromol Chem 264:60
217. Suto S, Hasegawa S (2002) J Mater Sci 37:4857
218. Suto S, Kobayashi T (1998) J Appl Polym Sci 70:1465
219. Wojciechowski P, Okrasa L, Ulanski J, Kryszewski M (1996) Adv Mater Opt Electron 6:383
220. Okrasa L, Ulanski J, Boiteux G (2002) Polymer 43:2417
221. Yamagishi T, Sixou P (1995) Polymer 36:2315
222. Hohn W, Tieke B (1997) Macromol Chem Phys 198:703
223. Shimamoto S, Uraki Y, Sano Y (2000) Cellulose 7:347
224. Marsano E, Costa C, Bianchi E (2002) Polymer 43:1021
225. Revol J-F, Godbout L, Dong XM, Gray DG, Chanzy H, Maret G (1994) Liq Cryst 16:127
226. Dong XM, Gray DG (1997) Langmuir 13:3029
227. Orts WJ, Godbout L, Machessault RH, Revol J-F (1998) Macromolecules 31:5717
228. Wang L, Huang Y (2000) Macromolecules 33:7062
229. Wang L, Huang Y (2001) Liq Cryst 28:1673
230. Chiba R, Nishio Y, Miyashita Y (2003) Macromolecules 36:1706
231. Nishio Y, Chiba R (2003) Ekisyo 7:218
232. Müller M, Zentel R, Keller H (1997) Adv Mater 9:159
233. Müller M, Zentel R (2000) Macromol Chem Phys 201:2055
234. Yue Z, Cowie JMG (2002) Polymer 43:4453
235. Yue Z, McEwen IJ, Cowie JMG (2002) J Mater Chem 12:2281
236. Sato T, Banno K, Maruo T, Nozu R (2005) J Power Sources 152:264
237. Arici E, Greiner A, Hou H, Reuning A, Wendorff JH (2000) Macromol Chem Phys 201:2083
238. Greiner A, Hou H, Reuning A, Thomas A, Wendorff JH, Zimmermann S (2003) Cellulose 10:37
239. Wegner G (2003) Macromol Chem Phys 204:347
240. Kobayashi S (2005) J Polym Sci Pt A Polym Chem 43:693
241. Karakawa M, Mikawa Y, Kamitakahara H, Nakatsubo F (2002) J Polym Sci Pt A Polym Chem 40:4167
242. Klemn D, Heublein B, Fink H-P, Bohn A (2005) Angew Chem Int Ed 44:3358

Adv Polym Sci (2006) 205: 153–197
DOI 10.1007/12_098
© Springer-Verlag Berlin Heidelberg 2006
Published online: 2 September 2006

Trapping of Reactive Intermediates
to Study Reaction Mechanisms in Cellulose Chemistry

Thomas Rosenau (✉) · Antje Potthast · Paul Kosma

Department of Chemistry, University of Natural Resources and Applied Life Sciences
Vienna (BOKU), Muthgasse 18, 1190 Vienna, Austria
thomas.rosenau@boku.ac.at

Abstract Reaction with specific scavenger agents, so-called *trapping*, is a direct way to prove the occurrence of reactive intermediates in a reaction system, and thus to elucidate the underlying reaction mechanisms. After an introduction on chemical intermediates, chemical trapping and the peculiarities of trapping methodology in the presence of cellulose, four chapters from cellulose chemistry were selected to illustrate how trapping can be used to determine reactive intermediates, to prove the occurrence of reactive intermediates in the respective reaction system, and to finally help establish the reaction mechanism. The four topics are: radical and ionic intermediates in Lyocell dopes that cause cellulose degradation, the chemistry of cellulose in DMAc/LiCl solutions, aging of alkali cellulose under steeping conditions, and cellulose degradation during carbanilation in dimethyl sulfoxide (DMSO) mixtures.

Keywords Cellulose · Trapping · Reactive intermediates · Reaction mechanism

Abbreviations

DMAc	N,N-dimethylacetamide
DMSO	dimethyl sulfoxide
DP	degree of polymerization
EPR	electron paramagnetic resonance
GPC	gel permeation chromatography
HPCE	high-performance capillary electrophoresis
IR	infrared
NMMO	N-methylmorpholine-N-oxide
NMR	nuclear magnetic resonance
SOMO	single occupied molecular orbital
THF	tetrahydrofuran
TLC	thin-layer chromatography

1
Reactive Intermediates and Trapping Reactions

1.1
Chemical Intermediates

A chemical intermediate is a species that is neither starting material nor product and occurs only in multi-step reactions. The term chemical intermediate should not be mixed up with the term transition state. While the latter portrays the geometry of highest potential energy along the reaction coordinate of an elementary reaction step, the former describes an individual, albeit short-lived, chemical compound with transition states leading to and from it. When generated in a chemical reaction, intermediates will quickly convert into another, more stable, molecule. Most chemical reactions consist of more than one elementary step, and thus inevitably involve chemical intermediates. The series of elemental steps makes up the reaction mechanism. The knowledge on each single step is thus necessary to formulate the complete mechanism of the overall reaction.

If the example reaction A + B → C is considered, it may proceed according to the separate steps A + B → X and X → C; in this case X is the chemical intermediate. The amount of an intermediate present in a reacting system at any instant of time will depend on the rate of the step by which it is formed and the rate of its subsequent reaction. A qualitative indication of the relationship between intermediate concentration and the kinetics of the reaction can be gained by comparing the rate constants of the reactions for intermediate production and consumption, as shown in the following for the most general two-step mechanism: reactants $\xrightarrow{k1}$ intermediate $\xrightarrow{k2}$ products. In some reactions, the situation $k1 > k2$ exists. Under these conditions, the concentration of the intermediate will build up, and the reactants are consumed faster than the products are formed. It will be possible to isolate, or at least to observe, the intermediate. In most reactions, the opposite case $k2 > k1$, or the case that both constants are very large, is found. Only a very small concentration of the intermediate will exist at any time. The reaction proceeds too rapidly to permit isolation of the intermediate, which will thus be a rather transient or short-lived species. Often, the term reactive intermediates – which is also preferred in the following – is used to distinguish such species from the isolable and more stable chemical intermediates in the case $k1 > k2$.

Examples for frequently encountered intermediates in organic reactions are: carbocations (carbenium ions, carbonium ions), carbanions, C-centered radicals, carbenes, O-centered radicals (hydroxyl, alkoxyl, peroxyl, superoxide anion radical etc.), nitrenes, N-centered radicals (aminium, iminium), arynes, to name but a few. Generally, with the exception of so-called persistent radicals which are stabilized by special steric or resonance effects, most radicals belong to the class of reactive intermediates.

1.2
Spectroscopic Detection of Intermediates

Identification of the intermediates in a multi-step reaction is the major objective of studies of reaction mechanisms. It is most useful to study intermediates present in low concentrations without chemical interference with the reacting system, i.e. by rapid spectroscopic methods. The most common methods in organic chemistry include ultraviolet–visible (UV-VIS), IR, and EPR spectroscopy. In principle, all other spectroscopic methods for the detection of reaction intermediates are also applicable provided that they are fast enough to monitor the intermediate and able to provide sufficient structural information to assist in the identification of the transient species.

UV-VIS and IR spectroscopy are often combined with the technique of fast cooling to detect and identify highly unstable intermediates. The quickly decreasing temperature drastically decreases reaction rates and mobility so that it becomes more likely to get a spectroscopic snapshot of the intermediate. An extreme example of this technique is called matrix isolation. In this method, the intermediate is trapped in a solid inert matrix, usually argon or another inert gas, at very low temperatures. Because each molecule is surrounded by inert gas atoms, there is no possibility for further intermolecular reactions and the rates of intramolecular reactions are slowed down by the low temperature. Matrix isolation is a very useful method for characterizing intermediates in photochemical or gas-phase reactions.

Free radicals and other intermediates with unpaired electrons can be detected in extremely low concentrations by electron paramagnetic resonance (EPR). This technique measures the energy absorbed to reorient an electronic spin in a magnetic field. It provides structural information on the basis of splitting of the signal by adjacent nuclei, much as in NMR. EPR is not only very sensitive but also very specific: as diamagnetic molecules present give no signals, the possibility for interference is greatly decreased. The method can only be applied to homolytic reactions, i.e. processes involving paramagnetic intermediates. Because of its sensitivity, it is important to demonstrate that any paramagnetic species detected are true intermediates rather than being involved only in minor pathways.

In the use of all spectroscopic methods, it must be remembered that the simple detection of a species does not prove that it is an intermediate. It must also be shown that the species is converted to the product. Therefore, it is necessary in most cases to determine the kinetics of the production and consumption of the intermediate and to demonstrate that this is consistent with the species being an intermediate, which in most cases is rather difficult.

1.3
Trapping of Reactive Intermediates

In contrast to the spectroscopic techniques, which in most cases provide only an indication for the occurrence of the intermediate but no unambiguous proof, trapping experiments – if successful – are capable of providing solid evidence for the occurrence of a reactive intermediate. In addition, the chemical structure of the intermediate will become evident. For trapping a reactive intermediate, a compound that is expected to react specifically with the intermediate in a well-defined manner is added to the reaction system. This reagent stops or slows down the reaction to the usual reaction product by competitively converting the intermediate into a different product. The intermediate is thus diverted from its normal reaction course, and evidence for the existence of the intermediate is obtained if the structure of the trapped product is consistent with expectation. Occurrence of the expected trapping product confirms the intermediacy, clarifies the structure of the intermediate, and proves the stepwise nature of the overall reaction.

The particular reasons for employing trapping methodology can be manifold:

- if it is not possible to isolate a chemical species or to detect it by spectroscopic means,
- if the concentration of the intermediate is very small and below the detection limit,
- if the occurrence of a species is assumed but not verified,
- if the observed outcome of a reaction needs to be related to an intermediate compound,
- if a reaction takes an unexpected course that must be clarified, or
- if the intermediate concentration at a certain point of time needs to be quantified.

The most prominent application of trapping methodology is to prove the occurrence of chemical species in a reacting system. From this knowledge, the processes leading to the intermediate and its subsequent reactions can be deduced, and the presence of the intermediate can be correlated with the experimental observations. Eventually, this will lead to clarification of the reaction mechanism of the sequence in which the intermediate is involved.

As the whole breadth of chemistry is found for reactive intermediates, the same must in principle be true for trapping agents. However, the well-defined action of trapping agents allows the definition of some general rules as to their chemical and physical properties. An ideal trapping agent should perform the following tasks and exhibit the following properties:

- selective reaction with the intermediate,
- formation of stable trapping products which do not undergo subsequent conversions,
- no parallel reactions with other species in the reaction system,
- much faster reaction rate of the trap with the intermediate than that of the intermediate to the usual products,
- formation of trapping products that unambiguously report the presence of the intermediate (no other reaction that the trapping reaction of the specific intermediate must lead to the trapping product),
- formation of a trapping product that can be separated from the reaction mixture.

1.4
Trapping Methodology in Cellulose Chemistry

In general, all the above characteristics of trapping agents also apply to reactions involving cellulose. However, there are some peculiarities if trapping methodology is to be applied in cellulose chemistry, which make the approach even more intricate than in conventional organic chemistry.

- Reactions of intermediates with cellulose will be polymer-analogous reactions. The reaction will occur more or less randomly along the polymer chain and will thus likely be difficult to prove or to relate to specific locations along the polymer chain.
- Typical reactions of cellulose, such as oxidation or chain cleavage, can be caused by many different reagents and conditions. It will thus be difficult to relate the reaction outcome to the presence of a specific intermediate.
- In the case of cellulose solutions, the reaction mixture will be rather viscous, which limits the intimate admixing of the trapping agent and quick reaction with the species to be trapped.
- In heterogeneous reactions involving cellulose, the accessibility for the trapping agent is a major problem.
- In both homogeneous and heterogeneous mixtures involving cellulose, the system is likely to be rather complex, so that separation and identification of trapping products might prove difficult.

In the following, four selected chapters from cellulose chemistry will illustrate how trapping was used in the hunt for reactive intermediates.

The first section covers the chemistry of cellulose solutions in an amine N-oxide solvent (NMMO), the so-called Lyocell chemistry, as encountered in the industrial production of cellulosic Lyocell material. The system is characterized by high reaction temperatures, the presence of a strong oxidant and high complexity by multiple (homolytic and heterolytic) parallel reactions. Trapping was used to address the questions that reactive intermediates are present in Lyocell solutions and are responsible for the observed side-reactions and degradation processes of both solvent and solute.

The second chapter deals with cellulose solutions in yet another solvent system for cellulose, namely DMAc/LiCl, which is not used on an industrial scale as is NMMO, but on the laboratory scale for analytical purposes. The presence of the somewhat exotic reaction medium poses special requirements on trapping methodology that was used to clarify the mechanisms of different degradation processes. This issue was of importance since maintenance of cellulose integrity is the key prerequisite for any analytical procedure which should report the polymer characteristics of the genuine cellulosic material.

The third chapter is concerned with the detection of a highly reactive intermediate, hydroxyl radicals, upon alkalization of cellulose. This processing step is frequently used in cellulose chemistry, especially in industrial-scale production of cellulose ethers and viscose. The presence of hydroxyl radicals with their inherent high reactivity and low selectivity would have implications as to the appropriate choice of reaction conditions and the underlying reaction mechanisms. However, the extreme conditions – working in concentrated lye – made application of trapping methodology especially challenging.

The final section addresses degradation and oxidation reactions in a commonly used derivatization system for cellulose, a mixture of DMSO and phenyl isocyanate to achieve cellulose carbanilation, e.g. for analytical purposes. Mechanistic studies were aimed at verifying the assumed oxidative action of this reaction system, and trapping methodology was employed to detect responsible intermediates.

2
Cellulose Solutions in NMMO (Lyocell)

2.1
Cellulose and NMMO Degradation in Lyocell Dopes

N-Methylmorpholine-N-oxide monohydrate, a tertiary, aliphatic amine N-oxide, is able to dissolve cellulose directly, i.e. without chemical derivatization, which is used on an industrial scale as the basis of the Lyocell process [1, 2]. This technology only requires a comparatively low number of process steps compared for instance to traditional viscose production. Cellulose material – mainly fibers – are directly obtained from the cellulose solution in NMMO; no chemical derivatization, such as alkalization and xanthation for rayon fibers, is required [3]. The main advantage of the Lyocell process lies in its environmental compatibility: very few process chemicals are applied, and in the idealized case NMMO and water are completely recycled, which is also an important economic factor. Even in industrial production systems NMMO recovery is greater than 99%. Thus, compared with cotton and viscose the Lyocell process pertains a significantly lower specific environmental challenge [4]. Today, Lyocell fibers are produced on an industrial scale, and other cellulosic products, such as films, beads, membranes and filaments, are also currently being developed or are already produced commercially.

As NMMO is a solid at room temperature, dissolution and processing of the spinning dope require elevated temperatures of about 100 °C. The dope is spun into an air gap and water, where cellulose is regenerated and NMMO is washed out. After purification and evaporation of the water, the amine N-oxide is reintroduced into the system and used again for cellulose dissolution.

Ideally, dissolution of cellulose in the amine N-oxide is supposed to be an entirely physical process without any chemical changes of pulp or solvent. However, in real-world processes there are several chemical processes observed, which cause formation of appreciable amounts of byproducts. A strong discoloration of the solution due to chromophore formation has been observed, which is accompanied by degradation of both the solute cellulose and the solvent NMMO at the elevated process temperatures, which in turn can provoke very severe effects, such as degradation of cellulose, temporary or permanent discoloration of the resulting fibers, decreased product performance, accelerated decomposition of NMMO and increased consumption of stabilizers. The most drastic effect of side-reactions are degradation processes that become uncontrollable, in the literature often denoted with the euphemistic terms "fast exothermic process" or "thermal runaway reaction" [5–7]. The question of the chemical stability of the system has always been a crucial issue for the development of Lyocell technology, and is still a major concern today as it is related to safety at work and integrity of the whole production line. Research on degradation processes in NMMO/cellulose mixtures in different industrial companies have therefore largely focused on the empirical search for new and better stabilizers for Lyocell dopes.

It was our aim to study the chemistry of NMMO and Lyocell solutions in order to put the prevention of undesired side-effects on a more scientific foundation [8, 9]. The improved understanding of the individual chemical processes in Lyocell solutions today allows an accurate deduction of the different tasks of optimum Lyocell stabilizers. The question of which reactive intermediates are involved in the degradation processes of both NMMO as the solvent and cellulose as the solute was a key issue in the studies of Lyocell chemistry.

2.2
NMMO-derived Radical Intermediates

2.2.1
Formation and Occurrence of NMMO-Derived Radicals

The positive effect of classical phenolic antioxidants on the stability of Lyocell solutions suggested that radicals were involved in the degradation processes. Analogous to tertiary amine oxides in general [10, 11], the primary radical species derived from NMMO was assumed to be the N-centered cation radical 3, which is generated by cleavage of the N – O bond. For this breakage to occur, activation of the *exo*-oxygen by protonation and concomitant one-electron reduction is required, finally producing 3 and a hydroxyl anion. The same species is obtained by one-electron oxidation of the corresponding tertiary amine, in this case N-methylmorpholine (2). Aminium cation radicals such as 3 are generally very labile and have a pronounced tendency towards mesolytic cleavage of the $C_\alpha - H_\beta$ bond. Mesolysis, in addition to the well-known terms homolysis and heterolysis, denotes a bond cleavage in radical ions that separates the charge from the single electron [12]. This extremely fast process results in uncharged C-centered α-amino radicals [13, 14]. Hence, also the primary cation radical 3 was expected to produce immediately neutral carbon-centered radicals by release of a β-proton, either from the exocyclic N-methyl group leading to 4, or from the ring N-methylene leading to 5. In accordance with general amine and amine N-oxide chemistry, cation

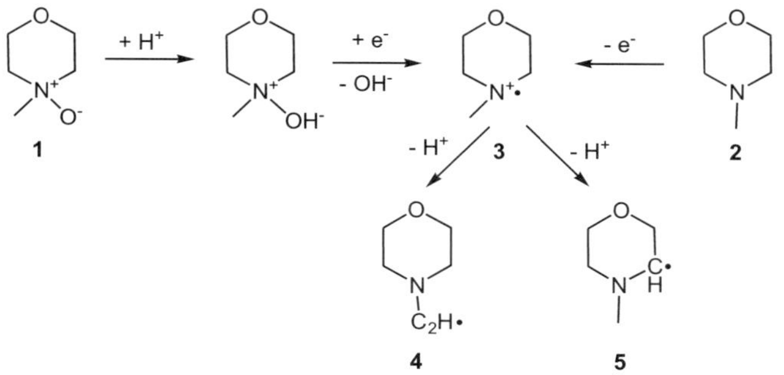

Scheme 1 Primary radical species derived from NMMO

radical **3**, the *exo*-centered radical **4**, and the ring-centered radical **5** are the three main initial intermediates in homolytic reactions of NMMO (Scheme 1).

Cation radicals of tertiary amines without steric hindrance around the nitrogen, such as **3**, are rather intricate to investigate by EPR spectroscopy. The detection mostly requires UV or γ-irradiation in inert matrices at low temperatures and provides only badly resolved spectra [15, 16]. EPR was therefore considered inappropriate for the identification of radicals present in NMMO reaction mixtures.

2.2.2
Trapping of NMMO-Derived Radicals

The application of conventional spin traps, such as substituted pyrroline-*N*-oxides, nitrones, or sterically hindered phenols, has been reported to fail in the presence of large amounts of tertiary amine oxides [10, 11]. For successful trapping of the radicals, γ-tocopherol (**6**), a component of natural vitamin E, was used as the specific trapping agent. Both the trapping agent and its products offer the advantage of supreme extractability into apolar solvents, such as *n*-hexane or petrol ether, due to the strongly lipophilic isoprenoid side-chain. Thus, even from very complex mixtures, they can readily be separated. From the chemical point of view, the well-defined coupling reaction was a clear benefit, as was the near absence of self-coupling reactions, in contrast to other phenolic spin traps.

The interaction of γ-tocopherol (**6**) with radicals generates the relatively stable γ-tocopheroxyl radical (**7**). While the *O*-centered form of the γ-tocopheroxyl radical has a lower affinity towards other radicals and forms relatively labile tocopheryl ether products, its *C*-centered resonance structure **7a**, with the radical being located at C-5, readily recombines with other radicals present in the mixture to give stable compounds. This way, trapping products of all three radicals **3**, **4**, and **5** were isolated and fully analytically characterized (Scheme 2) [17].

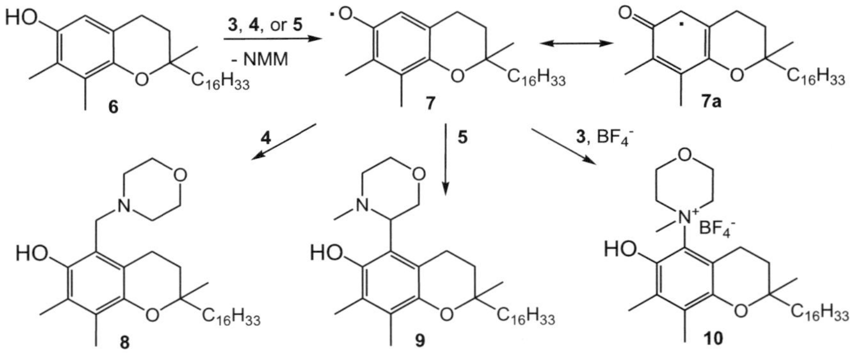

Scheme 2 Trapping of the NMMO-derived radicals 3, 4 and 5 with the trapping agent γ-tocopherol (6)

The production of the radicals was carried out by reduction of NMMO with Fe(II) or by oxidation of N-methylmorpholine with Fe(III), at different temperatures. Other reductants and oxidants gave nearly identical results. The similar outcome of NMMO reduction and NMM oxidation was already a clear indication that the structure of the radical intermediates was correctly proposed – apart from the definite proof by the chemical structure of the trapped radical intermediates.

Compound 8, the major trapping product, originated from coupling of radical 4 with 7a. Trapped 5 was also observed, however in significantly smaller amounts. This agreed very well with theoretical considerations based on computations on the ab initio level, which predict a lower stability of 5 compared to 4. Also the trapping product of the primary cation radical 3, the quaternary ammonium cation 10, was isolated. Only very small amounts of 10 were obtained, and purification was possible by precipitation as the corresponding tetrafluoroborate salt. The rather low yield of 10 was due to three reasons: first, the stability of 3, and thus its equilibrium concentration, is very low as it readily loses a β-proton to give the more stable 4 or 5. Second, tertiary amine cation radicals recombine rather slowly with γ-tocopheroxyl radicals: while the recombination of γ-tocopheroxyl radicals with carbon-centered radicals is a diffusion-limited process, with the rate being largely independent of the nature of the radical, the rate constant for the recombination with amine cation radicals is about three orders of magnitude smaller [18]. At third, the trapping product is extremely thermolabile and sensitive towards oxidation. Even short exposure to air at ambient conditions was sufficient to cause complete oxidative conversion into red 5,6-tocopheryldione, so-called α-tocored.

Care has been taken to rule out that the coupling products could have formed according to heterolytic (nonradical) pathways involving Mannich intermediates. In short, a modified trapping agent, γ-tocopherol methyl ether, was used which would exclusively react by electrophilic substitution, but not homolytically as from the blocked phenolic OH group neither the phenoxyl radical nor the *C*-centered resonance form could form. The fact that no trapping products were found in this case demonstrated that the formation of **8–10** was indeed caused by homolytic reactions.

The trapping reaction with γ-tocopherol (**6**) was also employed to prove the occurrence of the NMMO-derived radicals directly under Lyocell conditions, and not only in organic solutions of NMMO at ambient temperatures or below. While the trapping product of the *N*-centered radical was too labile under the prevailing conditions, trapping product **9** from radical **5** was indeed detected. For trapping *C*-centered radical **4**, the procedure was modified since trapping product **8** was thermally unstable above 60 °C, eliminating morpholine to produce the *ortho*-quinone methide **11** which immediately dimerized to spiro-dimer **12** (Scheme 3). Both **11** and **12** are typical products in the chemistry of α-tocopherol which differs from the employed trapping reagent γ-tocopherol by an additional methyl group in position 5. In the trapping product **8** as well as in *ortho*-quinone methide intermediate **11** and spiro-dimer **12** the additional carbon atoms originated from the *exo*-CH$_2$ group of the radical **4**. This was demonstrated by applying selectively deuterated NMMO (**1-D$_3$**) as the starting material [19]. Coupling of the corresponding radical **4-D$_2$** produced the bisdeuterated coupling product **8-D$_2$** which finally gave the tetradeuterated dimer **12-D$_4$**. Since the deuterium can only arise from the methyl group in the starting material NMMO, the occurrence of the labeled product is a direct proof of **4** as the intermediate.

Ab initio computations predicted the *exo*-centered radical **4** to be more stable by 2.3 kJ/mol (energy of formation) than the ring-centered radical **5**. With

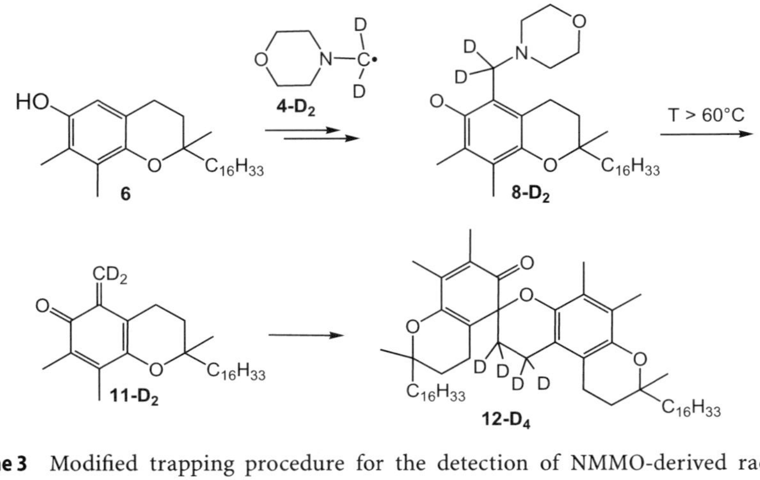

Scheme 3 Modified trapping procedure for the detection of NMMO-derived radical **4** under Lyocell conditions

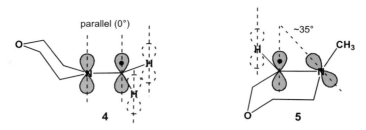

Scheme 4 Illustration of hyperconjugative stabilization and approximate geometries, leading to higher stability of radical **4** compared to radical **5**

the calculated activation energy difference $\Delta(\Delta E^{\#})$, the equilibrium ratios of **4** and **5** are given by $N_5/N_4 = \exp - (\Delta(\Delta E^{\#})/RT)$, assuming an irreversible formation reaction for **4** and **5**. The theoretical values calculated from this equation were in superb agreement with the experimental ones determined by measuring the ratios between these two radicals in the form of their trapping products at different temperatures.

Optimum hyperconjugative stabilization in **4** and the lack of steric strain account for its higher stability compared to **5**. Generally, aminyl radicals experience hyperconjugative stabilization of the SOMO by neighboring orbitals. The *p*-orbital SOMO in **4** can fully overlap with the two pseudo-*p* orbitals of the remaining two methylene protons and with the nitrogen *n–p* orbital, so that the hyperconjugative stabilization reaches its optimum. In contrast, in **5** there is only one pseudo-*p* orbital at the ring proton, and the nitrogen *n–p* orbital overlaps less effectively (34.6°) with the SOMO, so that the stabilization effect is much smaller. Furthermore, the spin-bearing carbon – as an sp^2 hybrid with a singly occupied *p* orbital – has nearly planar geometry. This means no steric strain at the *exo*-carbon of **4**, but severe twisting of the chair conformation for the ring-centered radical **5**, rendering the latter energetically unfavorable (Scheme 4). These effects account for the smaller stability of **5**, which is reflected experimentally by the fact that the trapping product of **4** dominated over that of **5** in all cases.

2.2.3
Radical Recombination of NMMO-Derived Radicals

The reaction of two radicals can be seen as a type of self-trapping, with one molecule of the radical species consuming another one in a recombination process. For this process to be observed, the radical concentration must be sufficiently high that the probability of two radicals encountering one another is increased and the recombination products are accumulated to detectable amounts. In the case of the *C*-centered NMMO-derived radicals **4** and **5**, two of the three possible recombination products were identified (Scheme 5).

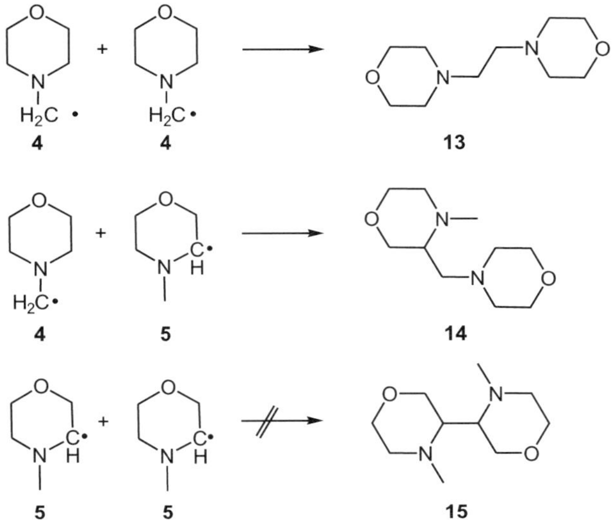

Scheme 5 Radical recombinations with C–C bond formation, involving the NMMO-derived C-centered radicals **4** and **5**

The proof of their structure was again provided by separation and analytical characterization of the pure compound. The symmetrical coupling product 1,2-bis(4-morpholino)ethane (**13**) was formed by recombination of two molecules of **4**. By analogy, **4** and **5** yielded 3-(4'-morpholinomethyl)-4-methylmorpholine (**14**). Compound **15**, as the self-trapping product of the ring-centered radical **5**, was not found [17]. Although likely to have formed, it was present only in concentrations too small for detection: first, the absolute concentration of **5** and thus the probability of two molecules of **5** to meet were rather low and, second, the steric conditions for the recombination of two secondary radicals were much more demanding than for the reaction between **4** and **4**, or **4** and **5**, respectively.

In the literature, the simultaneous formation of the major NMMO degradation products N-methylmorpholine, morpholine and formaldehyde [20] is always attributed to the disproportionation of the primary aminyl radical **3**, as a – not further defined – redox process between two molecules of **3**, in which one is reduced to N-methylmorpholine (**2**) and the other oxidized to N-(methylene)morpholinium cation (**17**), which upon addition of water, forms morpholine and formaldehyde. Also this disproportionation can be rationalized as a radical coupling reaction which proceeds through recombination of N-centered **3** and C-centered **4**, via the ammonium aminal intermediate **16** as the actual recombination product (Scheme 6). The intermediacy of this species was indeed confirmed by isolation from the reaction mixture and full characterization [17]. Compound **16** is quite labile and immediately de-

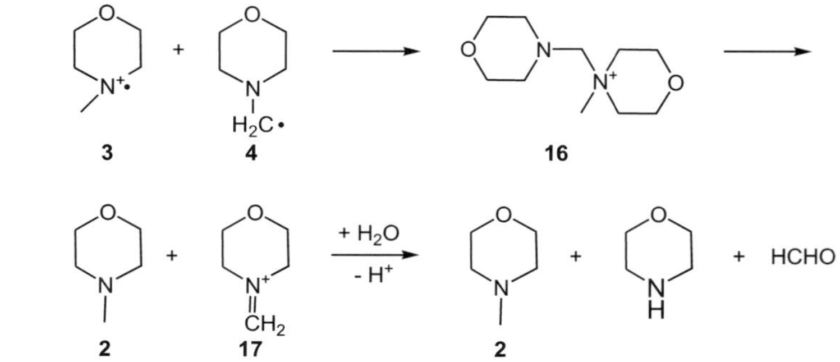

Scheme 6 Disproportionation of the NMMO-derived radicals

composed in quantitative yields into *N*-methylmorpholine, morpholine and formaldehyde when entering aqueous media – exactly the macroscopically observed outcome of the disproportionation.

2.2.4
Reactions of the NMMO-Derived Radicals with Cellulose

The NMMO-derived radicals are strongly electron-deficient and thus strongly oxidizing species. By analogy to hydroperoxyl radicals, they will react with electron-rich positions in cellulose or model compounds. The preferred positions for attack will be the CH-acidic groups in the α-position to carbonyl functions. Trialkylaminyl radicals react with α-hydroxyacids to give α-ketoacids, and with aldoses to furnish 2-ketoaldoses [21]. The main result of the action of NMMO-derived radical 3–5 on pulp will be the random introduction of keto groups into the 2-position of the anhydroglucose repeating

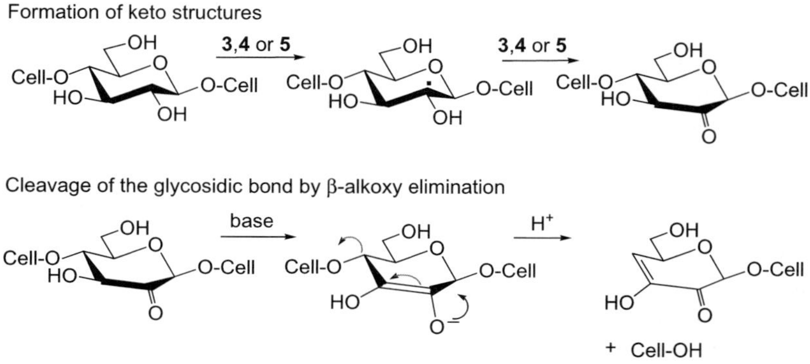

Scheme 7 Cellulose-chain scission as a consequence of homolytic reactions involving NMMO-derived radicals

units of cellulose, finally leading to chain cleavage by β-elimination and thus, a decreased DP (Scheme 7).

Typical values given in the literature for Lyocell dopes without stabilizers added are DP losses from 472 to 177 over 6 hours at 105 °C [22], or from 570 to 185 over 2 hours at 90 °C [23]. The activation energy for the cellulose chain cleavage in NMMO at temperatures above 115 °C was determined to be 69 kJ/mol. This observation strongly supported the hypothesis that the actual rate-determining step of the cellulose chain cleavage was a β-alkoxy elimination, with an activation energy of 67–72 kJ/mol [24, 25].

2.3
NMMO-Derived Ionic Intermediates

2.3.1
Formation and Trapping of Carbenium-iminium Ions

In the early phases of Lyocell research, degradation reactions of NMMO and cellulose solutions in NMMO were mainly thought to be homolytic (radical) processes, mostly sustained by the beneficial action of antioxidant additives, such as propyl gallate, which were assumed to act merely as a radical scavenger. However, the radical pathways could hardly account for uncontrollable degradation reactions, which on one hand would occur rather unpredictably also in the presence of stabilizers and on the other hand were not inducible by typical radical initiators. With the finding of nonradical, autocatalytic decomposition processes and *Polonowski*-type degradation reactions [26, 27], studies in Lyocell chemistry turned to those heterolytic (ionic) degradation processes, which were established as the cause of the observed exothermic processes – showing the system of side-reactions in the Lyocell system to be much more complex than commonly thought.

To prove the presence of formaldehyde and NMMO-derived carbenium-iminium ions under Lyocell mixtures was a key issue here, as the occurrence of the latter is critical with regard to autocatalytic degradation and instabilities of Lyocell solutions. Again, trapping methodology was used for this purpose, it served also to investigate the general chemistry of the NMMO-derived carbenium-iminium ions.

Formaldehyde is the aldehyde with the highest carbonyl reactivity: it reacts rather fast with any suitable coreactant present, for instance with cellulose, causing the formation of O-hydroxymethyl groups and methylene crosslinks. The critical process with regard to NMMO degradation is reaction with the NMMO-degradation product morpholine to form N-(methylene)morpholinium cations (17) via N-hydroxymethylmorpholine in neutral and acidic media, cf. Scheme 8. N-(Methylene)morpholinium cations (17) belong to the compound class of carbenium-iminium ions or so-called *Mannich* intermediates, they induce the autocatalytic decomposition of NMMO

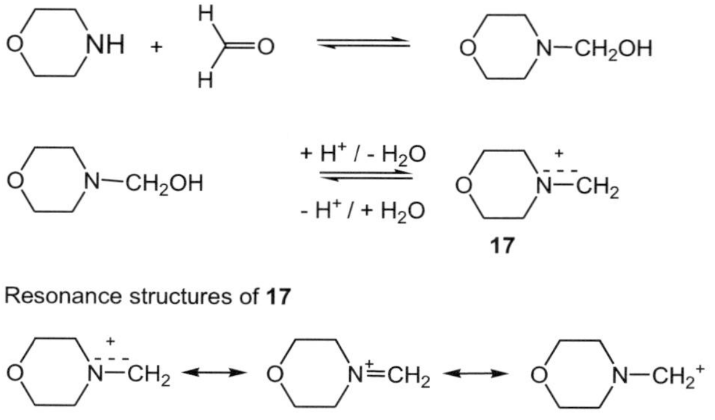

Scheme 8 The role of formaldehyde in the formation of carbenium-iminium ion 17 and resonance stabilization of the latter

and other tertiary amine N-oxides [26]. The detection of formaldehyde as one precursor of 17 was a prerequisite in confirming the decisive role of carbenium-iminium ions in Lyocell chemistry.

The determination of HCHO in the NMMO/water/cellulose reaction system was unusually difficult as all commonly employed reactions to determine aldehydes were unsuitable since either the trapping reagents were unstable under the prevailing reaction conditions, or they reacted with the carbohydrates present in the system. It should be recalled that working under Lyocell conditions meant performing the reactions in a viscous melt of a relatively strong oxidant in the presence of about 10% dissolved cellulose at temperatures of about 100 °C. Dilution of the system with water followed by determination of formaldehyde in the resulting aqueous mixture was not feasible, since determination of HCHO would not necessarily prove its presence in the Lyocell dope, but only in the aqueous extracts: there remained the possibility that the HCHO originated from hydrolytic reactions. A good trapping agent for HCHO had to react specifically with HCHO immediately upon its formation, to be sufficient stable under the drastic reaction conditions, and to form stable and separable products.

The trapping of formaldehyde in Lyocell dopes under processing conditions was performed using the trapping agent 5,5-dimethylcyclohexa-1,3-dione (dimedone, 18) in a two-phase system with o-dichlorobenzene as the organic phase and the cellulose/water/NMMO mixture as the "aqueous" phase [28]. Dimedone reacts with aldehydes in a well-defined reaction [29, 30]. It is insoluble in o-dichlorobenzene, whereas its reaction product with HCHO is soluble in this solvent and is thus continuously extracted and prevented from side reactions with NMMO. o-Dichlorobenzene has a sufficiently high boiling point (180 °C) to have a negligible vapor pressure at the Lyocell working temperature, and it is completely inert under the reaction conditions so that it does not interfere with the processes in the cellulose/water/NMMO layer in any way. Addition of dimedone into the Lyocell dope (0.5% relative to NMMO*H$_2$O) resulted in the formation of the dimedone–formaldehyde adduct 19, which upon formation was immediately extracted into the organic phase (Scheme 9). The solution characteristics were so favorable that only trace amounts of other compounds besides the desired product 19 were extracted. This was also reflected by the fact that the organic phase remained nearly colorless while the NMMO phase quickly turned dark in the reaction.

After having established the presence of HCHO in Lyocell dopes, trapping methodology was used once more to demonstrate the presence of N-(methylene)morpholinium cations (17). A Mannich-type conversion [31, 32] was certainly the most specific reaction of carbenium-iminium ions to be used for this task: the carbenium-iminium ions needed in this reaction would not be formed by separate reaction between secondary amine and aldehyde, but would already be present in the system as intermediates to be trapped. The actual trapping agent thus needed to be a suitable methylene-active carbonyl compound. However, the above described restrictions to the selection of a trapping reagent apply analogously, and first of all the trapping reagent must be stable under Lyocell conditions. This requirement eliminated the application of aliphatic aldehydes as trapping reagents. In addition, the products of the reaction, usually named Mannich bases, had to be stable and easily isolable. This turned out to be a major impediment as many Mannich bases readily eliminate the secondary amine group to form an α,β-unsaturated carbonyl compound.

After testing several arylmethylketones, 2-acetonaphthone (20) was found to be a very suitable reagent, which was used to demonstrate the in situ formation of N-(methylene)morpholinium cations in NMMO solutions of cellulose by formation of the Mannich base 3-(4-morpholino)propionaphthone (21), see Scheme 10 [28]. This trapping product did not eliminate morpholine at temperatures below 120 °C and was easily extractable into chloroform due to the relatively lipophilic naphthene structure. Although the yield of the trapping reaction was rather low depending on the respective conditions – between 0.1 and 5% relative to applied trapping agent – the ease of isolation of the compound made the trapping procedure quite reliable and reproducible so that different Lyocell dopes or different reaction conditions could be com-

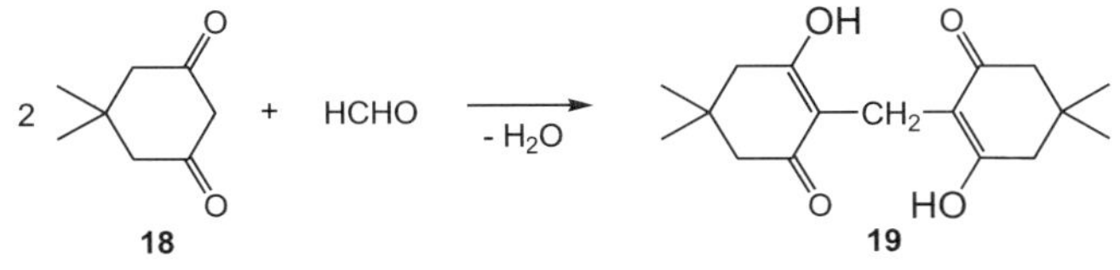

Scheme 9 Trapping of formaldehyde with dimedone (18) under Lyocell conditions in the two-phase system Lyocell dope/o-dichlorobenzene

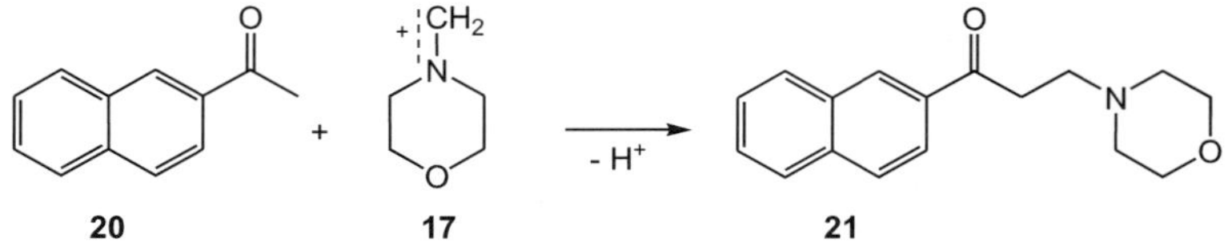

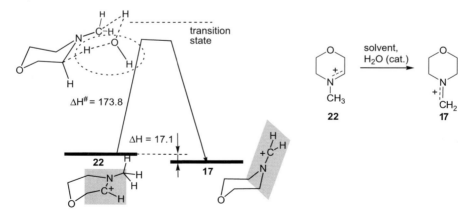

Scheme 10 Trapping of *N*-(methylene)morpholinium cations (**17**) in Lyocell dopes by 2-acetonaphthone (**20**)

pared with regard to the content of **6** by determining the amounts of trapping product formed.

2.3.2
Interconversion of NMMO-Derived Carbenium-iminium Ions

Trapping with 2-acetonaphthone was also used to clarify a peculiar intercon-version reaction between carbenium-iminium ions derived from NMMO [33]. According to theory, two such species were to be expected, the *exo*-centered carbenium–iminium ion **17** and the *ring*-centered carbenium–iminium ion **22** (Scheme 11). Even though **22** might appear more stable as its ring-centered positive charge seems better accommodated at a secondary position than the "exposed" charge in **17**, it was the latter which occurred predominantly in all different types of side-reactions starting from NMMO. In fact, the inter-mediate **17** was produced exclusively in NMMO reaction mixtures containing water – thus also under Lyocell conditions, whereas in nonaqueous solutions

Scheme 11 Carbenium-iminium ion conversion of **22** into **17**: schematic representation of the computed reactant and transition state geometries as well as reaction energies. The trigonal planar environments of the carbenoid carbons are *shaded in gray*. The water molecule participating in the transition state is *circled by a dotted line*

of NMMO smaller amounts of **22** were also found. Thus, both the presence of water and higher temperatures generally seemed to promote the consumption of **22** in favor of **17**.

Besides the trapping methodology, a combined approach consisting of isotopic labeling, kinetic studies, and computations on the density-functional theory (DFT) level was used to clarify the mechanism (Scheme 11) [33]. The NMMO-derived *ring*-centered carbenium-iminium ion **22** is energetically disfavored by ring strain imposed by the trigonal planar geometry of the carbenium carbon. In the presence of water it is rearranged into its counterpart, *N*-(methylene)morpholinium cation (**17**), which has no ring strain as the carbenium center is exocyclic. The reaction is an endothermic, bimolecular process proceeding via a highly organized transition state that involves one molecule of water. The reaction was the first example of a carbenium-iminium ion interconversion reported, which in addition was in complete agreement with the empirical data from Lyocell process chemistry as it explained the nearly complete absence of **22** and the observed predominance of **17**.

2.3.3
Reactions of NMMO-Derived Carbenium-iminium Ions with Cellulose

From the observation that *N*-(methylene)morpholinium cations (**17**) induced the autocatalytic decomposition of tertiary amine *N*-oxides in combination with the proof that these intermediates were actually present in Lyocell dopes, the question arose why Lyocell solutions were stable at all. The answer is found in the fact that carbenium-iminium ions generated are immediately consumed by reaction with water and carbohydrate structures as the most abundant and most reactive nucleophiles present. Only when the local concentration of **17** increases to such a high level that the consumption by reaction with water or cellulose cannot compensate for its production, does the reaction with NMMO become uncontrollable and lead to an exothermic event. The pulp used in the Lyocell process acts as a quasi-stabilizer for the solvent NMMO, albeit with the drawback of increased chromophore generation.

The strongly electrophilic *N*-(methylene)morpholinium cation attacks CH-acidic positions in cellulose, mainly neighboring a keto, acetal or eneamine structure, which basically constitutes a classical Mannich reaction. With α-hydroxy-carbonyl compounds, among them aldoses, ketoses and the anhydroglucose units in cellulose, Mannich intermediates react according to a multi-step sequence that formally exchanges the hydroxyl group with a formyl group to produce 1,3-dicarbonyl compounds, see Scheme 12 [34–36]. These highly reactive species will immediately undergo subsequent reactions, especially under the drastic conditions of Lyocell processing, which eventually cause formation of deeply colored compounds (**23–26**),

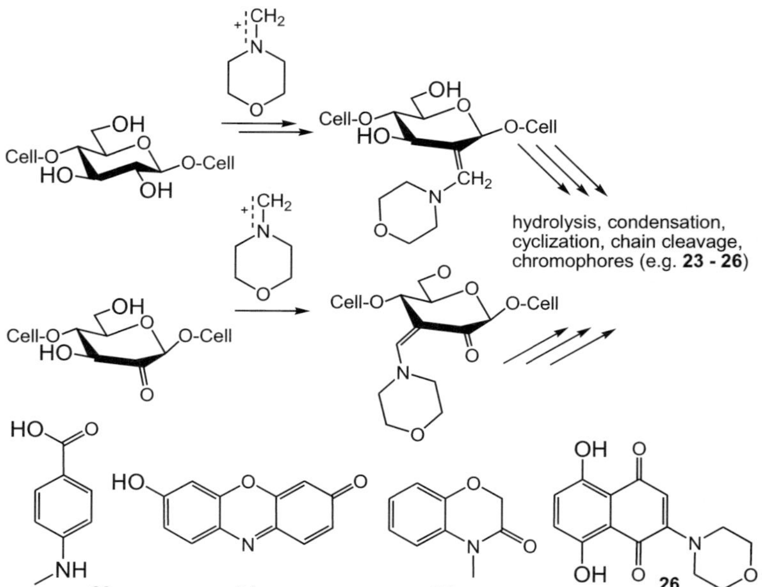

Scheme 12 Reactions of N-(methylene)morpholinium cations (**17**) with cellulose and oxidized structures in cellulose and formation of Lyocell-specific chromophores (**23–26**)

see Scheme 12. These chromophores are either released into the NMMO as the solvent or remain attached to the cellulose. Indeed, such chromophores as condensation products of NMMO degradation compounds and cellulose degradation compounds were isolated from Lyocell fiber material according to a novel chromophore analysis procedure [37, 38]. Compounds **23–26** proved generally the participation of NMMO or NMMO-derived intermediates in the formation of colored degradation products, since the latter contained nitrogen, of which NMMO was the only possible source. Naphthoquinone **26** even included an intact morpholine ring that indicated the likely participation of the carbenium-iminium ion **17** in the formation.

2.4
Trapping of Unsaturated Intermediates
Upon Thermal Degradation of NMMO

The mechanism of the thermal decomposition of NMMO and Lyocell solutions is extremely complex since the reaction, initiated by the action of carbenium-iminium ions, quickly enters an uncontrollable course. A central question was whether the heterocyclic ring of NMMO was cleaved during the reaction and whether products of this cleavage, having either vinyl ether or enamine structures, could be detected. Employing the trapping agents phencyclone (**27**) and tetracyclone (**28**) several vinyl ether and enamine-type compounds were trapped in situ in the reaction mixture during thermal degradation of Lyocell dopes (Scheme 13) [39]. These trapping reagents represent electron-deficient dienes, which would react with electron-rich dienophiles, such as the degradation products, in a Diels–Alder type [4 + 2]-cycloaddition. The reagents are relatively inert and suitable also for use under rather drastic conditions. The reaction products again offer good extractability into aromatic solvents, from which the products can be separated and identified. At the high temperatures present, the primary addition products undergo elimination and decarbonylative aromatization to a substituted triphenylene **29**, so that in the trapping mixture both the primary adducts and the aromatization products were found. The trapped olefinic intermediates originated from cleavage of the O – C and N – C bonds, respectively, of the morpholine ring and thus proved its destruction during exothermic events in Lyocell dopes.

Vinyl ether structures from the ring fragmentation of NMMO have also been identified in the gaseous phase from uncontrolled degradation reactions under Lyocell conditions, again by trapping methodology [8]. For this purpose, the hot reaction gases were introduced into an ethanolic solution of 5a-bromo-α-tocopherol (**30**). The local heating by the gas caused local formation of the *ortho*-quinone methide **11** as the actual trapping reagent. In the absence of vinyl-type structures, dimerization to the spiro-dimer **12** occurred. If vinyl ethers were present, corresponding trapping products were

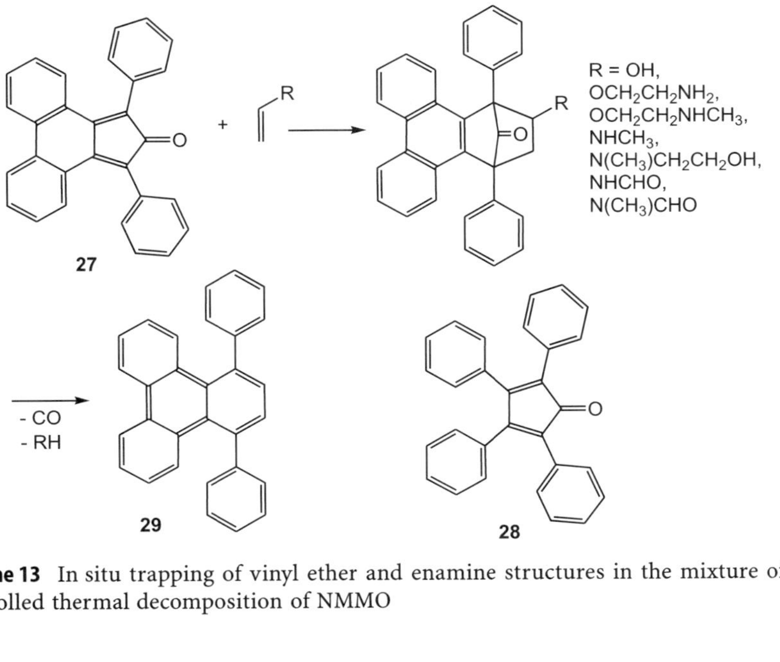

$$R = OH,$$
$$OCH_2CH_2NH_2,$$
$$OCH_2CH_2NHCH_3,$$
$$NHCH_3,$$
$$N(CH_3)CH_2CH_2OH,$$
$$NHCHO,$$
$$N(CH_3)CHO$$

Scheme 13 In situ trapping of vinyl ether and enamine structures in the mixture of uncontrolled thermal decomposition of NMMO

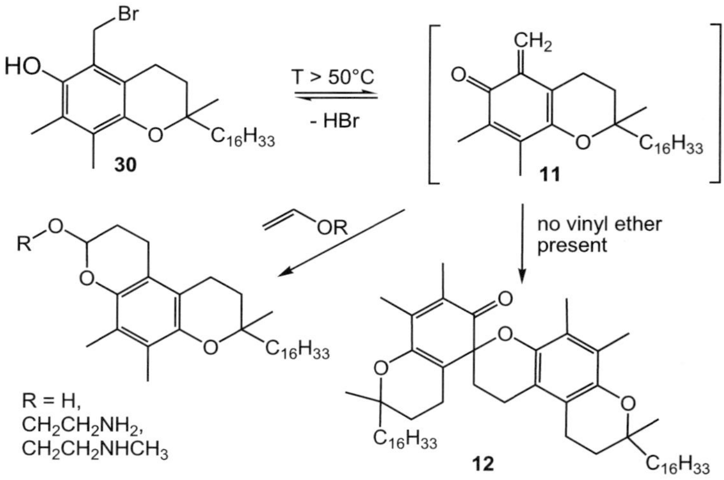

Scheme 14 Uncontrolled thermal decomposition of NMMO: in situ trapping of vinyl ether structures in the gas phase of the reaction

obtained (Scheme 14), which were readily extracted into *n*-hexane due to the lipophilic side chain. No reaction occurred with enamine double bonds. This way, the low-molecular-weight vinyl ethers in the gas phase of uncontrolled degradation reactions of NMMO were demonstrated to be similar to those found directly in the degradation mixture by trapping with the cyclone reagents. The application of 5a-bromo-α-tocopherol (**30**) directly in the reaction mixture instead of the phencyclone (**27**) or tetracyclone (**28**) traps – and not just for the gaseous phase – was not feasible as the reagent is degraded to the *ortho*-quinone methide intermediate **11** at temperatures above 50 °C, much lower than the temperatures of the degradation. Before eventually reaching the degradation temperatures, the trapping reagent would have been completely consumed and converted into the spiro-dimer **12**, which possessed no further trapping ability.

Origin and formation pathways of the different vinyl ether and enamine structures in mixtures from the uncontrolled NMMO degradation were difficult to assess, since the high temperatures during such processes also allowed disfavored decomposition processes to proceed, which are thermodynamically forbidden or disfavored under the usual Lyocell process conditions.

2.5
Summary

In summary, trapping methodology helped to confirm the presence of the three main NMMO-derived radical species, the nitrogen-centered radical cation **3** and the carbon-centered radicals **4** and **5** in Lyocell dopes. Trapping was not only performed by externally added trapping agent, but products from radical recombination reactions as a kind of internal trapping were also observed. The occurrence of NMMO-derived carbenium-iminium ion **17** and formaldehyde as its precursor compound were shown by trapping under Lyocell conditions. Trapping methodology was also used to study carbenium-iminium ion interconversions. Also intermediate vinyl ether and enamine-type degradation products in thermal decomposition mixtures of NMMO and the exhausted gases were detected by trapping.

3
Cellulose Solutions in DMAc/LiCl

3.1
Degradation of Solvent and Cellulose

N,N-Dimethylacetamide (DMAc, **31**) containing lithium chloride is a solvent system that is capable of dissolving cellulose within certain concentration ranges of LiCl and pulp. It has thus been very frequently used in polysaccharide chemistry [40, 41, 43]. First of all, it has become a standard solvent for gel permeation chromatography (GPC) measurements of cellulosics [44–49]. For dissolving pulps and low-molecular-weight pulps, dissolution occurs neatly. As paper pulps and some higher-molecular-weight dissolving pulps are largely insoluble in DMAc/LiCl, several activation procedures have been proposed, which aim to accelerate the dissolution in the case of soluble pulps, and to make dissolution possible at all in the case of initially insoluble pulps. All activation treatments are thought to cause intra- and intercrystallite swelling, breaking of hydrogen bonds and increased accessibility. Among those activation procedures are the treatment with liquid ammonia [50, 51], swelling in water followed by solvent exchange to ethanol and DMAc [52, 53], freeze-drying [54], or swelling in 0.1 M LiCl followed by a number of extraction steps [55].

A very common protocol is also heating or refluxing the cellulose samples in DMAc, or in DMAc containing low concentrations of LiCl [56, 57]. A yellowing of the heated mixture has been noticed, sometimes accompanied by a yellowish discoloration of the activated pulp sample. The chemistry of heated cellulose solutions in DMAc and DMAc/LiCl was investigated, and once more intensive use of trapping methodology was made. After having shown in previous work that the discoloration was caused by chromophores formed in LiCl-catalyzed condensation reactions from DMAc, such as dehydroacetic acid (**32**), isodehydroacetic acid (**33**), and 2,6-dimethyl-γ-pyrone (**34**) [58], see Scheme 15, the question of whether – and if so how – cellulose was degraded by the activation treatment became the key issue. If cellulose degradation indeed occurred, the common heating treatment would be

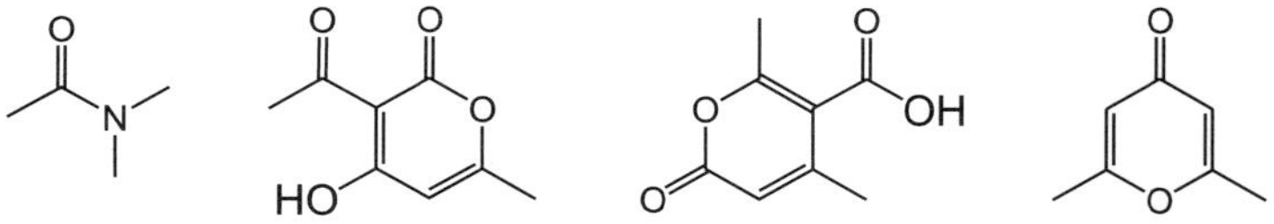

Scheme 15 Chromophoric condensation products of DMAc (**31**) formed by prolonged heating of DMAc or DMAc/LiCl

unsuitable for all processes which rely on an unchanged molecular weight distribution, such as GPC of pulps.

GPC measurements with initially soluble pulps demonstrated unambiguously that cellulosic material suffered a loss in average degree of polymerization (DP) upon heating in DMAc/LiCl. Already at temperatures as low as 80 °C there was a slow, linear, but noticeable decrease in molecular weight. At temperatures between 120 °C and the boiling point of DMAc (164 °C) the kinetics changed to an exponential decay, and the molecular weight degradation became rather severe and fast. The two degradation modes were observed for all pulps, although the degradation rate differed slightly for pulps of different providence.

The well-known fact that heating of initially insoluble pulps in DMAc/LiCl improves solubility or increases the dissolution rate of the material had thus to be attributed to pronounced cellulose degradation. The observed improved solubility was evidently accompanied by a progressive DP loss of the pulp. The solubility gain was thus not an activation of the pulp, but mainly a degradation process to material of lower molecular weight which naturally exhibited a higher solubility in the cellulose solvent.

From the GPC results it was clear that heating in DMAc/LiCl significantly influenced the molecular-weight distribution of the pulps. Consequently, GPC results from pulps which had undergone such an "activation" treatment did not reflect the data of the genuine cellulosic starting material. However, the actual degradation mechanism and the species causing this effect remained to be clarified. The two degradation processes that contribute to the DP loss have been identified: an *endwise peeling* reaction causing a rather slow decrease in the molecular weight at lower temperatures, and a random cleavage of glycosidic bonds, which is responsible for a much faster DP loss, at temperatures above 120 °C.

3.2
Slow "Thermal Endwise Peeling" of Cellulose in DMAc/LiCl

Thermal treatment of DMAc/LiCl caused the formation of *N,N*-dimethylacetoacetamide (**35**), independent of the presence of pulp [58]. This species is a highly reactive 1,3-dicarbonyl compound that readily undergoes subsequent conversions, such as aldol condensations. The first step in these processes is the addition of the highly activated, CH-acidic methylene group of 35 to a carbonyl function. To prove the presence of this reactive intermediate in heated DMAc/LiCl, glucose was used as a trapping reagent, which was neatly converted into UV-active furan derivative 36 (Scheme 16), which was isolated and comprehensively characterized. C1 and C2 of the aldose and C2 and C3 of the acetoacetamide were incorporated into the furan heterocycle.

Surprisingly, furan 36 was not just formed upon trapping with glucose, but also when cellulosic pulps were heated in DMAc or DMAc/LiCl, and the formation was continuous over time following zero-order kinetics. The formation of 36 from pulp was strongly accelerated when excess N,N-dimethylacetoacetamide (35) was added to the mixture. From these results it became evident that not only the reducing ends were derivatized forming furan structures, but the derivatized terminal anhydroglucose units were continuously cleaved and released in the form of condensation product 36. Thus, a cellulose chain with n anhydroglucose units formed the corresponding furan derivative, which was subsequently peeled off as 36 leaving behind the $(n - 1)$-cellulose chain with a "new" reducing end, which again reacted with acetoacetamide 35 and was once more cleaved. N,N-Dimethylacetoacetamide (35), the main thermal condensation product of DMAc, thus eventually caused a thermal endwise peeling of the pulp with concomitant release of furan 36.

This peeling mechanism was confirmed by reaction of 35 with cellotetraose. The cellotetraose acted here as a type of repetitive trapping agent as it produced aldose molecules capable of trapping 35 as the respective furan structures. While the formation of endwise furan structures started at temperatures as low as 80 °C, the subsequent peeling steps required higher temperatures of about 95 °C to proceed at a noticeable rate. Cellotetraose formed first the corresponding furan derivative, which was then cleaved to give cellotriose and furan 36, as followed by TLC and HPCE [59]. The cellotriose reacted immediately with 35 to give the corresponding end-derivatized furan, from which again 36 was peeled off under formation of cellobiose. Cellobiose, eventually, was converted into the corresponding endwise furan, and then finally cleaved into two molecules of 36. Thus, the starting oligosaccharide

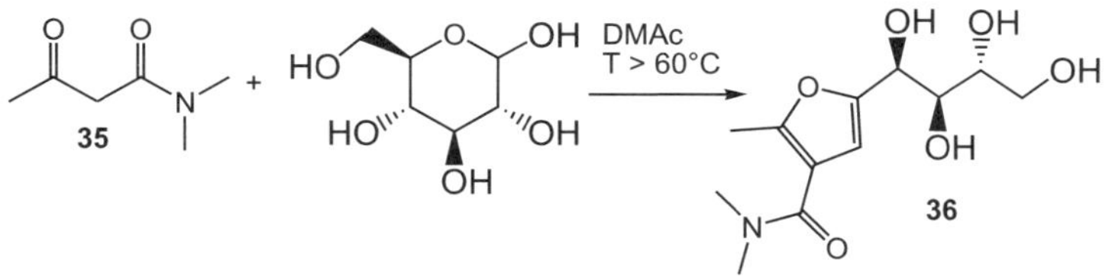

Scheme 16 Reaction of the main thermal condensation product of DMAc, N,N-dimethylacetoacetamide (35), with glucose to furan 36

reacted four times as a quasi-trap for N,N-dimethylacetoacetamide (35), and was neatly converted into four molecules of furan 36 without significant formation of byproducts. As pulp and also cellotetraose formed exclusively 36, but no other furan structures or condensation products, it was clear that the degradation had progressed from the reducing end. Cleavage of glycosidic bonds other than the respective proximal ones did not occur.

The mechanism of thermal endwise peeling of cellotetraose can be summarized as in Eq. 5, with the Eqs. 1–4 describing the consecutive, stepwise character of furan formation:

$$(\text{Glc})_4 + 35 \rightarrow (\text{Glc})_3 + 36 \tag{1}$$

$$(\text{Glc})_3 + 35 \rightarrow (\text{Glc})_2 + 36 \tag{2}$$

$$(\text{Glc})_2 + 35 \rightarrow \text{Glc} + 36 \tag{3}$$

$$\text{Glc} + 35 \rightarrow 36 \tag{4}$$

$$(\text{Glc})_4 + 35 \ (\text{Eq. 4}) \ \rightarrow 36 \ (\text{Eq. 4}) \ . \tag{5}$$

A plausible mechanism for the observed thermal endwise peeling reaction was proposed based on computational chemistry. In its minimum conformation, C-2 of the furan ring at the terminal anhydroglucose unit of a derivatized cellulose molecule 37 is placed in close proximity to the C-2 hydroxyl group of the $(n - 1)$ non-reducing anhydroglucose unit. This pre-organizational effect facilitates the addition of the hydroxyl group to the furan ring. The addition is an endothermic process and a reversible reaction; addition/elimination reactions of alcohols or hydroxyl ions to/from furans are well known in organic chemistry [60, 61]. The resulting seven-membered-ring structure 38, a perhydro-1,4-dioxepane derivative, will most likely fragment back into the starting material. However, ring opening by cleavage of the glycosidic bond between C-1′ and O-1′ is a competitive pathway with relatively low activation energy and high overall exothermicity. The intermediate, a resonance-stabilized oxonium cation, reacts immediately with water to a new reducing end, or directly with 35 to form 39 in which two molecules of acetoacetamide are bound to the cellulose at the same time. Elimination of O-2 releases furan 36 into the surrounding medium. After ring closure to tetrahydrofuran derivative 40 the whole reaction cycle starts again with the cellulose chain shortened by one glucose unit (Scheme 17).

Following from the computations, the formation of furan derivatives from the reducing end of cellulose thus represents a neighbor-group assisted reaction. According to the proposed mechanism, only cleavage of the glycosidic bond that is adjacent to the proximal furan structure occurs, in agreement with experimental data. All other glycosidic bonds are stable under the prevailing conditions as no neighboring, activating furan structures are available. The crucial effect of the neighboring C-2 hydroxyl group was experimentally confirmed. Selectively 2-O-methylated cellulose (41), obtained chemically by cationic ring-opening polymerization [62–64], was shown to

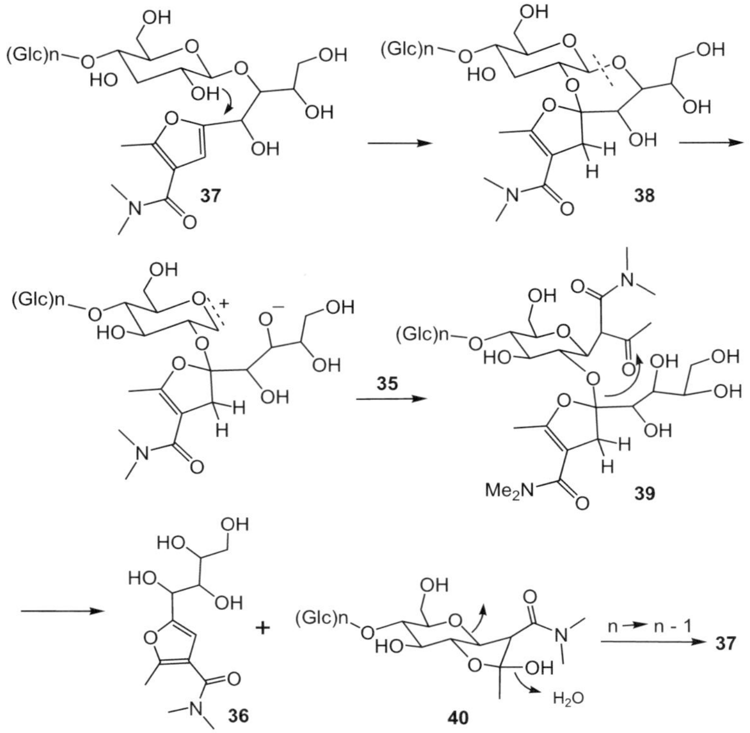

Scheme 17   Mechanism of the slow thermal endwise peeling of cellulose in heated DMAc/LiCl under release of furan **36**

be completely inert towards *N,N*-dimethylacetoacetamide (**35**), even upon prolonged reflux in DMAc. As this material contained no free 2-OH group, addition to the terminal furan structure was prevented, and further thermal endwise peeling could not proceed. In contrast, selectively 3-*O*-methylated cellulose (**42**) and 6-*O*-methylated cellulose (**43**) – similarly synthesized by ring-opening polymerization – underwent the furan peeling reaction by complete analogy to cellulose (Scheme 18). These cellulose derivatives produced the respective monomethoxy derivatives **44** and **45**, respectively, as the peeling products, instead of furan **36**.

A final conformation of the computationally predicted mechanism was provided by the use of isotopically labeled material. *N,N*-Dimethylaceto-acetamide-2-^{13}C was used to produce a snapshot of the reaction intermediates present during reaction with cellotetraose as a model compound, taken by ^{13}C NMR. A series of consecutive spectra produced a kinetic record how the positions with isotopic enrichment were distributed among different intermediates. From the prominent resonance in the starting material, the

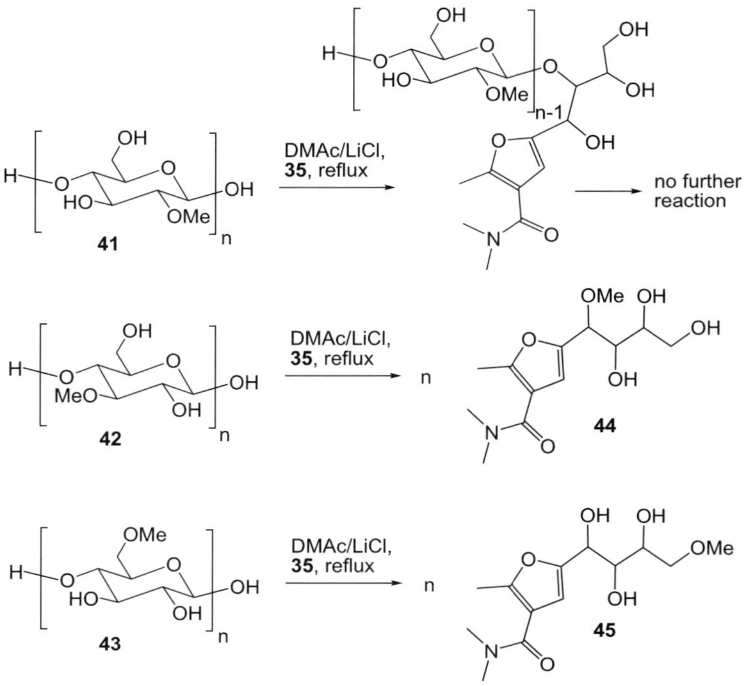

Scheme 18 Thermal behavior of selectively substituted methyl celluloses (41–43) in refluxing DMAc/LiCl

spectrum soon displayed four different resonances in agreement with the intermediates **38–40** in Scheme 17, and changed back to a single resonance of final furan product **36**.

3.3
Fast Degradation of Cellulose in DMAc/LiCl
Due to Random Cleavage of Glycosidic Bonds by Keteniminium Ions

The thermal endwise peeling mechanism could not explain the drastically increased degradation of cellulosic material in heated DMAc at temperatures above 120 °C. If the decrease in molecular weight was only due to endwise peeling, large amounts of peeled-off furan structures **36** should be present, but only minute amounts were found. Thermal endwise peeling by furan structures and cleavage of glycosidic bonds along the cellulose chain were thus two separate parallel processes, which both caused a decrease in molecular weight of the pulp heated in DMAc, the first one proceeding only slowly, the latter one rather fast.

As N,N-dimethylacetoacetamide (35) alone induced only the slow end-wise peeling pathway, and the other thermally formed byproducts of DMAc (32–34) did not effect a chain cleavage at all, an even more reactive species, which was responsible for the chain cleavage, was likely to be present. This intermediate was identified as N,N-dimethylketeniminium cations (46) by trapping reactions (Scheme 19) [65, 66].

Keteniminium salts are extremely reactive compounds; they are more electrophilic than ketenes and do not dimerize as most ketenes do [67, 68]. To prove the intermediacy of this species we used a typical reaction of keten-iminium ions (and ketenes), the reaction with non-activated olefins to cyclobutanes in a thermal, i.e. suprafacial–antarafacial, [2 + 2]-cycloaddition. In order to be able to separate the trapping products from the DMAc medium with its high solvation power, strongly lipophilic trapping reagents were chosen that allowed extraction into apolar solvents. 1-Decene (47) and 3,4-dehydro-α-tocopheryl acetate (48) were chosen as the olefinic traps, the latter having the additional advantage of a high boiling point so that it can be used also in refluxing DMAc without yield penalty. The intermediacy of 46

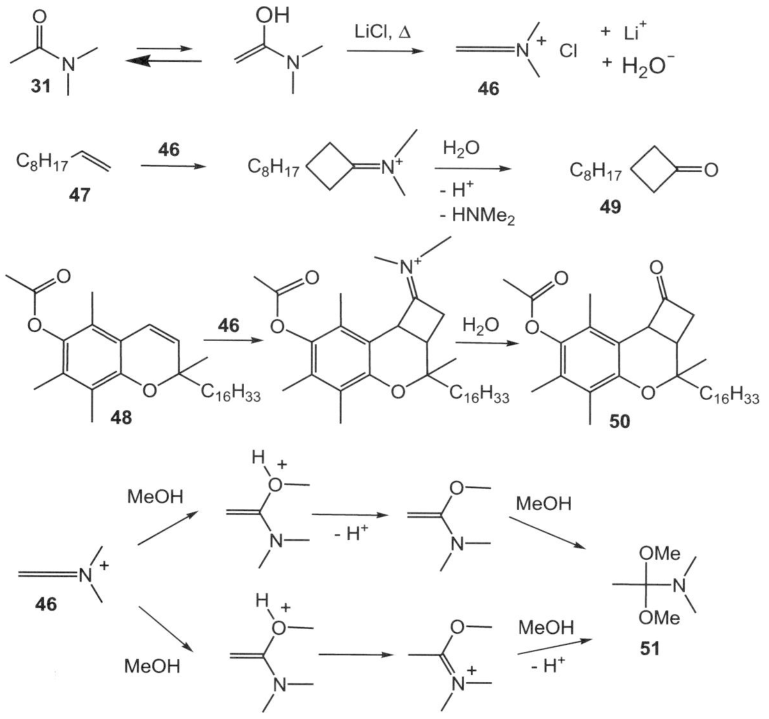

Scheme 19 Formation of N,N-dimethylketeniminium ions (46) from DMAc (31) and their trapping in [2 + 2]-cycloadditions and by consumption with methanol

in DMAc/LiCl of varying composition at temperatures above 120 °C was then unambiguously proven by isolation of the trapping products 3-octylcyclobutanone (49) from 47, and cyclobutanone 50 from the tocopherol trap 48, respectively. The regioselectivity of the addition reaction, e.g., the selective formation of the 3-octylcyclobutanone as compared to the 2-octyl derivative starting from 47, is exclusively determined by steric factors as usual in thermal [2 + 2]-cycloadditions. The trapping of keteniminium 46 succeeded in DMAc/LiCl solutions independent of the presence of dissolved cellulose. In its presence, however, the amount of trapping product obtained was significantly lower, which indicated that also the pulp consumed the reactive intermediate, thereby being degraded.

The equilibrium concentration of 46 is determined by the endothermicity of the formation process and the sum of strong ionic interactions in the solution, and is usually very low. However, if keteniminiums are removed from the reaction system, for instance by reaction with pulp or by trapping, they are continuously regenerated according to the equilibrium constant. Addition of methanol to the respective DMAc/LiCl mixture was the most simple way of consuming N,N-dimethylketeniminium ions, which were converted into 1-dimethylamino-1,1-dimethoxy-ethane (51) (Scheme 19). Alternatively, the reaction with methanol can be seen as a trapping of the reactive intermediate 46 with methanol as the trapping agent. The formation reaction proceeded in a neat way, and the working procedure was quite convenient: product 51 was simply distilled off together with excess methanol. Keteniminium 46 was continuously regenerated in the reaction mixture and converted to the trapping product 51 by methanol. The capacity of DMAc/LiCl to generate 46 was quite large, ranging in the 1–5 percent range upon refluxing for 3 hours, depending on the amount of LiCl present. From the mechanistic point of view, the trapping started with methanol nucleophilically attacking the keteniminium carbon. This was followed by two competitive reaction sequences that both lead to the same trapping product 51: first, proton loss produced an enol ether intermediate (an O-methylated ketene-semiaminal) which added a second equivalent of methanol to produce 51, or second, a [1,3]-sigmatropic proton shift gave an iminium salt that reacted with methanol to provide the same product after proton release (Scheme 18). Both pathways differed in whether the second molecule of methanol is added to a C = C double bond or to the C = N double bond.

The N,N-dimethylketeniminium cation (46) is not formed directly from DMAc, but via its enol form (52), see Scheme 19. The formation of the enol form (imine form) is strongly facilitated by the presence of lithium ions, which coordinate to the amide oxygen. Also the intermediacy of enol precursor 52 was proven by means of trapping reactions: in the presence of allyl alcohol as the trap, the enol 52 formed an allyl enol ether, which at the prevailing elevated temperatures immediately underwent a Claisen-type rearrangement to produce 4-pentenoic acid N,N-dimethylamide (53) (Scheme 20). This

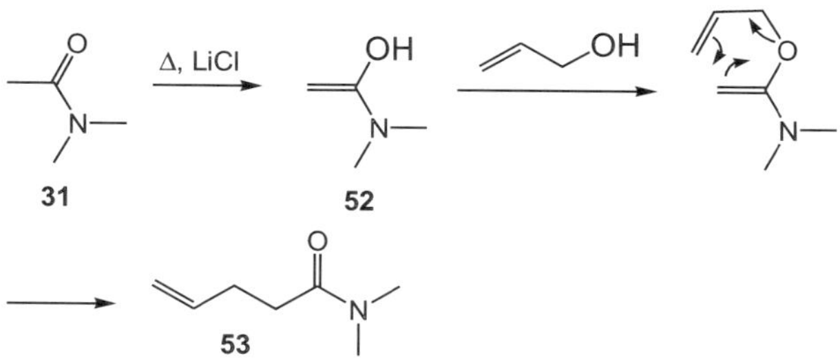

Scheme 20 Trapping of the enol form of DMAc (52) as the precursor of *N,N*-dimethyl-keteniminium cations (46)

amide, which was again detected both in the presence and absence of pulp, was isolated from the reaction mixture by extraction with ethyl acetate after addition of water. The rearrangement was quite convenient as a trapping reaction as it proceeded neatly without byproduct formation.

3.4
Reactions of the DMAc-Derived Keteniminium Ions with Cellulose

The absolute amount of keteniminium ions present in heated DMAc/LiCl solutions is rather low. However, due to the extreme high reactivity of this intermediate and due to its continuous regeneration upon consumption, these small amounts were evidently sufficient to cause the observed pronounced cellulose degradation. Keteniminium salts are known as reagents to effect mild cleavage of ethers, even nonreactive diaryl ethers, acetals, and ketals [67, 69, 70]. In the first step, the keteniminium ion attacks the ether oxygen electrophilically, followed by ether bond cleavage. In solutions containing electrolytes, such as ammonium salts or soluble alkali salts, hydroxyl groups and amino groups do not significantly interfere with the reaction, as these structures in ionic solutions are surrounded by a solvent shell (*ion cloud*) and blocked by strong hydrogen bonds, thus being shielded from attack.

An analogous mechanism was assumed for the cleavage of glycosidic bonds in cellulosic pulp. The strongly electrophilic *N,N*-dimethylketeniminium ion (46) attacked the glycosidic oxygen, followed by cleavage of the glycosidic bond under formation of a ketene-semiaminal derivative, which underwent secondary hydrolytic reactions, shown in Scheme 21 for a cellotetraose section of a cellulose molecule.

In heated DMAc/LiCl, the keteniminium ions are generated in situ from the solvent. When the degrading agent was added externally in the form of a keteniminium precursor, cellulose degradation was even more pronounced. The thermolabile keteniminium precursor compound 4-(1-chloro-2-methylpropenyl)morpholine (54) immediately releases keteniminium ions 55 upon thermal treatment (Scheme 22) [71]. Addition of the precursor in very small

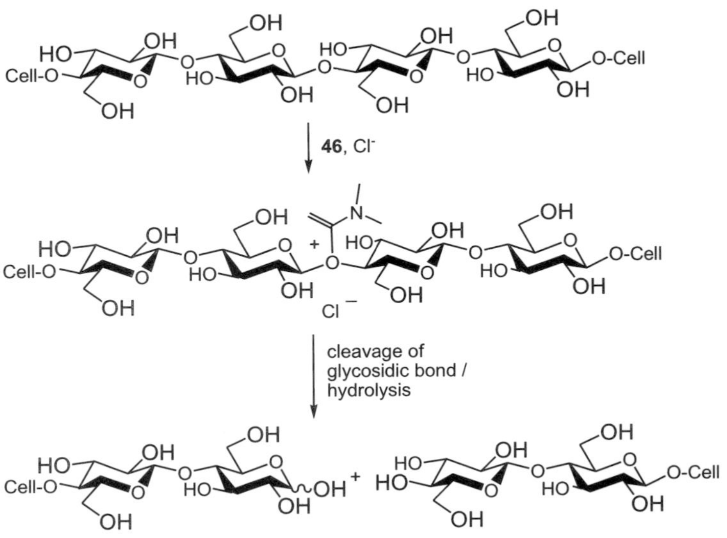

Scheme 21 Cleavage of glycosidic bonds in cellulose by *N,N*-dimethylketeniminium ions (**46**)

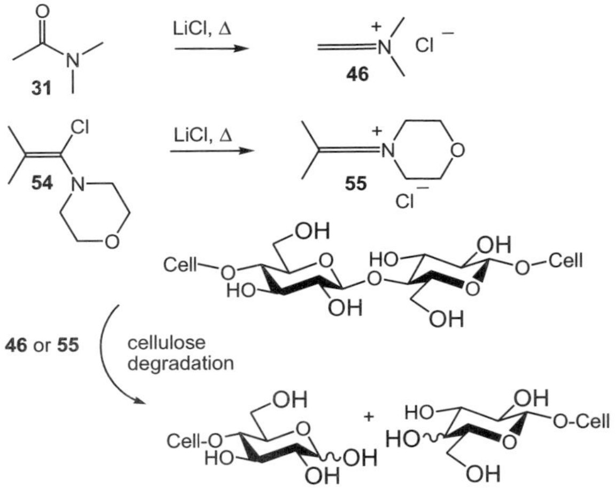

Scheme 22 Cellulose degradation by keteniminium ions being either generated in the reaction mixture or added externally

amounts to a solution of pulp in heated DMAc/LiCl at 125 °C caused degradation of the cellulose down to oligomers (DP < 30) within less than 1 h. This was an additional illustrative proof of the detrimental action of keteniminium species on cellulose.

3.5
Summary

Two thermal degradations mechanisms of cellulose dissolved in DMAc/LiCl have been clarified. The lower-temperature process consists of an endwise peeling under release of furan structures induced by the thermal condensation product of DMAc, N,N-dimethylacetoacetamide (35). The high-temperature mechanism for cellulose degradation is based on DMAc-derived N,N-dimethylketeniminium ions (46), the presence of which in heated DMAc/LiCl, independent of the presence of pulp, was confirmed by trapping. Also the occurrence of a precursor enol intermediate (52) was demonstrated by trapping methodology. The general degrading effect of keteniminium ions on cellulose was experimentally shown in addition.

4
Cellulose in NaOH (Alkali Cellulose)

4.1
Aging of Alkali Cellulose

Alkalization of cellulose – often also called alkaline steeping – is a central process step in the production of many cellulose derivatives, most notably xanthation in the viscose process [72] or etherification in the production of carboxymethyl celluloses or alkyl celluloses [73]. In viscose rayon production, alkalization is not only used for activation of the hydroxyl groups towards xanthation, but also to free the pulp from impurities, such as hemicelluloses and cello-oligosaccharides. The alkalization step involves treatment of dissolving pulp with strong alkali hydroxides – mostly 18% NaOH – which converts cellulose into sodium cellulose I. The alkalization step is followed by the (pre-)ripening procedure: excess sodium hydroxide is pressed off and the press cake is left at temperatures slightly above ambient in the presence of air for several hours. In this stage, the appropriate pulp viscosity, i.e. the cellulose DP, is adjusted for further processing to viscose. It is known that the changes in the molecular-weight distribution are brought about by oxidative processes [74, 75], which involve introduction and conversion of oxidized functions, such as keto and aldehyde groups [76], in addition to chain shortening.

Mechanistic aspects of the alkaline degradation of cellulosic materials have mainly been addressed by research groups in the 1950s and 1960s [77–79]. Chain cleavage was shown to be predominantly caused by introduction of keto groups followed by β-elimination. As early as 1949, a radical chain mechanism for the aging of alkali cellulose was proposed [80, 81]. The formation of hydrogen peroxide during aging of alkali cellulose under oxygen was demonstrated [82], as well as the formation of a number of low-molecular-weight acids. Thus, reactive oxygen species, namely hydroperoxyl, alkylperoxyl and hydrogen peroxide along with the respective radical anions, play a key role in the aging of alkali cellulose. The question of whether also the extremely reactive hydroxyl radicals were involved was controversially discussed. On one hand, the formation of hydroxyl radicals from peroxides and other reactive oxygen species should proceed relatively easily, on the other hand, hydroxyl radicals – if present in reasonably high concentrations – should cause much more pronounced chain degradation than usually observed.

Problems concerning the presence of a distinct radical species are usually addressed by specific trapping [83], employing spin traps that react with transient single-electron species to form stable radicals, which are subsequently analyzed by EPR or UV spectrometry. Such approaches have been widely and successfully used for a large variety of structurally different radicals, employing numerous special spin traps. Almost all of these applications have been concerned with the detection of radicals under physiological conditions or conditions coming close to those in living systems. The number of spin traps remaining operational under extreme conditions, i.e. at high temperatures or in the presence of concentrated acids, bases or other aggressive chemicals, is extremely limited. This was the reason, for instance, why a nonconventional spin trap had to be used to clarify the chemistry in Lyocell dopes (see above). For the detection of OH radicals during alkalization of celluloses the extreme conditions posed a similar problem.

4.2
Development of a Hydroxyl Radical Trap Working in Alkali Cellulose

There have been reports on spin traps working under alkaline bleaching conditions [84, 85], and a literature report on a colorimetric determination of OH radicals in weakly alkalized cellulose samples using N,N'-(5-nitro-1,3-phenylene)bisglutaramide [86], but these approaches were not viable for application in the case of alkali cellulose. Generally, there was no spin trap known to work under the drastic conditions of cellulose alkalization, a reaction essentially carried out in concentrated aqueous NaOH. Recovery of the spin traps and separation from the alkalization byproducts remained as additional obstacles. In most cases, both the starting trap and the trapping product cannot withstand the concentrated alkaline medium. Therefore, aromatic hydroxylation was chosen as another means to report the presence of hydroxyl radicals. Aromatic systems are converted into hydroxyaromatics (phenols) in the presence of hydroxyl radicals. Usually, salicylic acid or phenylalanine are used as aromatic traps in hydroxylation assays under physiological conditions [87]. However, application to alkali cellulose posed the problem of retrieving spin trap and trapping products, as the mixture was rather complex and contained a nearly inseparable blend of aromatic degradation products from low-molecular-weight celluloses and hemicelluloses.

γ-Tocopheryl-*iso*-propyl ether (56) was introduced as a new hydroxyl-selective spin trap to work under conditions of cellulose alkalization [88]. The reagent was synthesized according to a facile Williamson etherification procedure starting from γ-tocopherol (6) and 2-bromo-propane. The reagent is stable under ambient conditions, but should be kept in the dark under inert gas for long-term storage. The trapping reagent offered the advantage of only one aromatic position free for hydroxylation, which limited the number of possible hydroxylation products and facilitated subsequent analysis. Furthermore, the spin trap and its reaction products are strongly lipophilic and thus readily extractable into apolar solvents, a fact that had already been exploited and mentioned repeatedly in the case of tocopherol-based trapping agents. Even from the semi-solid cellulose alkalization mixture, retrieval of trapping agent and problems posed no problems.

The selectivity of the trap towards hydroxyl radicals was demonstrated by several control experiments using different radicals, showing that the formation of the respective hydroxylation product, 5-hydroxy-6-*O*-*iso*-propyl-γ-tocopherol (57), was caused exclusively by hydroxyl radicals, but not by hydroperoxyl, alkylperoxyl, alkoxyl, nitroxyl, or superoxide anion radicals. These radicals caused the formation of spin adducts from standard nitrone- and pyrroline-based spin traps, whereas a chemical change of spin trap 56 was only observed in the case of hydroxyl radicals. This result was independent of the use of monophasic, biphasic, or micellar reaction systems: in all OH radical generating test systems, the trapping product 57 was found. For quantitation, compound 57 was extracted with petrol ether, separated by adsorption onto basic alumina and subsequently oxidized in a quantitative reaction to α-tocored, the deeply red-colored 5,6-tocopheryldione, which was subsequently determined by UV spectrophotometry (Scheme 23).

The trapping selectivity of γ-tocopheryl-*iso*-propyl ether (56) was due to a combination of both electronic and steric effects. The aromatic part in tocopherols is electron-rich, which is the prerequisite for a high trapping efficiency towards electrophilic radicals. The electron density is increased by the mesomeric effects of the two oxygens at C-6 and C-8a and the inductive influence of the alkyl groups. γ-Tocopheryl-methyl ether (58) can be regarded as a model trap in which only the electronic contributions are active, since the methyl substituent is sterically innocent, exerting no influence on the neighboring non-substituted aromatic position. Compound 58 reacted also with hydroperoxyl, alkylperoxyl and alkoxyl radicals besides hydroxyl radicals: the high selectivity for hydroxyl radicals as found in 56 was evidently lost by exchanging the isopropyl for a methyl group. On the other hand, the iso-propyl group was still small enough to allow hydroxyl radicals approaching the neighboring aromatic position. A *tert*-butyl structure as in γ-tocopheryl-*tert*-butyl ether (59), carrying only one CH_3 group more than trap 56, had

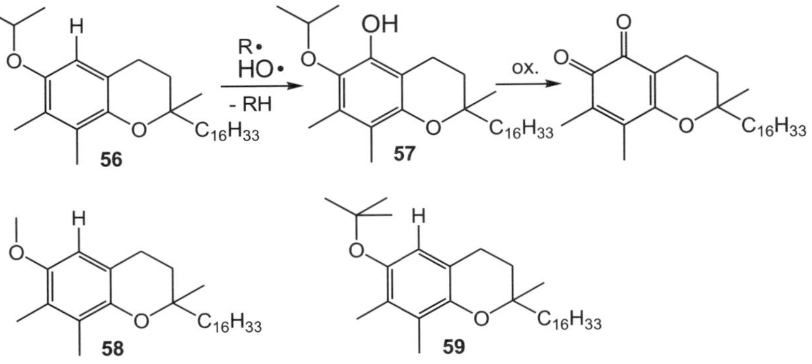

Scheme 23 Hydroxyl radical-selective trapping reaction with γ-tocopheryl-*iso*-propyl ether (**56**) and γ-tocopheryl ether derivatives (**58, 59**) showing different reactivity towards radicals

apparently lost this property, as hydroxyl radicals attacked this molecule almost randomly, with only a slight preference remaining for the reactive, but now largely inaccessible position 5.

4.3
Hydroxyl Radical Trapping in Alkali Cellulose During Aging

By application of hydroxyl radical trap **56**, the presence of hydroxyl radicals during ripening of alkali cellulose was proven for the first time. Aging of alkali cellulose was performed under conditions of industrial relevance, over a time of 19 h at two temperatures, 35 °C and 40 °C. The spin trap was added during the initial shredding of the alkali cellulose, and samples were taken over the reaction time. The samples were extracted with petrol ether to remove all tocopherol products, and the trapping product quantified as described above. The obtained data corresponded to the amount of hydroxyl radicals trapped during the respective time, which is an integral value, reflecting the amount of hydroxyl radicals trapped from the beginning of the aging process to a specific time rather than a snapshot of the radical concentration at the respective point of sampling time.

OH-radical production was quite intense in the initial phase of alkali cellulose aging, whereas in the later stages radical production gradually slowed down. The leveling off at longer reaction times could not be attributed to a diminished trap reservoir, as approximately 70% of the trap remained unchanged after completion of alkalization. About 15% of the added reagent was converted within the first three hours, nearly 30% at the end of the aging procedure after 19 °C. Thus, about one half of the trapped OH radicals were generated in the first sixth of the overall aging time. This allowed the interesting conclusion that hydroxyl radicals were mainly active in the initial phases of cellulose aging, but involved little in later stages. This appeared reasonable as alkalization of cellulose under oxygen is well-known to produce a variety of low-molecular-weight (aromatic) compounds and semi-stable radicals from cello-oligosaccharide and xylo-oligosaccharide degradation products, all of which act as competitive substrates trapping hydroxyl radicals [89–93]. The concentrations of these species are low in early aging phases, so that hydroxyl radicals predominantly react with the tocopherol-derived trap **56**. In later stages, the reaction with the condensation products becomes increasingly competitive as their concentration increases, so that less hydroxyl radicals react with spin traps or with cellulose.

4.4
Cellulose Degradation During Steeping of Alkali Cellulose

The course of hydroxyl radical generation agreed very well with the observed cellulose chain degradation. The DP loss was pronounced in the early alkalization stage, and slowed down with longer reaction times, parallel to the activity of hydroxyl radicals as detected through the trapping approach. For instance, cellulose of an initial DP of approximately 1400 suffered a DP loss of 400–500 within the first three hours, whereas the DP loss in the remaining 16 h was only about 300–400 DP units more. The hydroxyl radicals present – this applies to other radicals as well – appeared to attack the alkali cellulose mainly in the early phases of the aging procedure; in later phases competitive reactions of the radicals with low-molecular-weight products became increasingly dominant, so that aging cellulose degradation could be assumed to be increasingly caused by ionic rather than radical processes.

Also with trap **56** no quantification of the total amount of generated hydroxyl radicals could be performed – as is the case with any other trapping method – since hydroxyl radicals are overly aggressive chemical species, reacting with any structure in their immediate chemical environment. However, aging of alkali cellulose in the absence and presence of the radical trap was compared and showed that the presence of γ-tocopheryl isopropyl ether (**56**) in aging alkali cellulose evidently slowed down the chain cleavage of cellulose and thus impeded the rate of the DP loss. The differences in the aging kinetics were related to the trapped hydroxyl radicals, with the missing amount of OH radicals corresponding to the amount of hydroxylation product isolated. However, the contribution of OH radicals to the overall cellulose degradation, approximately 100 DP units, was comparatively small, the larger part was attributable to the action of different radicals, reactive oxygen species, and heterolytic processes. Also OH radicals, which hydroxylated the trap, would in the absence of the trap not necessarily attack and cleave the cellulose chain directly, but rather initiate a complex set of reactions, which as a whole resulted in chain degradation. These processes, for instance, might consist of the generation of secondary radicals which attack the cellulose, or the introduction of oxidized functionalities into the cellulose chain followed by base-induced cleavage (β-elimination). The aging temperature had the expected strong effect; the cellulose degradation proceeded faster with increasing temperature.

4.5
Summary

The presence of OH radicals during aging of alkali cellulose – as performed in the ripening step of viscose production – was proven for the first time by application of a newly developed, tocopherol-derived trapping agent **56**. The action of OH radicals was shown to be especially pronounced in the earlier phase of the aging reaction, in later stages the effect of OH radicals was increasingly attenuated by competitive reactions with (aromatic) carbohydrate degradation products.

5
Cellulose Carbanilation in DMSO

5.1
Cellulose Carbanilation in Different Solvents

Cellulose tricarbanilate, obtained by reaction of cellulose with phenyl isocyanate – mostly in DMSO or pyridine as the solvents – has been used widely for the determination of analytical parameters of celluloses by gel permeation chromatography (GPC) in organic solvents, such as THF.

Carbanilation offers some advantages over direct dissolution, e.g. in DMAc/LiCl, or over preparation of other derivatives for analytical purposes, such as cellulose nitrates. For instance, carbanilates are quite stable and can be stored over extended periods of time [94], and the eluants used in GPC of carbanilates – mostly THF – are much less exotic than, for instance, DMAc/LiCl or transition-metal complexes. The increase of the polymer mass by carbanilation in combination with the large refractive index increment dn/dc accounts for increased sensitivity upon light-scattering and refractive-index detection, respectively. In addition, the introduction of aromatic moieties upon derivatization allows for monitoring by UV detection. Carbanilates are also suitable derivatives for the analysis of partially substituted celluloses, such as cellulose ethers or esters. After derivatization of the remaining free hydroxyl groups by carbanilation the degree of substitution in relation to the molecular weight can be assessed without a cleavage of the primary substituents [94, 95].

One obstacle to a broad application in cellulose analytics is a certain discrimination effect: low-molecular-weight parts are lost in precipitation and purification steps. The main disadvantage, however, is cellulose degradation upon carbanilation, and the lack of suitable means to prevent this negative effect, so that a reliable polymer-analogous derivatization could not be performed so far. Many derivatization procedures for cellulose tricarbanilates have been reported [96–105]. In most protocols, DMSO was used as the derivatization solvent, sometimes pyridine or mixtures of both solvents. Cellulose degradation was found to occur in both solvents, but was described to be more pronounced in DMSO [102–104]. The degradation in DMSO was assumed to be due to an oxidizing effect of the solvent in combination with isocyanate, similar to Moffat oxidation systems [106], which would introduce carbonyl structures that subsequently cause chain cleavage by β-elimination procedures. The latter was also assumed to be the cause of cellulose degradation in pyridine as the carbanilation solvent.

It was the main goal of our studies to verify the existence of an oxidizing effect of DMSO-based carbanilation mixtures on cellulose, and – if this effect existed – to clarify the responsible reactive intermediates. Trapping reactions were again the key elements in these mechanistic studies.

5.2
Oxidizing Effects of Carbanilation Mixtures Containing DMSO

The oxidative power of carbanilation mixtures containing DMSO and isocyanates was concluded from cellulose degradation in these mixtures. The mechanism of cellulose degradation was proposed to be a combination of oxidation and subsequent chain cleavage at the oxidized positions according to a β-elimination mechanism triggered by auxiliary bases [106].

Three carbinol model compounds with a $\alpha,\alpha,\alpha',\alpha'$-tetraalkyl substitution pattern were applied to screen for oxidizing DMSO-derived species: 2,2,4,4-tetramethylpentan-3-ol (60), 2,2,6,6-tetramethylcyclohexanol (62), and fenchyl alcohol (64) the latter being closer to the structure of cellulosic anhydroglucose units by having a similar six-membered ring in chair conformation. The main advantage of the model compounds was their inability to form carbanilates due to steric crowding. For instance, 2,2,4,4-tetramethylpentan-3-ol (60) can also be termed di-*tert*-butyl-carbinol – a name that better reflects the steric hindrance. Using classical Swern oxidation conditions, it was confirmed that small molecules, such as sulfonium ylides, could still approach this site to perform oxidation, whereas steric hindrance around the hydroxyl groups would prevent larger molecules – such as isocyanates – from reacting. Thus, the chosen model compounds would make oxidation observable without carbanilation as a competitive process. Furthermore, the three product ketones were inert towards the carbanilation mixtures and thus were not further consumed by side-reactions.

Indeed, the oxidizing effect of a DMSO/PhNCO mixture under standard cellulose carbanilation conditions, i.e. reaction times of 2–3 d at tem-

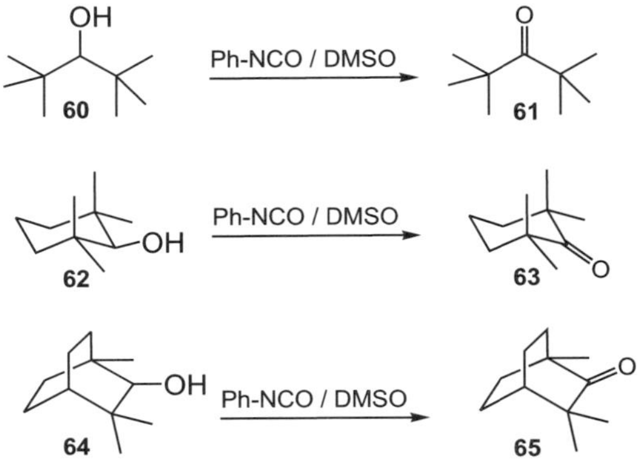

Scheme 24 Proving the oxidizing effect of DMSO/phenyl isocyanate mixtures under conditions used for carbanilation of cellulose by means of sterically hindered alcohol model compounds

peratures between 40 °C and 90 °C and a molar DMSO/isocyanate ratio of about 100 : 1, was demonstrated by means of the model compounds. All three alcohols were converted into the corresponding ketones, 2,2,4,4-tetramethylpentan-3-one (**61**), 2,2,6,6-tetramethylcyclohexanone (**63**) and fenchone (**65**), and no alcohol-derived byproducts, especially no carbanilates, were found (Scheme 24). In the presence of cellulose as a competitive substrate, the model alcohols were still oxidized, but at a much slower rate. This confirmed that the oxidative effect of the DMSO/PhNCO mixture was still active and not suppressed by the cellulose, and implied that cellulosic hydroxyls were oxidized to keto functions similar to the hydroxyls of the model alcohol probes. As the model compounds used were strongly sterically hindered, it was moreover likely that the oxidizing effect on the relatively easily accessible cellulosic hydroxyls was even more pronounced.

5.3
Carbanilation in the Presence of Trapping Agents

The proof of the oxidizing power of the DMSO-based carbanilation did not answer the question of which chemical species were actually causing the oxidation. It was very likely that these species were sulfonium ylides (**66**) by analogy to Moffat-type and Swern-type oxidations. Apart from their action mode as oxidants, sulfonium ylides are able to react with double bonds to cyclopropanes or epoxides, depending on whether the coreactant is an olefin, an α,β-unsaturated ketone, or an α,β-unsaturated ester. This chemical behavior was employed in a trapping approach to prove the presence of sulfonium ylide species.

From the apparent wealth of unsaturated compounds as coreactants, very few seemed appropriate as trapping agent for sulfonium ions or ylides, since the trapping approach once more sets special requirements: the reaction products – possibly formed in minute amounts only – had to be separated from the cellulose carbanilation mixture, purified, and identified. For this reason, once more a tocopherol derivative, 3,4-dehydro-α-tocopheryl acetate (**48**), was chosen due to its strong lipophilicity which guaranteed supreme extractability into apolar media, such as *n*-hexane or petrol ether, also from complex and heterogeneous carbanilation mixtures. The apolar extraction medium contained only the trapping agent and its products; all other components were too hydrophilic to enter the alkane phase.

Carbanilation in the presence of the trapping agent and subsequent extraction into *n*-hexane provided a mixture of non-reacted trap and a main product, the corresponding cyclopropa[c]chromanol derivative **67**, indicating that the double bond of the trapping agent had undergone cyclopropanation (Scheme 25). The cyclopropane derivative was separated, purified and fully analytically characterized, and its structure additionally confirmed by comparison with an authentic sample.

The trapping product constituted unambiguous evidence for the presence of dimethylsulfonium ions and the derived sulfonium ylide in the carbanilation mixture consisting of DMSO/PhNCO/cellulose. These sulphur-containing species – known to be strongly oxidizing as the actual oxidants in Moffat-type and Swern-type procedures – cause the oxidative effect of the

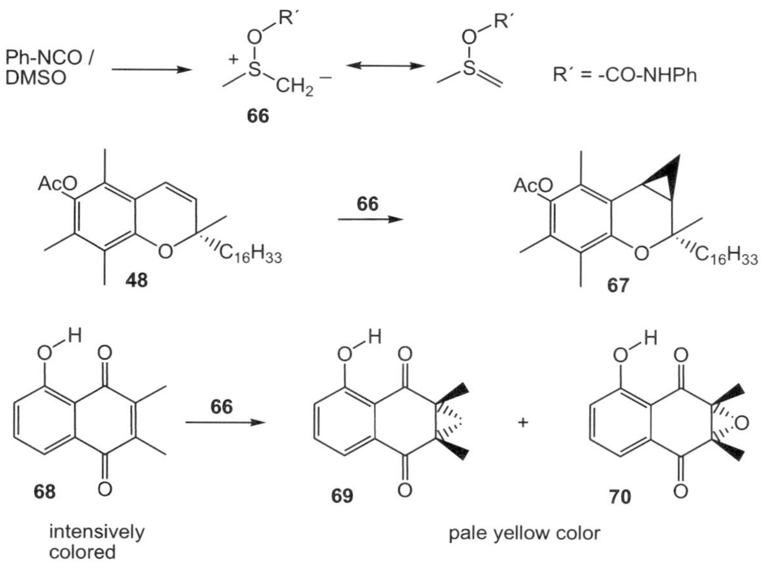

Scheme 25 Generation of methylsulfonium ylides in DMSO-based cellulose carbanilation mixtures, trapping by 3,4-dehydro-α-tocopheryl acetate (**67**), and quick color testing for their presence by naphthoquinone derivative **69**

DMSO/isocyanate carbanilation medium on cellulose. It should be noted that the oxidation per se does not cause chain cleavage and cellulose degradation, but only the introduction of carbonyl functionalities along the cellulose chain. However, these groups constitute points of pronounced chemical instability where subsequent cleavage, mainly under basic conditions in β-elimination processes, will readily occur.

A second proof of the presence of the cellulose-damaging sulfonium ylide **66** was provided by employing 5-hydroxy-2,3-dimethyl-[1,4]-naphthoquinone (**68**) as the trapping agent. This compound, which dissolved with a dark-blue, almost black, color – was similarly attacked at the quinoid double bond, and converted into a mixture of the cyclopropane derivative 1a,7a-dihydro-1*H*-cyclopropa[b]naphthalene-2,7-dione (**69**) and the epoxide derivative 1a,7a-dihydro-1-oxa-cyclopropa[b]naphthalene-2,7-dione (**70**) in an approximate 3 : 1 ratio (Scheme 25). As both of the products lacked the strongly chromophoric benzoquinoid system, the solution turned increasingly light during the reaction, finally reaching a pale-greenish hue.

Besides the direct proof of the presence of the oxidizing sulfonium species, this reaction offered a surprisingly easy and convenient way to check whether certain carbanilation conditions will cause oxidative damage of cellulose or whether they are safe to perform with regard to cellulose integrity. For this purpose, the naphthoquinone **68** was used as a quasi-color indicator in very low concentrations, just enough to cause a visible coloration of the solution. If the reaction mixture exerted an oxidizing effect due to dimethylsulfonium ions, the color faded; if no oxidizing sulfonium species were present, the color remained unchanged. With the finding of this probe, it became readily possible to monitor the effect of different influencing factors, such as temperature, isocyanate type, concentration, cosolvents, auxiliaries and others, on the oxidative power of the carbanilation system, in order to minimize oxidative damage to the cellulose and subsequent DP losses.

5.4
Summary

The oxidative effect of the DMSO/PhNCO system on cellulose was confirmed by means of alcoholic model compounds (**60, 62, 64**) that were neatly oxidized into the corresponding ketones. The presence of the active species, the oxidatively acting sulfonium ylide **66**, in the cellulose carbanilation mixture was proven by trapping with two reagents, a tocopherol-based compound (**48**) and a naphthoquinone (**68**) that was also used in a facile color test to estimate the degrading effect of certain carbanilation mixtures and conditions on cellulose.

Acknowledgements The financial support of the work by the Austrian Fonds zur Förderung der Wissenschaftlichen Forschung (projects P-14687, P-17426 and P-17428), by the Austrian Christian Doppler Research Society, by Lenzing AG, Austria, and Roche Pharmaceuticals, Switzerland, are gratefully acknowledged. We are grateful to all students, co-workers and cooperation partners that were involved in the studies described.

References

1. Chanzy H (1980) J Polym Sci, Polym Phys Ed 18:1137
2. Chanzy H, Nawrot S, Peguy A, Smith P (1982) J Polymer Sci 20:1909
3. Marini I, Brauneis F (1996) Textilveredelung 31:182
4. Firgo H, Eibl M, Eichinger D (1995) Lenzinger Ber 75:47
5. Brandner A, Zengel HG (1980) German Patent DE-OS 3,034,685; Chem Abstr 97 7727d CA
6. Ringel C (1969) Z Chem 9:188
7. Buijtenhuis FA, Abbas M, Witteveen AJ (1986) Papier 40:615
8. Rosenau T, Potthast A, Sixta H, Kosma P (2001) Progr Polym Sci 26:1763
9. Rosenau T, Potthast A, Adorjan I, Hofinger A, Sixta H, Firgo H, Kosma P (2002) Cellulose 9:283
10. Ferris JP, Gerwe RD, Gapski GR (1968) J Org Chem 33:3493
11. Ferris JP, Gerwe RD, Gapski GR (1967) J Am Chem Soc 89:5269
12. Malask P (1993) Top Curr Chem 168:1
13. Schmittel M, Burghart A (1997) Ang Chem 36:2550
14. Linker T, Schmittel M (1998) Radikale und Radikalionen in der Organischen Synthese. Wiley, Weinheim New York
15. Eastland GW, Rao DNR, Symons MCR (1984) J Chem Soc, Perkin Trans II 1551
16. De Meijere A, Chaplinski V, Gerson F, Merstetter P, Haselbach E (1999) J Org Chem 64:6951
17. Rosenau T, Potthast A, Sixta H, Kosma P (2002) Tetrahedron 58:3073
18. Yamauchi R, Matsui T, Kato K, Ueno Y (1990) Agric Biol Chem 54:2703
19. Rosenau T, Potthast A, Kosma P (1999) Synlett 1972
20. Taeger E, Michels C, Nechtawal A (1991) Papier 12:784
21. Chow YL, Danen WC, Nelsen SF, Rosenblatt DH (1978) Chem Rev 78:243g
22. Guthrie JT, Mannings CS (1990) The cellulose/N-methylmorpholine-N-oxide/H_2O system; degradation aspects. In: Kennedy JF, Phillips GO, Williams PA (eds) Cellulose sources and exploitation. Ellis Horwood Ltd., Chichester, p 49
23. Loubinoux D, Chaunis S (1987) Textile Res J 2:61
24. Novoselov NP, Tret'yak VM, Sinel'nikov EV, Sashina ES (1997) Russ J Gen Chem 67:430
25. Ioleva MM, Goikhman AS, Banduryan SI, Papkov SP (1983) Vysokomol Soedin Ser B 25:803
26. Rosenau T, Potthast A, Kosma P, Chen CL, Gratzl JS (1999) J Org Chem 64:2166
27. Rosenau T, Potthast A, Hofinger A, Sixta H, Kosma P (2002) Holzforschung 56:199
28. Potthast A, Rosenau T, Kosma P, Chen CL, Gratzl JS (2000) Holzforschung 54:101
29. Hopkin A, Williams E (1950) Organic Reagents for Organic Analysis, 2nd edn. The Chemical Society, Chadwell Heath, London, p 61
30. Cremlyn RJ, Osborne AG, Warmsley JF (1996) Spectrochim Acta A 52:1433
31. Tramontini M (1973) Synthesis 703
32. Tramontini M, Angiolinin L (1990) Tetrahedron 46:1791
33. Rosenau T, Potthast A, Kosma P (2003) Tetrahedron 60:301
34. Gevorgyan GA, Agababyan AG, Mndzhoyan OL (1984) Russ Chem Rev 53:561

35. Kuhn C, Florent JC (1995) Tetrahedron Lett 36:3137
36. Grumbach HJ, Arend M, Risch N (1996) Synthesis 7:883
37. Rosenau T, Potthast A, Milacher W, Hofinger A, Kosma P (2004) Polymer 45:6437
38. Rosenau T, Potthast A, Milacher W, Adorjan I, Hofinger A, Kosma P (2005) Cellulose 12:197
39. Potthast A, Rosenau T, Kosma P, Schelosky N, Baldinger T (2000) Holzforschung 54:641
40. Dawsey TR, McCormick CL (1990) J Macromol Sci, Rev Macromol Chem Phys C30:405
41. Morgenstern B, Kammer HW (1996) TRIP 4:87
42. Heinze T (2001) Prog Polym Sci 26:1689
43. Rahn K, Diamantoglou M, Klemm D, Berghmans H, Heinze T (1996) Angew Makromol Chem 238:143
44. Kennedy JF, Rivera ZS, White CA, Lloyd LL, Warner FP (1990) Cellulose Chem Technol 24:319
45. Striegel AM (1997) Carbohydr Polym 34:267
46. Scheloski N, Röder T, Baldinger T (1999) Papier 53:728
47. Potthast A, Rosenau T, Buchner R, Röder T, Ebner G, Bruglachner H, Sixta H, Kosma P (2002) Cellulose 9:41
48. Röder T, Potthast A, Rosenau T, Kosma P, Baldinger T, Morgenstern B, Glatter O (2002) Macromol Symp 190:151-160
49. Chrapava S, Touraud D, Rosenau T, Potthast A, Kunz W (2003) Phys Chem Chem Phys 5:1842
50. Morgenstern B, Berger W (1993) Acta Polymerica 44:100
51. Röder T, Morgenstern B, Scheloski N, Glatter O (2001) Polymer 42:6765
52. Kennedy JF, Rivera ZS, White CA, Lloyd LL, Warner FP (1990) Cellulose Chem Technol 24:319
53. Kvernheim AL, Lystad E (1989) Acta Chem Scand 43:209
54. Röhrling J, Potthast A, Rosenau T, Lange T, Ebner G, Sixta H, Kosma P (2002) Biomacromol 3:959
55. Schult T, Hjerde T, Optun OI, Kleppe PJ, Moe S (2002) Cellulose 9:149
56. Klemm D, Philipp B, Heinze T, Heinze U, Wagenknecht W (1998) Comprehensive Cellulose Chemistry, vol 2. Wiley, Weinheim, p 331
57. Tosh B, Saikia CN, Dass NN (2000) Carb Res 327:345
58. Rosenau T, Potthast A, Hofinger A, Kosma P (2001) Holzforschung 55:661
59. Sjöberg J, Adorjan I, Rosenau T, Sixta H, Kosma P (2004) Carbohydr Res 339:2037
60. Iovel IG, Lukevics E (1998) Chem Heterocyclic Comp 34:1
61. Panda H (2001) Chem Weekly 46:155
62. Nakatsubo F, Kamitakahara H, Hori M (1996) J Am Chem Soc 118:1677
63. Nishio N, Takano T, Kamitakahara H, Nakatsubo F (2005) Cellulose Chem Technol 39:377
64. Karakawa M, Mikawa Y, Kamitakahara H, Nakatsubo F (2002) J Polym Sci A: Pol Chem 40:4167
65. Potthast A, Rosenau T, Sartori J, Sixta H, Kosma P (2002) Polymer 44:7
66. Potthast A, Rosenau T, Sixta H, Kosma P (2002) Tetrahedron Lett 43:7757
67. Ghosez L, Marchand-Brynaert J (1976) In: Böhme J, Viehe HJ (eds) Iminium Salts in Organic Chemistry, Part I. Wiley, New York
68. Falmagne JB, Escudero J, Talbe-Sahraoui S, Ghosez L (1981) Angew Chem Int Ed Engl 20:879
69. Arcelli A, Cecchi R, Porzi G, Rinaldi S, Sandri S (2001) Tetrahedron 57:6843

70. Penn JH, Deng DL (1992) Tetrahedron 48:4823
71. Marchand-Brynaert J, Ghosez LJ (1972) J Am Chem Soc 94:2869
72. Götze K (1940) Kunstseide und Zellwolle nach dem Viskose-Verfahren. Springer, Berlin
73. Fengel D (1980) Papier 34:428
74. Sihtola H, Neimo L (1963) Tappi J 46:730
75. Barthel P, Philipp B (1967) Faserforsch Textiltechn 18:266
76. Potthast A, Röhrling J, Rosenau T, Borgards A, Sixta H, Kosma P (2003) Biomacromol 4:743
77. Göransson S (1968) Svensk Papperstid 71:131
78. MacDonald DM (1965) Tappi J 48:708
79. Majdanac L, Galogaza V, Theodorovic M (1983) Cellulose Chem Technol 17:333
80. Entwistle D, Cole EH, Wooding NS (1949) Textile Res J 527
81. Entwistle D, Cole EH, Wooding NS (1949) Textile Res J 609
82. Lindgren BO, Sundin S (1978) Svensk Papperstid 81:485
83. For the chemistry of oxygen radicals and oxygen derived species see: Halliwell B, Gutteridge JMC (1989) Free Radicals in Biology and Medicine, 2nd edn. Clarendon, Oxford, p 22
84. Smith K, Argyropoulos DS (2003) Int Symp Wood Pulping Chem 1:167
85. Smith K, Argyropoulos DS (2002) Int Pulp Bleaching Conference, Portland, USA, p 97
86. Kolar J, Strlic M, Pihlar B (2001) Anal Chim Acta 431:313
87. Halliwell B, Gutteridge JMC (1989) Free Radicals in Biology and Medicine, 2nd edn. Clarendon, Oxford, p 53
88. Rosenau T, Potthast A, Möslinger R, Kosma P (2005) J Wood Chem Technol 26:1
89. Popoff T, Theander O (1976) Acta Chem Scand B 30:397
90. Popoff T, Theander O (1976) Acta Chem Scand B 30:705
91. Popoff T, Theander O, Westerlund E (1978) Acta Chem Scand B 32:1
92. Olsson K, Pernemalm PA, Theander O (1978) Acta Chem Scand B 32:249
93. Theander O, Westerlund E (1980) Acta Chem Scand B 34:701
94. Hearon WM, Hiatt GD, Fordyce CR (1943) J Am Chem Soc 65:833
95. Hearon WM, Hiatt GD, Fordyce CR (1943) J Am Chem Soc 65:829
96. Hall DM, Horne JR (1973) J Appl Polym Sci 17:3727
97. Valtasaari L, Saarela K (1975) Paperi ja Puu 57:5
98. Schroeder LR, Haigh FC (1979) Tappi J 62:103
99. Wood BF, Conner AH, Hill CG (1986) J Appl Polym Sci 32:3702
100. Rantanen T, Färm P, Sundquist J (1986) Paperi ja Puu 57:634
101. Lauriol JM, Froment P, Pla F, Robert A (1987) Holzforschung 41:109
102. Evans R, Wearne RH, Wallis AFA (1991) J Appl Polym Sci 42:821
103. Evans R, Wearne RH, Wallis AFA (1991) J Appl Polym Sci 42:813
104. Evans R, Wearne RH, Wallis AFA (1989) J Appl Polym Sci 37:3291
105. Terbojevich M, Cosani A, Camilot M, Focher B (1995) J Appl Polym Sci 55:1663
106. Fischer M (2005) PhD thesis, Dresden University of Technology

Adv Polym Sci (2006) 205: 199–291
DOI 10.1007/12_100
© Springer-Verlag Berlin Heidelberg 2006
Published online: 8 September 2006

Functional Polymers Based on Dextran

Thomas Heinze (✉) · Tim Liebert · Brigitte Heublein · Stephanie Hornig

Kompetenzzentrum Polysaccharidforschung, Friedrich-Schiller-Universität Jena,
Humboldtstraße 10, 07743 Jena, Germany
thomas.heinze@uni-jena.de

Abstract Dextran, comprising a family of neutral polysaccharides consisting of an α-(1→6) linked D-glucose main chain with varying proportions of linkages and branches, depending on the bacteria used, is an interesting starting material for chemical modification reactions for the design of new functional polymers with promising properties. The review summarises recent results on structure characterisation of dextran including some comments on biosynthesis of this important class of biopolymers. Applications of dextran are discussed as well. Chemical modification reactions of dextran are increasingly studied for the structure and hence property design. The review highlights recent progress in esterification of dextran, both inorganic and organic polysaccharide esters, etherification reactions towards ionic and non-ionic ethers, and the huge variety of different conversions mainly developed for the binding of drugs. It summarises recent developments in the application of dextran derivatives with a focus on the chemical structures behind these materials such as prodrugs, bioactivity of inorganic dextran esters, heparin sulfate mimics, hydrogels, nanoparticles and self assembly structures for surface modification.

Keywords Functionalised dextran · Structural analysis · Prodrugs · Nanostructures · Bioactivity

Abbreviations

AFM	Atomic force microscopy
AGU	Anhydroglucose units
AT	Antithrombin
AZT	Azidothymidine
BSH	$Na_2B_{12}H_{11}SH$
BSA	Bovine serum albumin
CAC	Critical association concentration
CDI	*N,N′*-Carbonyldiimidazole
CMC	Critical micelle concentration
CMD	Carboxymethyl dextran
CMDB	Carboxymethyl dextran benzylamide
CMDBSSu	Carboxymethyl dextran benzylamidesulfonate sulfate
CMDBSu	Carboxymethyl dextran benzylamide sulfate
CMDEE	Carboxymethyl dextran ethyl ester
CMDSu	Carboxymethyl dextran sulfate
COSY	Two-dimensional correlated spectroscopy
CsA	Cyclosporin A
2D	Two-dimensional
Dach-Pt	*cis*-Dihydroxo(cyclohexane-*trans*-L-1,2-diamine)platinum II
DCC	*N,N′*-Dicyclohexylcarbodiimide

DCMD	Dicarboxyethyl dextran
DEAE	2-(Diethylamino)ethyl
DexC$_n$	2-Hydroxyoctyl- and 2-hydroxydodecyl
DexPEG$_{10}$C$_n$	Poly(ethylene glycol)-alkyl dextran ether
DexP	2-Hydroxy-3-phenoxy propyl dextran ethers
DLS	Dynamic light scattering
DMA	*N,N*-Dimethylacetamide
DMAP	4-*N,N*-Dimethylaminopyridine
DPA	Dextran propionate acetate
DS	Degree of substitution (DS is the average number of functional groups per anhydroglucose repeating unit arriving at a level of three after complete functionalisation of all hydroxyl groups)
EC	Endothelial cells
EDC	1-Ethyl-3-[3-(dimethylamino)propyl]carbodiimide
EEDQ	*N*-Ethoxycarbonyl-2-ethoxy-1,2-dihydroquinoline
ELISA	Enzyme-linked immunosorbent assay
EPR	Electron paramagnetic resonance
FITC	Fluorescein isothiocyanate
GI	Gastrointestinal
GLC-MS	Gas–liquid chromatography/mass spectrometry
Glc*p*	Glucopyranose
HCII	Heparin cofactor II
HEMA	2-Hydroxyethyl methacrylate
HIV	Human immunodeficiency virus
HMDS	1,1,1,3,3,3,-Hexamethyldisilazane
HMPC	Hydrophobically modified polycationic dextran
HMQC	Heteronuclear multiple quantum correlation
HPLC	High performance liquid chromatography
HSV	Herpes simplex virus
HTLV-III	Human T-cell lymphotropic virus type III
LbL	Layer-by-layer
LD$_{50}$	Lethal dose
β-LG	*β*-Lactoglobulin
Lm	*Leuconostoc mesenteroides*
M_n	Number average molecular weight
M_w	Weight average molecular weight
2M2B	2-Methyl-2-butene
MCA	Monochloroacetic acid
MesCl	Methanesulfonic acid chloride
MRI	Magnetic resonance imaging
MWD	Molecular weight distribution
NHS	*N*-Hydroxysuccinimide
NMP	*N*-Methyl-2-pyrrolidone
NMR	Nuclear magnetic resonance
PCL	Poly(*ε*-caprolactone)
PEC	Polyelectrolyte complex
PEG	Poly(ethylene glycol)
PLA	Poly(lactide)
PrPres	Protease-resistant isoform of a host-encoded protein
PSDA	Particle size distribution analyser

PS-DVB	Polystyrene-divinylbenzene
Py	Pyridine
QCM	Quartz crystal microbalance
R_G	Radius of gyration
ROP	Ring-opening polymerisation
RT	Room temperature
S	*Streptococcus*
S_N	Nucleophilic substitution
SEC	Size exclusion chromatography
SEM	Scanning electron microscopy
SMC	Smooth muscle cells
SPR	Surface plasmon resonance
SRA	Scavenger receptor class A
T-2513	7-Ethyl-10-(3-aminopropyloxy)-camptothecin
TBAH	Tetrabutylammonium hydroxide
TEA	Triethylamine
TMSCl	Chlorotrimethylsilane
TosCl	*p*-Toluenesulfonyl chloride
TOCSY	Total correlated spectroscopy
Trityl	Triphenylmethyl
TTd	Tetanus toxoid
UV/Vis	Ultraviolet/visible light

1
Introduction

Polysaccharides are fascinating macromolecules possessing structural diversity and functional versatility. Focusing on polyglucans (i.e. polysaccharides containing glucose repeating units only) a broad variety of structures appear, resulting from the stereochemistry of the anomeric C-atom, from the regiochemistry of the glycosidic linkage and from the pattern of branching. Cellulose, the β-(1 → 4) linked polyglucan of a very uniform molecular structure, is by far the most important polysaccharide. Cellulose is not only the most abundant organic compound but also most intensively used as raw material for the commercial production of fibres, beads and sponges as well as for the commercial synthesis of various esters and ethers [1]. Next to cellulose, the polyglucan starch has to be mentioned. As very well known, starch consists of two primary polymers containing D-glucose, namely the linear α-(1 → 4) linked amylose and the amylopectin composing additional α-(1 → 6) linked branches. Amylose and amylopectin occur in varying ratios depending on the plant species [2]. Starch is also used to a high extent as raw material for chemical modification reactions, though the commercially produced derivatives possess a low degree of substitution (DS). In addition to these polysaccharides mainly isolated from different plant sources, various fungi and bacteria synthesise polyglucans, e.g. curdlan (β-(1 → 3) linked), scleroglucan and schizophyllan (β-(1 → 3) linked main chain and β-(1 → 6) linked branches) and pullulan (α-(1 → 4) and α-(1 → 6) linked).

However, the most important polysaccharide for medical and industrial applications produced by bacterial strains is dextran, a family of neutral polysaccharides consisting of an α-(1 → 6) linked D-glucose main chain with varying proportions of linkages and branches depending on the bacteria used (Fig. 1). The α-(1 → 6) linkages in dextran may vary from 97 to 50% of total glycosidic bonds. The balance represents α-(1 → 2), α-(1 → 3) and α-(1 → 4) linkages usually bound as branches [3]. Different bacterial strains are able to synthesise dextran mainly from sucrose. In 1861, Pasteur found slime-producing bacteria, which were later named *Leuconostoc mesenteroides* by van Tieghem in 1878 [4, 5]. Scheibler named the segregated carbohydrate

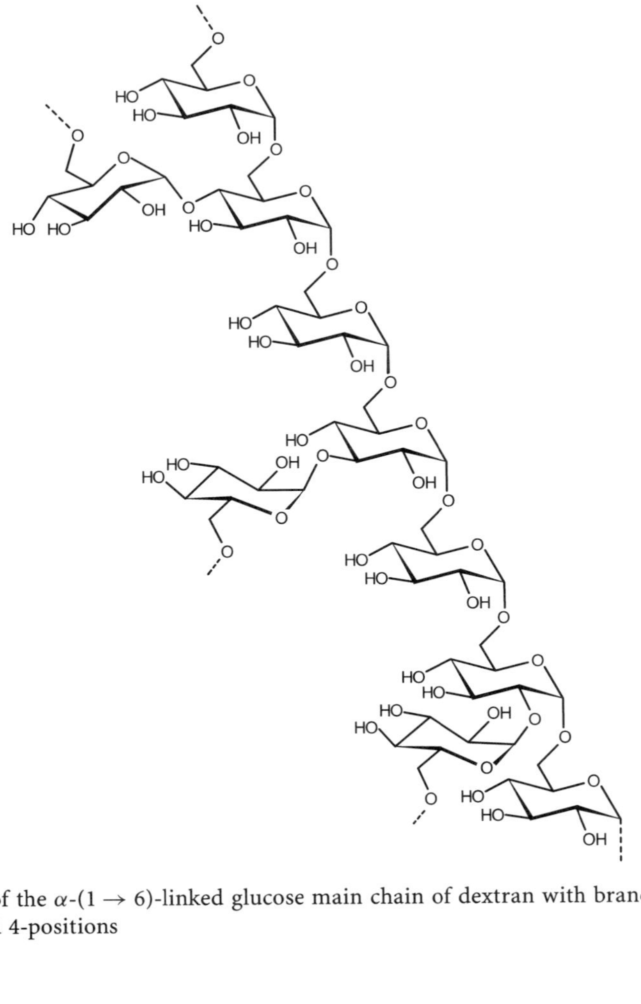

Fig. 1 Part of the α-(1 → 6)-linked glucose main chain of dextran with branching points in 2-, 3- and 4-positions

"dextran". Subsequent investigations have shown that dextran can be formed by several bacterial strains mostly gram-positive, facultatively anaerobe cocci, e.g. *Leuconostoc* and *Streptococcus* strains [6].

The commercial production carried out by various companies is estimated to be ca. 2000 t year^{-1} worldwide [7]. Due to the common solubility in water and various other solvents (e.g. DMSO, formamide), the biocompatibility, and the ability of degrading in certain physical environments, dextran is already successfully applied in the medical and biomedical field [8]. The physiological activity of dextran and its derivatives, indicated also by a very large number of publications in this area of research, is in contrast to inadequate structural analysis of both dextran and their semi-synthetic products. Only a few publications, in contrast to extensive studies in cellulose and starch chemistry [9, 10], deal with the defined functionalisation and characterisation of dextran for adjusting desired features.

From the chemist's point of view, dextran is really an interesting polysaccharide as starting biopolymer for the design of structures and hence of properties by polymeranalogous reactions. The homopolymer structure of dextran composed of D-glucose units without any relevant imperfection and the availability of a number of dextran samples with a narrow molecular weight distribution are advantageous for the chemical modification. In addition to the products obtained from cellulose and starch, dextran derivatives may show different properties due to the differences in the structures of the polysaccharide chain and the three secondary hydroxyl groups, even if the same functional groups are introduced. Multifunctional dextran derivatives can be prepared with tunable properties depending on the introduced substituent. The main goal of this article is to review some aspects of the chemistry of dextran. Emphasis is placed on ethers and esters, which have been a recent focus of interest.

2
Sources, Structure and Properties of Dextran

2.1
Occurrence of Dextran

Leuconostoc mesenteroides and *Leuconostoc* ssp. are found in fermented foods of plant origin [11]. The occurrence of these bacteria in sugar refineries is responsible for problems in filtration processes because of increased viscosity by the presence of soluble dextran [12, 13]. Furthermore, dextran retards the rate of crystallisation of sucrose and adversely affects the crystal shape. The occurrence of dextran in the matrix of dental plaque results from certain *Streptococcus* strains [14]. The principle organism, *Streptococcus mutans*, is able to produce water-soluble glucan (named dextran) and water-insoluble glucan (mutan) from sucrose [15, 16]. These polysaccharides provide a protective matrix for the colonisation of cariogenic bacteria on the tooth surface and attend as reserve carbohydrates [12]. Consequently, dental plaque promotes caries and can thus be reduced by application of dextranases, enzymes which hydrolyse specific glycosidic bonds [17]. One of the major industrial applications of dextranases is the reduction of sliming in sugar production processes [18].

For industrial, medical and scientific interests, a variety of dextran samples from different origins are commercially available. Depending on the fermentation conditions, dextran with particular features can be produced as described in Sect. 3.2.

2.2
Structure Characterisation

Dextran is a homopolymer of glucose with predominantly α-($1 \rightarrow 6$) linkages (50–97%) [12]. Figure 1 shows a part of the dextran main chain with branching points in the 2, 3 and 4 positions. The degree and nature of branching units is dependent on the dextran-producing bacterial strain [6].

Preliminary examinations of dextran structures were conducted by optical rotation, infrared spectroscopy and periodate–oxidation reactions. More detailed results can be achieved by methylation analysis [19]. The hydroxyl groups are methylated with methyl iodide after activation with sodium methylsulfinyl carbanion (Fig. 2). The methyl dextran is hydrolysed to the corresponding different methylated monosaccharides, which are furthermore reduced and peracetylated. The resulting alditol acetates of methylated sugars are separated by gas chromatography and identified by their retention times. In particular, a combined capillary gas–liquid chromatography/mass

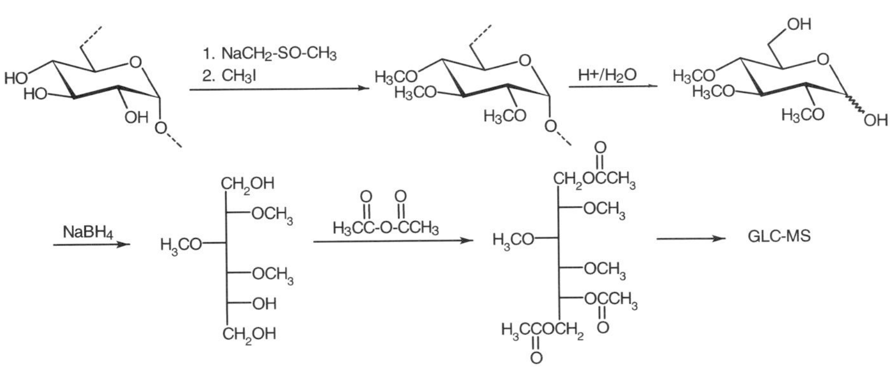

Fig. 2 Methylation analysis with subsequent GLC-MS for structure determination of dextran demonstrated for an α-($1 \rightarrow 6$) linked glucose unit

Table 1 Percentage of different glycosidic linkages in dextran of different bacterial strains obtained by methylation analysis

Strain number[a]	Solubility in water	α-Linkages (%)				Refs.
		1 → 6	1 → 2	1 → 3	1 → 4	
Lm NRRL B-512F	+	95		5		[21, 23]
Lm NRRL B-1355 fraction 1	+	54		46		[20]
Lm NRRL B-1355 fraction 2	−	95		5		[24]
Lm NRRL B-1299 fraction 1	+	68	29	3		[25]
Lm NRRL B-1299 fraction 2	−	63	27	8		[25]
Lm NRRL B-742 fraction 1	+	50		50		[24]
Lm NRRL B-742 fraction 2	−	87			13	[24]
S mutans 6715 fraction 1	+	64		36		[16]
S mutans 6715 fraction 2	−	4		96		[26]
S mutans GS5	+	70		30		[27]
S downei	+	90		10		[28]

[a] *Lm Leuconostoc mesenteroides, S Streptococcus*

spectrometry (GLC-MS) is an effective method for structure determination of dextran [20, 21].

Furthermore, structural investigations were carried out by the use of degradative enzymes of known specificity followed by means of thin-layer chromatography, HPLC and ^{13}C NMR spectroscopy [12, 16]. Table 1 shows examples of the linkage analysis of several dextran fractions produced by different bacterial strains.

The length of the side chains in dextran produced by *Lm* NRRL B-512F has been studied by sequential alkaline degradation [22]. The procedure is based on the substitution of the terminal non-reducing glucopyranose (Glc*p*) at position 6 with *p*-toluenesulfonylmethyl groups. Analysis by GLC-MS reveals that about 40% of the side chains contain one glucose residue, 45% are two glucose units long and the remaining 15% are longer than two. These results confirm HPLC studies of enzymatically hydrolysed dextran from *Lm* NRRL B-512F [3].

2.2.1
NMR Spectroscopic Characterisation of Dextran

Beside chemical modification and degradation of the polymer backbone, NMR spectroscopy is a capable method for structure determination of dextran [29–32]. Moreover, from the experiences of the authors in the field of structure characterisation of polysaccharides and their derivatives, NMR spectroscopy is one of the most important methods for obtaining detailed structure information [33]. The chemical shifts of the hydrogen and carbon atoms (NMR spectra measured in DMSO-d_6) of the glucose unit in the α-$(1 \rightarrow 6)$ linked main chain are summarised in Table 2. In addition to the ring protons, the protons of the hydroxyl moieties are located at 4.10–4.12 ppm (OH2), 4.51–4.52 ppm (OH3) and 4.63–4.64 ppm (OH4). Anomeric proton resonances of reducing end groups are down-field shifted for α- and high-field for β-anomers [31]. In the ^{13}C NMR spectrum, six signals for the corresponding C-atoms of the dextran main chain are found (Table 2).

In addition to the resonances of the main chain, signals corresponding to certain branches can be observed. Strain *Lm* NRRL-B 512F is known to produce dextran with about 5% α-$(1 \rightarrow 3)$ linkages, determined by methylation analysis. Beside the resonances of the α-$(1 \rightarrow 6)$ linked glucose unit, the signal at 100.5 ppm indicates C1 corresponding to an α-$(1 \rightarrow 3)$ linkage (Fig. 3). The signal at 84.4 ppm indicates C3 of the α-$(1 \rightarrow 3, 6)$ linked branching unit. The C6 atom of the non-reducing end group (indicated as $C6_s$, where s is substructure) is assigned to the signal at 61.7 ppm. The α-$(1 \rightarrow 3)$ linkage content can be estimated as 5% by integration of the C1 signals in α-$(1 \rightarrow 6)$ and α-$(1 \rightarrow 3)$ linkages.

Table 2 Chemical shifts (ppm) for α-$(1 \rightarrow 6)$ linked dextran in DMSO-d_6

	Position						
	1	2	3	4	5	6	6′
^{1}H	4.70–4.69	3.19–3.28	3.43–3.47	3.19–3.28	3.63–3.65	3.73–3.77	3.55–3.59
^{13}C	98.9	72.5	74.1	71.0	71.1	67.1	

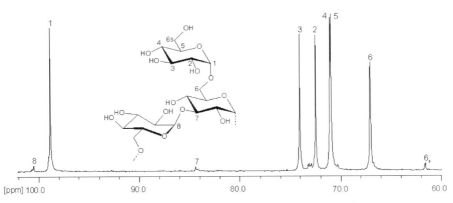

Fig. 3 ^{13}C NMR analysis (in DMSO-d_6) of dextran (M_w 60 000 g mol^{-1}) produced by *Lm* NRRL B-512F in DMSO-d_6

More complex structures need highly resolved NMR spectra applying two-dimensional (2D) spectroscopy for peak assignment. Extensive NMR studies have been carried out for dextran produced by *Leuconostoc ssp.* strain number 10817 (M_w 5400 g mol^{-1}) [34]. A variety of signals beside the resonances for the main chain indicates the structural diversity. A slight signal corresponding to the α-(1 → 3) linkage at 100.5 ppm can be observed. Branching resulting from α-(1 → 2) linkages is predominant. At 97.1 ppm, the anomeric carbon atom appears to participate in the α-(1 → 2) branch [25]. The C1 involved in an α-(1 → 6) linkage and substituted on C2 is situated at 96.3 ppm whereas the C2 is situated at 77.4 ppm. Furthermore, no cross-peak in the HMQC-TOCSY spectrum was observed that corresponded to H1 of the α-(1 → 2) branch at 4.28–4.30 ppm and C6 of the glucopyranosyl residue in the side chain, but one with the C6 of the non-reducing end group. These results indicate branching lengths of one Glc$p$ unit. In accord with the low M_w, the amount of reducing end groups is comparatively high and can be detected in the HMQC-TOCSY spectrum. The signal of the α-anomeric (β-anomeric) proton of the reducing end group at 4.91–4.92 ppm (4.28–4.30 ppm) gives a cross-peak with the corresponding carbon at 92.8 ppm (97.5 ppm). A part of the HMQC-TOCSY spectrum is shown in Fig. 4.

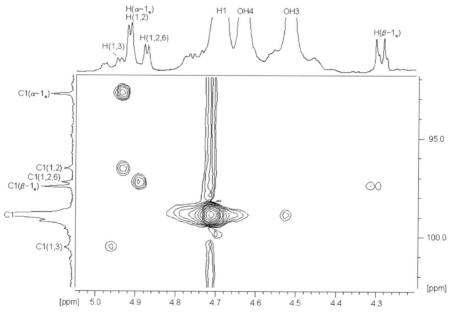

Fig. 4 HMQC-TOCSY spectrum in the region of anomeric atoms of dextran produced by *Leuconostoc ssp.* strain no. 10817 (M_w 5400 g mol^{-1}) in DMSO-d_6 (type of linkage of corresponding AGU and anomeric form of reducing end group, respectively, in *brackets*)

Even with the use of modern NMR instruments, the exact assignment of all resonances beside the α-(1 $\rightarrow$ 6) glycosidic linked main chain is not feasible because of the weak mobility with increasing molecular weight of the polymer. Additionally, a calculation of the degree of branching by integration of relevant signals can not satisfactorily be carried out because of overlay and noise.

2.2.2
Molecular Weight

The weight average molecular weight (M_w) can be determined by light scattering, ultracentrifugation, small-angle neutron scattering and viscometry [35]. Membrane osmometry and end group analysis give information about the number average molecular weight (M_n). Native dextran is generally of a high average molecular weight ranging from 9×10^6 to 5×10^8 g mol^{-1} with a high polydispersity [36–38]. The polydispersity of dextran increases with the molecular weight as a result of increasing branch density [39]. However, defined molecular weight fractions are of interest for many current applications. Beside the fractional precipitation with subsequent molecular weight determination, size exclusion chromatography (SEC) is a useful tool for analysis of the molecular weight distribution (MWD). Figure 5 shows MWD curves obtained by SEC from four dextran samples (a–d), which are used as standards for aqueous SEC, and five further commercially available dextrans (e–i). Even the lower molecular weight dextran samples (e–h)

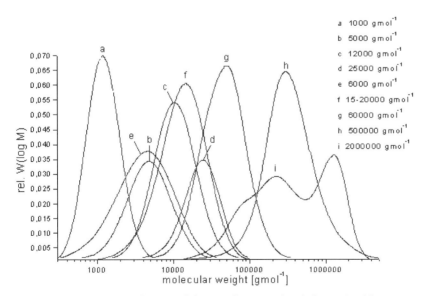

Fig. 5 Molecular weight distribution of dextran fractions (*a–g*) determined by SEC

for technical and scientific use have a low polydispersity and thus a narrow MWD.

2.2.3
Physico-Chemical Characteristics

Being more rod-like below M_w 2000 g mol^{-1}, dextran molecules behave as a random coil in aqueous, low concentration solution, as examined by small angle X-ray scattering (SAXS) measurements [40–42]. If a certain overlap concentration is reached, the individual chains interpenetrate each other and form a transient network of entangled chains resulting in more compact coil geometry and thus a non-Newtonian behaviour [43, 44]. The radius of gyration, R_G, is a useful parameter for estimating the size of dextran molecules in solution. With increasing M_w the R_G increases, whereas higher concentrations or poor solvents lead to decreased R_G values (Table 3) [41, 45].

Naturally occurring dextran is basically amorphous. However, single crystals with lath-like shape can be grown in a mixture of water/polyethylene glycol at temperatures ranging from 120 to 200 °C [47]. Combined electron and X-ray diffraction studies indicate that the unit cell contains two antiparallel dextran chains of two glucopyranosyl residues each [48].

Table 3 Molecular dimensions of dextran from *Lm* NRRL B-512F as radius of gyration (R_G) in nm depending on M_w, concentration and solvent

Solvent/ concentration (mg mL^{-1})	R_G(nm) $M_w \times 10^{-3}$ (g mol^{-1})						Refs.
	40	70	100	500	1000	2000	
Water/1.25	8.5	9	–	–	–	–	[41]
Water/5	6	8	9.5	20	27.5	38	[46]
Water/50	30	–	–	–	–	–	[42]
Water/100	–	10	–	–	–	–	[41]
10% Ethanol/50	25	–	–	–	–	–	[42]
1M Urea/50	30	–	–	–	–	–	[42]

2.3
Properties

Depending on the dextran-producing bacterial strain and thus the structural diversities, the properties of dextran may be different for different samples. In general, the occurrence of the α-(1 → 6) glycosidic bond provides an increase of chain mobility and is responsible for the solubility in a variety of solvents including water, DMSO, DMA/LiCl, formamide, ethylene glycole, glycerol, 6 M aqueous urea, 2 M aqueous glycine and 4-methylmorpholine-4-oxide [13, 45, 49]. The molecular weight of the dextran fraction can influence the time of the dissolution. A 10% aqueous dextran solution (M_w 40 000 g mol^{-1}) will develop precipitates during storage, indicating that dextran solutions are not stable [42, 50]. The adsorption of dextran molecules on the air–liquid interface is the first step of insolubilisation. The precipitates can be resolved in boiling water or DMSO. Even hydrogel formation occurs from concentrated solutions (50–60%) of low molecular weight dextran (M_w 6000 g mol^{-1}) caused by sol–gel conversion, which leads to crystallisation [51].

The colloid osmotic pressure of aqueous dextran solutions can be regulated by molecular weight and concentration of the solute [52]. Dissolved dextran in low concentrations possesses Newtonian flow characteristics [45]. The relationship between viscosity and concentration is shown in Fig. 6 for different dextran fractions [52]. The molecular weight dependence of the intrinsic viscosity can be estimated by several equations [37, 46, 53].

The specific rotation of dextran differs with the solvent and the structural features. Rotations at 25 °C in water and formamide range from + 195 to + 201° and from + 208 to + 233°, respectively [6, 52].

Dextran is a physiologically harmless biopolymer because of its biocompatible, biodegradable, non-immunogenic and non-antigenic properties [54, 55]. It can be depolymerised by different α-1-glycosidases (dextranases) occurring in liver, spleen, kidney and lower part of the gastrointestinal tract.

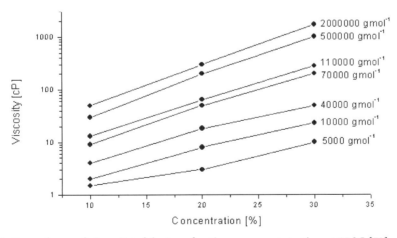

Fig. 6 Dependence of viscosity of dextran fractions on concentration at 25 °C [52]

3
Production and Application of Dextran

3.1
Biosynthesis of Dextran

The majority of dextrans in nature are produced extracellularly via dextran-sucrase from sucrose by several lactic acid bacteria, principally *Leuconostoc* and *Streptococcus* species [13]. Dextran is also synthesised by dextrinase of different *Gluconobacter* species [56]. Referring to this enzyme, fermentation of maltodextrins leads to α-(1 $\rightarrow$ 4) branched dextrans with comparatively lower M_w. However, dextransucrase from *Leuconostoc mesenteroides* NRRL B-512F has attracted most interest because of commercial use.

Dextransucrase is the active enzyme, which catalyses the transfer of D-glucopyranosyl residues from sucrose to dextran, circumstantially named 1,6-α-D-glucan-6-α-glycosyl tranferase [45, 57]. The relatively high binding energy of the glycosidic bond in sucrose is utilised by the enzyme to synthesise the α-(1 $\rightarrow$ 6) linkages of the polymer backbone [13]. Therefore, no adenosine triphosphate or cofactors are required. Beside sucrose, a number of natural, as well as synthetic, donors yield dextran in the presence of dextransucrase. Lactulosucrose [58], α-D-glucopyranosyl fluoride [59], *p*-nitrophenyl-α-D-glucopyranoside [60] and even dextran [61,62] can act as donor substrates. For the investigation of the mechanism, pulse-chase studies with ^{14}C-labelled sucrose were conducted [63]. The proposed insertion mechanism involves two nucleophiles at the active site, which attack sucrose and yield two β-glycosyl intermediates by displacing fructose [64]. The primary hydroxyl group of one glycosyl residue attacks nucleophilically the C1 of the second glycosyl intermediate, resulting in the formation of an α-(1 $\rightarrow$ 6) linkage and a free enzyme nucleophile. The unoccupied nucleophile site attacks another sucrose molecule forming a new β-glycosyl intermediate. The primary hydroxyl group of the latter attacks the C1 hydroxyl group of the isomaltosyl unit and is thus conveyed to the glycosyl residue. The glycosyl and dextranosyl groups are alternately transferred between the nucleophiles while a growth of the dextran chain proceeds (Fig. 7). If an acceptor displaces the dextran from the active site, termination of chain extension occurs. Acceptors are polyhydroxyl compounds, for instance maltose, nigerose, methyl-α-D-glycoside and dextran itself [65]. In the latter case, branching results from an attack of the C3 hydroxyl group of a dextran acceptor chain at the glucosyl or dextranosyl unit of the enzyme (Fig. 7, [66]).

Imidazolium groups of histidine are essential for the transfer of a hydrogen ion to the displaced fructosyl moiety [12]. The reprotonation of the imidazole takes place by abstracting a proton from the C6 hydroxyl moiety and facilitating the formation of the glycosidic linkage.

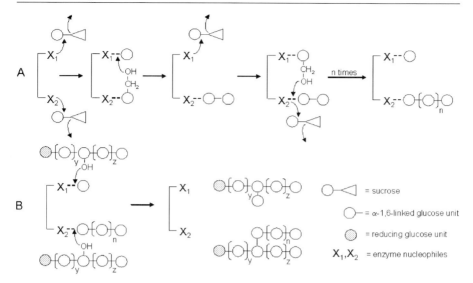

Fig. 7 Mechanism proposed for the synthesis of the α-(1 → 6) glycosidic linked dextran backbone (A) and α-(1 → 3) branches (B) by B-512F dextransucrase (adapted from [12])

3.2
Industrial Production

Recently, most major producers of dextran apply a method based on the batch-wise culture of *Leuconostoc mesenteroides* NRRL B512(F) or B512 strains in the presence of sucrose [67]. Aside from serving as energy source for the microorganism, sucrose induces the dextransucrase production [68]. Special nutritional requirements are satisfied by a combination of complex medium components, for instance yeast extracts, acid hydrolysed casein, corn steep liquors or malt extracts with the addition of peptone or tryptone broth [67, 69]. Low levels of calcium and phosphate are necessary for optimal enzyme and dextran yields. Initial pH values for the fermentation media generally lie between 6.7 and 7.2 where maximum enzyme production takes place [70]. Due to the liberation of organic acids, e.g. lactic acid, the pH decreases to approximate 5, the value for maximum enzyme activity [71]. The stability of the enzyme can be improved by the addition of low concentrations of high M_w dextran, methyl cellulose, poly(ethylene glycol) (M_w 20 000 g mol^{-1}) or non-ionic detergents. For practical purpose, the fermentation process for reaching high M_w dextran takes place at a temperature of 25 °C [69, 72]. At lower temperatures, the amount of low M_w dextran increases, while over 25 °C higher branching occurs [73, 74]. Another influence on branching and weight average molecular weight is the concentration of sucrose. With increasing sucrose content, both the degree of branching and

the yield of high M_w dextran decreases; thus optimal conditions are concentrations of 2–10% and continuous feeding with sucrose [74, 75]. After
24–48 h of fermentation, the viscous culture fluid is precipitated in ethanol
or methanol [76]. The obtained native dextran has a M_w up to 5×10^8 g mol^{-1}
together with a broad distribution [36–38].

Dextran can also be produced enzymatically using cell-free culture su-
pernatants that contain dextransucrase [68]. Low levels of high molecular
weight dextran, poly(ethylene glycol) (PEG) or non-ionic detergents stabilise
the enzyme [77]. The highest activity of dextransucrase after purification
is obtained by the phase partitioning method using aqueous solutions of
dextran and PEG 6000 [78]. This procedure is simple, inexpensive and less
time-consuming than other purification methods involving ultrafiltration and
chromatography.

For clinical and technical dextran, the partial hydrolysis and further
fractionation of native dextran gives products with the desired molecular
sizes [79, 80]. Variation of the concentration of the hydrochloric or sulfuric
acid, the time and temperature of hydrolysis, and the nature and ratio of the
precipitant (alcohol/water) for phase separation permits the control of the
resultant average molecular weights. Recently, a new fractionation method,
called continuous spin fractionation, was developed to improve the efficiency
of the separation process [81]. The use of spinning nozzles facilitates the
transfer of the soluble, low molecular weight polymer species into the extract-
ing agent to such an extent that successful fractionation becomes possible
even with concentrated polymer solutions. The molecular weight distribution
and therefore the polydispersity can be significantly narrowed. Attractive alternative methods for producing defined dextran fractions include the use
of chain degrading enzymes, dextranases and chain-terminating acceptor
reactions [18, 82].

3.3
Chemical Synthesis of Dextran

Beside the natural fermentation process, dextran can be synthesised chemically via a cationic ring-opening polymerisation (ROP) of levoglucosan (1,6-
anhydro-β-D-glucose), a pyrolysis product of polyglucans (Fig. 8, [83]).

The polymerisation of 1,6-anhydro-2,3,4-tri-O-benzyl-β-D-glucopyranose
with phosphorous pentafluoride as catalyst under high vacuum and subsequent cleavage of the benzyl ether linkage by Birch reduction leads to dextrans with M_n 41 800–75 750 g mol^{-1} [84, 86]. Furthermore, tri-O-methyl- glucopyranose and tri-O-ethyl-glucopyranose can serve as monomer but yield
a low molecular weight or rather polydisperse polymer (Table 4). The synthetic dextran has features similar to the natural product concerning solubility, optical rotation and spectroscopic characteristics. X-ray powder patterns
show higher crystallinity compared to native dextran [87]. A more convenient

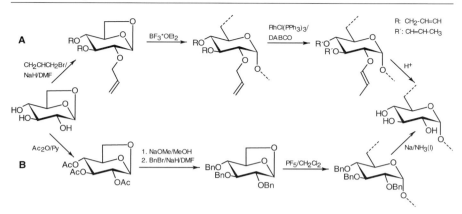

Fig. 8 Synthesis of dextran via a cationic ring-opening polymerisation of 1,6-anhydro-2,3,4-tri-O-allyl-β-D-glucopyranose (*path A*) and 1,6-anhydro-2,3,4-tri-O-benzyl-β-D-glucopyranose (*path B*) [84, 85]

synthesis strategy is the protection of the relevant hydroxyl groups with allyl ether moieties [85]. After the cationic ROP to the stereoregular polymer of 2,3,4-tri-O-allyl-α-(1,6)-D-glucopyranan, an isomerisation with RhCl(PPh₃)₃ as catalyst results in propenyl ether groups, which can be cleaved under acidic conditions. Polymers with a M_n 21 700 g mol^{-1} and a narrow molecular weight distribution (M_w/M_n 2.1) were obtained (Table 4).

Table 4 Ring-opening polymerisation of 1,6-anhydro-2,3,4-tri-O-substituted β-D-glucopyranose in CH_2Cl_2

Monomer[a]	Initiator	[I]/[M][b]	Temp (°C)	Time (h)	Yield (%)	M_n[c] (g mol^{-1})	M_w/M_n	Refs.
Me	$BF_3 \cdot OEt_2$	0.030	+ 25	157	6	1400	–	[84]
Me	PF_5	0.037	– 28	144	85	19 950	–	[84]
Et	PF_5	0.118	– 78	167	70	14 800	–	[84]
Et	PF_5	0.190	– 78	12	51	47 200	62.5	[84]
Bz	PF_5	0.202	– 78	87	86	56 100	–	[84]
Bz	PF_5	0.106	– 78	91	95	75 750	–	[84]
Allyl	$BF_3 \cdot OEt_2$	0.480	0	144	84	21 700[d]	2.1	[85]
Allyl	$CF_3SO_3CH_3$	0.480	0	80	25	11 000[d]	1.2	[85]

[a] Substituent of 1,6-anhydro-β-D-glucopyranose
[b] Molar ratio of monomer (M) to initiator (I)
[c] Determined by membrane osmometry in toluene
[d] Determined by SEC in $CHCl_3$

3.4
Application of Dextran

Clinical grade dextran with molecular weights of 40 000, 60 000 and 70 000 g mol^{-1} (designated dextran 40, 60 and 70) in 6 or 10% aqueous solutions are at present available for replacing moderate blood losses [13, 45]. The polymer essentially substitutes blood proteins, e.g. albumins, in providing colloid osmotic pressure to pull fluid from the interstitial space into the plasma. Because of the low antigenicity and high water solubility, dextran produced by *Lm* NRRL B-512(F) is the material of choice for clinical dextran. Furthermore, the high percentage of α-(1 → 6) glycosidic linkages is responsible for the biological stability in the human bloodstream. The antithrombotic effect of clinical dextran fractions provides a prophylactic treatment for deep venous thrombosis and postoperative pulmonary emboli. Dextran 40 has the special feature of improving the blood flow, presumably caused by reduction of blood viscosity and inhibition of erythrocyte aggregation. Hypersensitivity reactions are limited by a preinjection of low molecular weight dextran (M_w 1000 g mol^{-1}) as a monovalent hapten.

Technical dextran fractions are common in various industries [67, 70]. Because of the narrow molecular weight distribution, specific fractions are used as SEC standard for molecular weight determination. Dextran in X-ray and other photographic emulsions improves the efficiency of silver without loss of fineness of grain [12]. The polysaccharide is used as an ingredient for cosmetics and in bakery products due to superior moisturising properties. Advanced heat shock stability and an increase of the melting temperature can be observed after addition of native dextran to frozen dairy products [43]. The presence of dextran in aqueous solutions helps to confer a favourable physiological environment due to the colloid osmotic pressure. Thus, the use of dextran is a benefit in the preservation of viable organs and as an ingredient for ophthalmic formulations such as artificial tears and eye drops. Dextran is able to prevent protein opsonisation and hence is used for surface modification, e.g. coating of ferrite particles for hyperthermia in cancer therapy [88–90]. Dextran and PEG are known to separate in water [91]. This aqueous two-phase system has proved successful for separating mixtures of biomolecules and subcellular particles, for instance in enzyme-catalysed peptide synthesis [92, 93].

3.4.1
Cross-linked Dextran

Cross-linking can be achieved by physical interactions and chemical reactions. Concentrated solutions of low molecular weight dextran (M_w 6000 g mol^{-1}) lead to the formation of hydrogels by crystallisation [51]. Cross-linking with epichlorhydrin gives commercial products known as Sephadex for gel filtration and Debrisan, a wound-cleansing agent that shortens the healing time [94]. Another matrix for the separation of macromolecules is Sephacryl, an allyl dextran cross-linked by N,N'-methylene bisacrylamide [95, 96]. Furthermore, hydrogels can be obtained by reaction of the polymer with various bifunctional reagents like diisocyanates and phosphorus oxychloride, or by acylation (Sect. 4.2.1), e.g. with methacrylate groups and subsequent UV irradiation [97–99]. Cross-linking of carboxymethyl dextran (CMD) leads to stimuli-responsive hydrogel membranes (Sect. 5.2.2.6). In general, dextran hydrogels have received increased attention due to their biodegradable and biocompatible properties as an efficient matrix system for industrial polymer separation (gel filtration) [54, 55].

4
Esterification of the Polysaccharide

4.1
Inorganic Esters of Dextran

Among the inorganic esters of dextran, only the sulfuric acid half esters (sulfates) and the phosphoric acid esters (phosphates) have gained significant interest. The introduction of sulfate or phosphate groups leads to polyelectrolytes with an improved water solubility, giving aqueous solutions with defined rheological properties, which are valuable as viscosity-regulating agents. Nevertheless, the pronounced biological activity is the most important feature of such dextran derivatives. The anticoagulating properties of the heparin analogue inorganic dextran esters were revealed as early as the 1940s [100].

4.1.1
Dextran Phosphates

The preparation of dextran phosphate (Fig. 9) is simply achieved by treatment of the polysaccharide with polyphosphoric acid in formamide yielding products with up to 1.7% phosphorus.

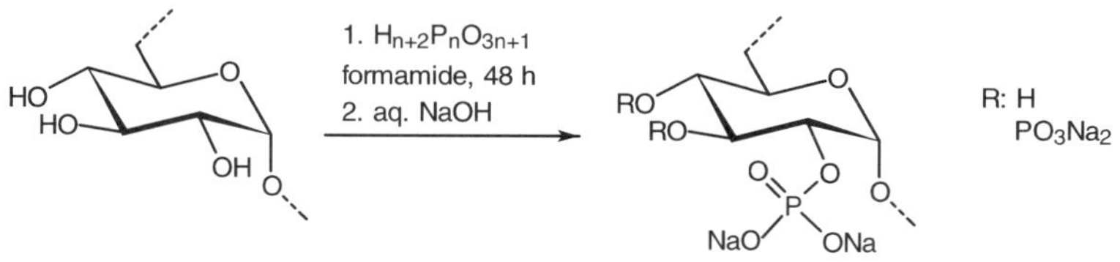

Fig. 9 Preparation of dextran phosphoric acid esters (phosphates) with polyphosphoric acid

These pure dextran phosphates exhibit immunostimulatory effects independent of the molecular weight. It was shown that the mitogenic response of murine splenocytes can be enhanced [101]. Moreover, dextran phosphate (M_w 40 000 g mol^{-1}) increases the survival rate of mice infected with influenza A2 virus (H2N2). Intraperitoneal administration of dextran phosphate, an interferon inducer, shows a 1-day delay in the virus growth in lung, and production of HAI antibody, when compared to the non-treated or dextran-treated mice as controls. More significant was the 2-day delay in the development of lung consolidation, which led to 40% survival of the treated mice [102].

4.1.1.1
Palmitoyldextran Phosphates

Dextran phosphates with additional long chain aliphatic ester functions such as palmitoyldextran phosphates exhibit antitumour activity (Fig. 10).

A growth regression of Sarcoma 183 ascites tumour up to 82% is described [103]. Comparable results were obtained for dextran modified by palmitoylation and/or phosphorylation, which yields three derivatives, i.e. palmitoyldextran phosphate, dextran phosphate and palmitoyldextran. Of these compounds, only palmitoyldextran phosphate showed a growth-inhibitory effect on Ehrlich solid tumour in mice. In combination therapy with mitomycin C, bleomycin, cyclophosphamide and 5-fluorouracil, palmitoyldextran phosphate manifested strong synergistic effects in case of Sarcoma 180 ascites and L1210 leukemic tumours. The compound is not directly cytocidal to Sarcoma 180 ascites tumour, but it appears to act via activation of peritoneal macrophage. The antitumour activity of palmitoyldextran phosphate apparently is mainly due to immunological host-mediated mechanisms [104].

In addition, esters of dextran-bearing palmitoyl groups and phosphate functions can be used for the preparation of specifically modified liposomes, which are exploited for the entrapment of peptides such as hirudin, the most potent inhibitor of thrombin [105].

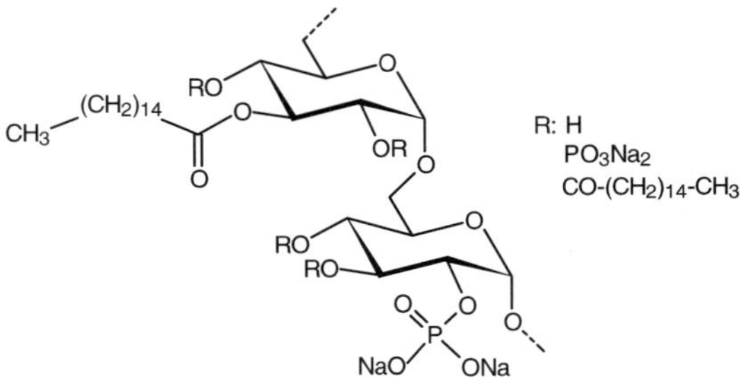

Fig. 10 Typical structural element of palmitoyldextran phosphates

4.1.2
Dextran Sulfuric Acid Half Esters (Dextran Sulfates)

In contrast to dextran phosphate, which is only of scientific interest up to now, the dextran sulfuric acid half ester and its sodium salt, usually simple referred to as dextran sulfate, is a commercially available product today. Its high purity, water solubility and reproducible quality commend it for many applications in molecular biology and the health care sector.

Different approaches for the synthesis of dextran sulfates are known [106]. First attempts exploited treatment of the polysaccharide with concentrated or slightly diluted H_2SO_4 in order to achieve sulfation. A remarkable depolymerisation occurs under these conditions. H_2SO_4 can also be applied in combination with alkyl alcohols yielding alkyl sulfates as reactive species. Here the polymer degradation is comparably low. Powerful sulfating agents are chlorosulfonic acid and sulfur trioxide [107, 108]. A major drawback of these reagents is the sensitivity against moisture. Both compounds strongly react with water. A convenient method to reduce the risk during synthesis is the application of the complexes of $ClSO_3H$ and SO_3 with organic bases, e.g. triethylamine (TEA) and pyridine (Py) or aprotic dipolar solvents, e.g. N,N-dimethylformamide (DMF, Fig. 11). The commercially available SO_3-DMF- and SO_3-Py complexes are white solids that are easy to use. Homogeneous reactions of dextran with these complexes are established methods [109].

A remarkable new method for the conversion of dextran and dextran derivatives, mainly towards mimetics of heparan sulfate uses comparable sulfation but applies 2-methyl-2-butene (2M2B) as an acid scavenger of neutral character [110]. This procedure shows a more efficient reaction combined with diminished chain degradation, as can be seen in Table 5. The method was used for the sulfation of carboxymethylated dextran.

Although sulfation of dextran is an easy synthetic step, most of the studies directed towards biological activity of dextran sulfate are carried out with commercially available products. The DS of these derivatives is usually rather high. The DS of the most widely applied dextran sulfate from Sigma is 2.3. A typical ^{13}C NMR spectrum of a commercial dextran sulfate is shown in Fig. 12. The spectrum confirms the high degree of sulfation and reveals com-

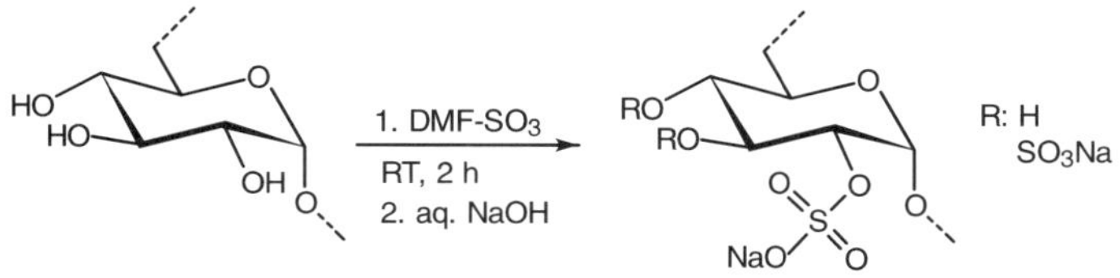

Fig. 11 Preparation of dextran sulfuric acid halfesters with N,N'-dimethylformamide (DMF)-SO_3 complex

Table 5 Results for the sulfation of carboxymethyl dextran with different sulfating agents applying 2-methyl-2-butene (2M2B) as an acid scavenger (reaction temperature 30 °C, [110])

Sulfating agent	Molar ratio reagent/AGU	DS_{CM}	DS_S	Reducing sugars $(nmol\,Glu\,mg^{-1})$
ClSO$_3$H	2 : 1[a]	0.37	0.35	12.71
SO$_3$-Me$_3$N	2 : 1	0.49	0.05	1.98
SO$_3$-Et$_3$N	2 : 1	0.29	0.40	2.61
SO$_3$-pyridine	2 : 1	0.44	0.17	1.84
SO$_3$-DMF	2 : 1	0.37	0.23	1.57
SO$_3$-DMF/2M2B	2 : 1	0.56	0.42	0.82
SO$_3$-DMF/2M2B	3 : 1	0.57	1.00	0.71
SO$_3$-DMF/2M2B	5 : 1	0.55	1.20	0.50

[a] At 22 °C

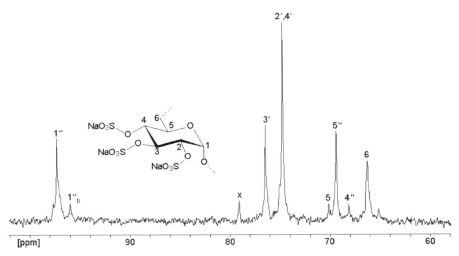

Fig. 12 Typical ^{13}C NMR spectrum of a commercial dextran sulphate; $'$ indicates functionalisation at the corresponding position, $''$ indicates sulfation at the neighbouring position, b indicates a C-atom of a branching structure, X corresponds to a signal caused by a 1,3 branching and substitution at C2 and C4 or a 1,2 branching and substitution at C3

plete functionalisation of position 2. The molecular weight of the commercial dextran sulfates is in the range 5000 to 500 000 g mol^{-1}. Besides the use of dextran sulfates with different M_w, the DS and the structure of the derivatives have not usually been considered in most of the studies discussed below.

4.1.2.1
Bioactivity

Both pure dextran sulfates and mixed derivatives of dextran containing sulfate groups exhibit a broad spectrum of biological activities. The structural features of so-called mimetics of heparan sulfate are discussed in Sect. 5.2.2.1. The pure dextran sulfates used are almost exclusively commercial products with DS values above 2.0.

Among the first biological effects recognised for dextran sulfate was the anticoagulating activity [100, 111]. Thus, dextran sulfates with different molecular weight (7000 up to $458\,000\,\mathrm{g\,mol^{-1}}$) have been tested as potential substitutes for heparin in anticoagulant therapy, showing that products with the lowest molecular weight display the highest anticoagulating properties [112–116]. Investigations on sulfates of cellulose and dextran suggest that the anticoagulant activities of these compounds are at least partially mediated through antithrombin III [117]. However, at best this is only 15% of heparin's activity. Nevertheless, the dextran derivative can be applied as mimic or substitute for heparan sulfate for specific interactions. For instance, it was shown that it protects the kidney against endothelial damage in a model of thrombotic microangiopathy [118]. The beneficial effect of dextran sulfate could be attributed to its ability to protect endothelial cells from coagulation and complement activation, i.e. it acts as a "repair coat" of injured glomerular endothelium.

It is also a well established fact that dextran sulfate modulates the immune response with effects on macrophages [119], proliferation of B-lymphocytes [120], and helper T-lymphocytes [121]. It has been implicated as an activator of C3 via the alternative pathway of complement [122]. Numerous reports claim that dextran sulfate is a potent inhibitor of human immunodeficiency virus (HIV), Herpes simplex virus (HSV) and other pathogens [123–125].

Comparison of the activity of sulfated homopolysaccharides such as dextran and cellulose esters with that of neutral homopolysaccharides and sulfated heteropolysaccharides such as heparin and heparan sulfuric acid half esters shows potent virucidal activity against human T-cell lymphotropic virus type III (HTLV-III) for the sulfated homopolysaccharides. In contrast, neutral homopolysaccharides have no effect and sulfated heteropolysaccharides exhibit only a little effect on HTLV-III activities. This suggests that the sulfate moiety and the type of polysaccharide are most important in inhibiting growth of HTLV-III [126].

In vitro studies have shown that dextran sulfate inhibits HIV binding, replication and syncytium formation, probably because they interfere with the ionic interaction between cell surface components such as CD4 or sulfated polysaccharides and positively charged amino acids concentrated in the V3 region of HIV gp120 [127–129]. Again, low molecular weight derivatives with

$M_w < 10\,000$ g mol^{-1} are the most efficient [130]. Dextran sulfate was shown to be a potent inhibitor of, e.g. HSV infectivity in vitro (Table 6, [131]).

In contrast, it has not been possible to provide any protection in murine models of HSV infection. In addition, it was found in preclinical trials that dextran sulfate administered orally [132] and intravenously [133] did not lead to a decrease in circulating HIV antigen. Reasons may be the poor absorption of the polysaccharide derivative from the intestine and desulfation processes [134, 135]. Nevertheless, the treatment with dextran sulfate and derivatives of dextran sulfate of optimised DS for short intervals using high dosages may overcome these problems. In this case, the sulfated polysaccharide is not substantially endocytosed or degraded by cell receptor binding in the mammal and thereby retains antiviral activity in vivo [136]. Additionally, the combination of azidothymidine (AZT), which has shown promise in prolonging life in patients with AIDS but exhibits significant blood toxicities, and dextran sulfate leads to a higher effectivity than either alone [137].

Recent research and development are also concentrated on the cancerostatic effects of dextran sulfate. A number of mechanisms for this specific biological interaction were established. Previous studies centre on the ability of dextran sulfate to reduce peritoneal- [138] and pulmonary metastasis [139]. These studies used derivatives of both low (7000 g mol^{-1}) and high molecular weight (500 000 g mol^{-1}) to reduce adherence of melanoma or lung tumour cells to endothelium of lung and peritoneum. The dextran sulfate was not toxic to cancer cells, therefore the anti-metastatic capacities were thought to be due to the polyanionic nature of the polymer. It was suggested that the dextran sulfate binding to both tumour and endothelial cells rendered them more negatively charged. This increasing electrostatic repulsion altered the adhesiveness between the tumour cells and the vascular endothelium. The altered adhesion properties were thought to reduce or inhibit the settlement of the tumour cells in the capillaries of organs, resulting in a subsequent reduction

Table 6 Infectivity of HSV-1 (Herpes simplex virus) and HSV-2 strains pretreated with different concentrations of dextran sulfate for 1 h at 37 °C

Concentration (nM)	Plaque forming unit (% of control) for	
	HSV-1 (F)[a]	HSV-2 (22)[a]
0.10	88.4 ± 23.7[b]	69.2 ± 1.4
0.25	72.6 ± 30.2	63.1 ± 6.2
0.50	58.8 ± 21.2	55.2 ± 3.1
1.00	49.3 ± 22.7	43.5 ± 3.1
10.00	15.6 ± 9.1	3.2 ± 1.5
50.00	0	0

[a] Wild-type strain
[b] Results are the mean (± the SD) of four independent experiments

in metastasis. Recent findings show that the ability of dextran sulfate to completely inhibit hyaluronidase needs to be invoked [140, 141]. Hyaluronidases have been previously implicated in tumour progression and metastasis. The ability of the dextran derivative to inhibit enzymatic degradation of hyaluronan into biologically active fragments was documented. Consequently, the finding of reduced metastasis after intravenous or intraperitoneal administration of dextran sulfate could be easily attributed to the hyaluronidase inhibitory action as well as an anti-adhesive mechanism of action caused by the polyelectrolyte.

Interestingly, the polysaccharide sulfate is also considered as an antiscapie drug [142]. Thus, it was included in an experimental setting studying the mode and the site of action of the major antiscrapie drugs, investigating their effects on the abnormal protease-resistant isoform of a host-encoded protein (PrPres) and on its accumulation in mouse spleen. Dextran sulfate delayed the beginning of the clearance phase but then blocked PrPres synthesis for a long period of time, probably because of its immunological effects on the spleen.

Although dextran sulfate exhibits such promising biological properties its application is limited because of various side effects. In early studies, clinical trials were discouraging with reports of, inter alia, stiff and painful joints and loss of hair [143]. Likewise chronic toxicity studies revealed retardation in weight gain and osteoporosis [133, 144]. More serious is the fact that dextran sulfate may cause colitis. Today, treatment of mice with the polysaccharide ester has even become a standard method to experimentally induce colitis. It is used both to study the mechanism of inflammatory reactions and to elucidate the reasons for colonic cancer. Thus, in recent investigations the "dextran sulfate-induced colitis" was induced by oral administration for 3 and 7 days to clarify the microvascular changes and the effector sites of lansoprazole during the formation of colitis [145]. Moreover, changes in expression or a redistribution of intercellular tight junction proteins were investigated using acute and chronic dextran sulfate-induced colitis in mice [146]. In acute inflammation, the changes are more pronounced than in chronic inflammation. In addition, studies applying dextran sulfate-induced colitis in mice have suggested that the enhanced release of reactive oxygen species plays an important role in the pathogenesis of inflammatory bowel disease, such as ulcerative colitis and Crohn's disease. It was shown that free radical scavengers (edaravone and tempol) suppressed the colonic shortening and the damage score. A clinical effect for edaravone and tempol in inflammatory bowel disease patients is strongly expected [147].

4.1.2.2
Dextran Sulfates in Supramolecular Assemblies

Besides the bioactivity, dextran sulfate was studied for a broad variety of applications using its polyelectrolyte nature and its ability to form polyelectrolyte complexes (PEC) with polycations such as chitosan. The process is based on the sequential deposition of interactive polymers from their solutions by electrostatic, van der Waals, hydrogen bonding and charge transfer interactions [148]. These interactions can be applied to create layer-by-layer (LbL, Fig. 13) assemblies of functional material surfaces with defined biodegradability or bioactivity.

Alternate anti-vs procoagulant activity of human whole blood on a LbL assembly between chitosan and dextran sulfate has been achieved [149, 150]. Furthermore, the technique permits the formation of biodegradable nanostructures with nanometer-order thickness on surfaces, which is an important requirement for biomedical applications. The alternating enzymatic hydrolysis of a LbL assembly formed from chitosan and dextran sulfate by chitosanase was demonstrated via measurements with a quartz crystal microbalance (QCM) [151]. The hydrolysis of the assembly was clearly dependent on the surface component. The hydrolysis of the assembly with the dextran sulfate surface was saturated within 10 min and was much faster than the hydrolysis of the assembly with the chitosan surface, although chitosanase can hydrolyse chitosan (Fig. 14).

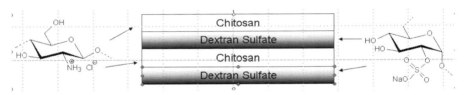

Fig. 13 Polyelectrolyte complex formed by layer-by-layer assembly of dextran sulfate and chitosan

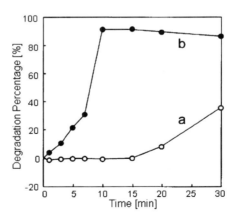

Fig. 14 Enzymatic hydrolysis with chitosanase of (a) the layer-by-layer (LbL) assembly of dextran sulfate/chitosan (five steps, total thickness 22 nm) with a chitosan surface and (b) a comparable LbL assembly (dextran sulfate/chitosan, six steps, total thickness 55 nm) with a dextran sulfate surface

This unique hydrolysis may appear from the electrostatic condensation of the enzyme with its cationic surface on the outermost surface of dextran sulfate, leading to subsequent hydrolysis of chitosan in the underlayer of the assembly.

In addition, the complexes based on the system dextran sulfate/chitosan show remarkable pH selectivity. Complex gels with approximately equivalent concentrations of amino and sulfate groups are highly sensitive to the change in external pH in a narrow range. The maximum volume of the complex gel was observed at pH 10.5 and was approximately 300 times as large as the initial one. Compared with the swelling behaviour of a complex gel prepared from carrageenan, it seems to be the high density of the ionisable functional groups, as well as the flexibility of acidic polymer chains, which contribute to the high pH sensitivity [152].

The strong interaction of dextran sulfates with cationic functions in porous support materials is exploited to create new highly charged surfaces for adsorption of proteins. It was revealed that new and strong ionic exchange resins are accessible by simple and rapid deposition of dextran sulfates on commercial DEAE- or MANAE-agarose. The material is characterised by an increased charge density on the porous surface of the support, which can perfectly bind protein material, as demonstrated in Fig. 15 [153].

PEC can adsorb the majority (80%) of the proteins contained in crude extracts from *Escherichia coli* and *Acetobacter turbidans* at pH 7 and can be used to immobilise industrially relevant enzymes, e.g. *Candida antarctica* A and B with very high activity recoveries and immobilisation rates. In contrast to covalent binding, the proteins can be recovered and do not undergo irreversible structural alteration.

A comparable strategy is applied to obtain compounds with a pronounced biological selectivity. By treatment of a porous NH_3^+-containing polypropylene membrane with the sulfate of dextran, a material for convenient removal

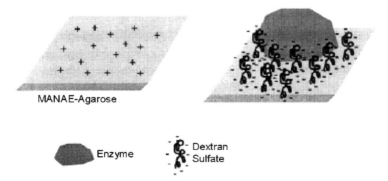

MANAE-Agarose

Enzyme   - Dextran
                - Sulfate

Fig. 15 Strong and non-distorting reversible binding of enzymes on polyelectrolyte complexes [153]

of HIV and related substances from blood, plasma or other body fluids is obtained. Filtration of HIV-containing human plasma results in 99.2% removal of HIV [154].

The "chitosan-dextran sulphate PEC" is widely used for the preparation of colloidal systems as well. Slow drop-wise addition of the components is generally used for the formation of PECs, which allows elaborating both cationic and anionic particles with an excess of chitosan or dextran sulfate, respectively. The PEC particles show a core/shell structure. The hydrophobic core results from the segregation of complexed segments, whereas excess component in the outer shell ensured the colloidal stabilisation against further coagulation. Considering the host/guest concept for the formation of PECs, the influence of the molecular weight of components on particle sizes could be well explained by the chain length ratios of the two polymers and is schematically demonstrated in Fig. 16.

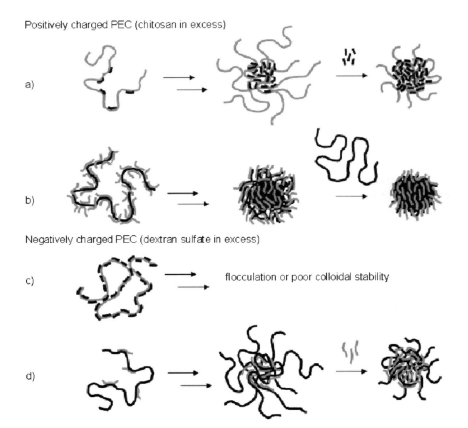

Fig. 16 Preparation of differently charged microparticles, dependent on the chain length of the polymers applied [155]

As an irreversible flocculation occurred with a drop-wise approach for both cationic and anionic PEC particles when the mixing ratio was close to unity, a more versatile and simpler method was found to be the one-shot addition of one solution to the other. Characterisation of particles obtained via this path revealed very similar properties to those obtained by a slow drop-wise approach [155].

Polyelectrolyte multilayer microspheres, prepared by alternating adsorption of dextran sulfate and protamine on melamine formaldehyde cores followed by the partial decomposition of the core, were used to immobilise the peroxidase and glucose oxidase. Retention of enzymic activity of the peroxidase/glucose oxidase system incorporated into the microspheres was demonstrated. These bienzyme system immobilised in the microspheres can be applied for kinetic glucose assays [156].

Highly engineered nanoparticles containing dextran sulfate are prepared for applications in the field of medical diagnostics, e.g. beginning atherosclerosis, which is an inflammatory disease of the arterial walls. A targeted magnetic resonance imaging (MRI) contrast agent for in vivo imaging of early stage atherosclerosis was designed. Early plaque development is characterised by the influx of macrophages, which expresses a class of surface receptors known collectively as the scavenger receptors. The macrophage scavenger receptor class A (SRA) is highly expressed during early atherosclerosis. The macrophage SRA therefore presents itself as an ideal target for labelling of lesion formation. By coupling a known ligand for the scavenger receptor (dextran sulfate) to a MRI contrast agent, early plaque formation can be detected in vivo [157].

It should be mentioned that the defined interaction of dextran sulfate with amino functions is not only applied for the design of structures on the supermolecular level but also on the molecular level. Thus, a preferred handed helical structure was induced into the polyaniline main chains by chemical polymerisation of achiral aniline in the presence of dextran sulfate as a molecular template. This affords a novel chemical route for the synthesis of chiral conducting polymers [158].

4.2
Organic Esters of Dextran

In contrast to the broad variety of applications of $1 \rightarrow 4$ and $1 \rightarrow 3$ linked glucans after reaction with (C_2 to C_4) carboxylic acid anhydrides and chlorides [159], the use of dextran esters of short chain aliphatic acids such as acetates or propionates is rather limited. According to the present knowledge, the commonly applied acetylation of polysaccharides with the acetic acid anhydrides or acetyl chlorides in the presence of triethylamine or pyridine as base does not lead to pure and soluble dextran acetate with significant DS values. In contrast, dextran propionates [160] and butyrates [161] can be easily prepared in heterogeneous reactions of the polymer suspended in pyridine using the carboxylic acid anhydride. The acylation of dextran yields derivatives with defined hydrophobic character. Thus, the solubility of dextran esters in water varies with the DS and the chain length of the substituent. It was shown that the maximum DS values while maintaining water solubility are 0.26 for C_6 carboxylic acid esters and 0.50 for C_4 carboxylic acid esters of dextran [162]. These values are comparable with data for solubility of ethyl (DS 0.81) and butyl (DS 0.69) carbonate substituted dextran [163]. Such hydrophobically modified dextran derivatives form aqueous biphase systems in combination with dextran [162] or with PEG [164, 165], which are exploited for the separation of biological material. Using dextran esters with C_3, C_4 and C_6 acid groups, a family of biphase systems is accessible by varying the properties of the hydrophobically modified dextran. This approach allows control of the phase boundary defining the onset of biphase formation as well as the solvency properties of the more hydrophobic phase. Dextran esters of aliphatic carboxylic acids are also exploited for pharmaceutical coatings [166].

Acylation reactions of dextran in combination with NMR spectroscopy are tools for the elucidation of the reactivity of dextran [30, 161, 167] and for the analysis of structural features of the polymers. The acetyl group distribution in the Glc$p$ units in acetylated amylose, dextran and cellulose was studied by ^{1}H and ^{13}C NMR spectroscopy. It was concluded that the reactivity of OH groups decreased in the order C2 > C3 > C4 for dextran.

Subsequent conversion of dextran derivatives with acetic acid or propionic acid anhydride is an effective method for revealing structural features on the molecular level. This is illustrated on a dextran propionate (M_w 5430 g mol^{-1}) which can be completely acetylated with acetic anhydride/pyridine in a separate step yielding a peracetylated sample (dextran propionate acetate, DPA). The assignment of the chemical shifts of DPA is carried out via 2D NMR (Fig. 17).

DS_{Prop} can be determined precisely from highly resolved polymer-^{1}H NMR spectra using the spectral integrals of the anhydroglucose unit's (AGU) proton region at 3.1–5.3 ppm and the integrals of the acetyl proton signals at 1.8–2.1 ppm. The appearance of three signals for the acetyl protons and the complexity of the COSY signals for H2, H3 and H4 suggest a statistic pattern of propionylation.

The introduction of longer aliphatic acids can also be achieved by acylation with the so-called impeller method. The carboxylic acids or their anhydrides are converted in situ to reactive mixtures of symmetric and mixed anhydrides (Fig. 18).

Chloroacetyl, methoxyacetyl and most importantly trifluoroacetyl moieties are used as impellers. Carboxylic acid esters of polysaccharides with almost complete functionalisation can be obtained. Thus, chloroform-soluble dextran stearates and dextran myristates with DS 2.9 are prepared by

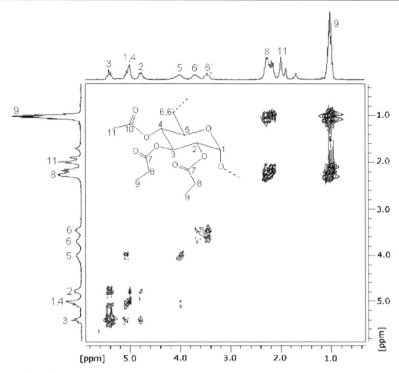

Fig. 17 ^{1}H, ^{1}H-COSY NMR spectrum (in CDCl$_3$) of a dextran acetate propionate (DS$_{Prop}$ 1.70) [160]

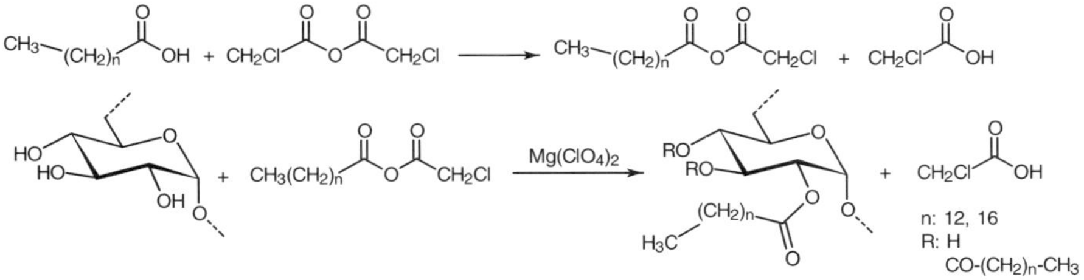

Fig. 18 Formation of mixed anhydrides during esterification of dextran using the impeller method

treating dextran in chloroacetic anhydride with the corresponding acids at 70 °C for 1 h. The presence of magnesium perchlorate as catalyst is necessary [168].

4.2.1
Homogeneous Esterification Reactions

Dextran usually dissolves in water, DMSO, DMA/LiCl, DMF/LiCl, formamid, aqueous urea and glycerin. The solubility may be restricted by a high crystallinity of the polymer (Sect. 2.3). The most versatile solvents for homogeneous acylation reactions are DMSO or mixtures of a polar aprotic solvent and a salt. The broadest application found the combination of substituted amide with LiCl. Thus, dextran dissolves easily in the mixture DMF/LiCl upon heating to 90–100 °C. This solvent can be exploited for the esterification of dextran as displayed in Fig. 19.

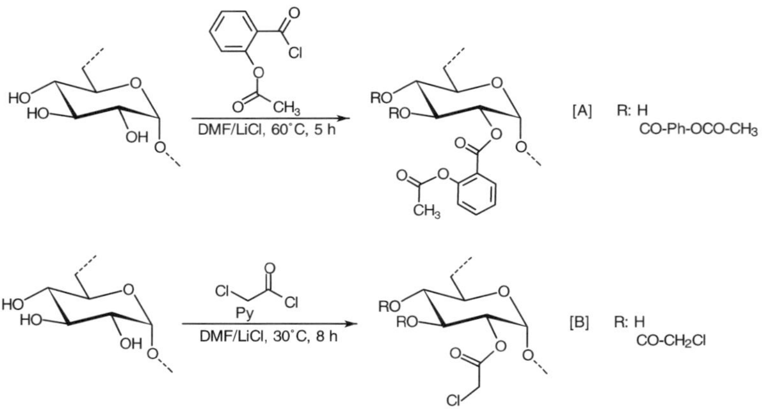

Fig. 19 Dextran esters synthesised homogeneously in DMF/LiCl [169, 170]

Conversion of dextran in DMF/LiCl with succinic anhydride is a popular method for the introduction of spacer-bound large molecules [171, 172]. Moreover, homogeneous esterification is exploited for the binding of bioactive compounds yielding prodrugs and for the introduction of unsaturated moieties giving precursors for hydrogels as discussed in the next sections.

4.2.1.1
Hydrogels Based on Unsaturated Dextran Esters

Hydrogels prepared via homogeneous esterification of dextran with unsaturated carboxylic acids are advanced polysaccharide-based products useful for drug delivery systems and protective encapsulants, e.g. of viruses used in gene therapy [173]. Very promising in this regard is the dextran maleic acid monoester [174], which can be obtained by conversion of dextran in DMF/LiCl with the maleic anhydride in the presence of TEA. The DS of the products can be easily controlled with the amount of anhydride applied but is also influenced by temperature, amount of catalyst and reaction time, as displayed in Table 7.

The dextran maleates are easily soluble in various common organic solvents such as DMSO, DMF, N-methyl-2-pyrrolidone (NMP) and DMA. The hydrogels are manufactured by irradiation of dextran maleate with long-wave UV light (365 nm). The minimum DS required for proper UV cross-linking of the derivatives is 0.60. The hydrogels show a high swelling capacity (swelling ratio up to 1489% at pH 7) depending on DS and the pH of the medium, i.e. highest swelling ratio in neutral pH, followed by acidic (pH 3) and alkaline conditions (pH 10). The swelling ratio increases with increasing DS. The surface and interior structure of a dextran methacrylate hydrogel (DS values up to 0.75), prepared in a comparable manner (Fig. 20), is investigated

Table 7 Influence of the temperature, reaction time and amount of maleic anhydride on the degree of substitution for the esterification of dextran in DMF/LiCl with maleic anhydride

Temp. (°C)	Time (h)	Molar ratio [a]	DS
20	20	1 : 1.0	0.30
40	20	1 : 1.0	0.51
60	20	1 : 1.0	0.99
60	1	1 : 1.0	0.36
60	5	1 : 1.0	0.57
60	10	1 : 1.0	0.84
60	20	1 : 1.0	0.99
60	8	1 : 0.5	0.60
60	8	1 : 1.0	0.90
60	8	1 : 2.0	1.47
60	8	1 : 3.0	1.53

[a] Mol OH group of dextran/mol maleic anhydride

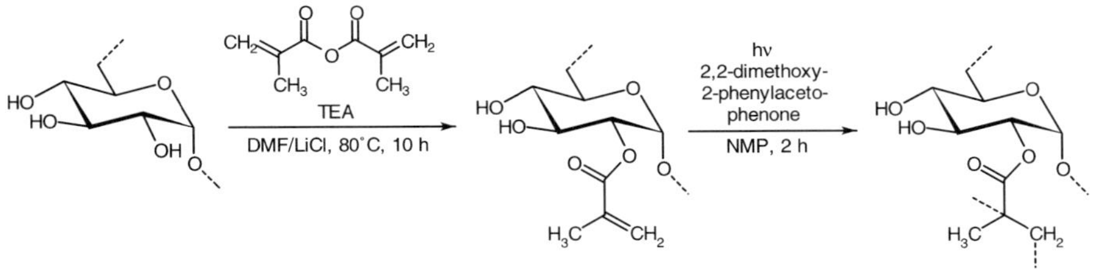

Fig. 20 Synthesis of dextran methacrylate usable for the formation of hydrogels after photo-initiated cross-linking [175]

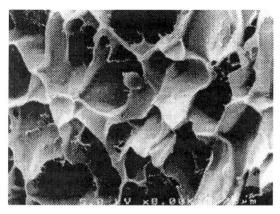

Fig. 21 Three-dimensional porous structure of a dextran methacrylate hydrogel observed by means of SEM

by means of SEM after application of special cryofixation and cryofracturing techniques.

A unique 3D porous structure is observed in the swollen hydrogel (Fig. 21), which is not determined in the unswollen state. Different pore sizes and morphologies between the surface and the interior of swollen hydrogels are visible [175].

Transparent hydrogels useful for adhesion inhibitors, tissue adhesives, wound dressings, hemostatics and embolisation materials are obtained from dextran methacrylates via polymerisation with *N*-isopropylacrylamide in DMSO in the presence of azobisisobutyronitrile [176]. A broad variety of new hydrogels with different sensitivities and tunable degradation behaviour is accessible by grafting L-lactide onto 2-hydroxyethyl methacrylate (HEMA) and binding this polymerisable group on dextran via activation with *N,N'*-carbonyldiimidazole (CDI, Sect. 4.2.2) [177].

4.2.2
Synthesis of Dextran Esters with Bioactive Moieties (Prodrugs) via In Situ Activation Reactions

The majority of acylation reactions on dextran are exploited to prepare complex esters of the polymer such as fluorescent-labelled dextran or polymers forming defined superstructures. By far the most important class of esters in this regard are still conjugates, i.e. dextran coupled to drugs, proteins, or hormones [178, 179]. Conjugation to dextran may overcome many problems in drug design including limited solubility of the drugs, short plasma half life, toxicity of the drugs and unspecific interaction with tissue [8]. In addition, binding of pharmaceutical products to the polymer can lead to controlled-release compounds. Polysaccharide esters are well suited for this concept because of the ease of deesterification by simple hydrolysis or enzymatic attack. A broad variety of drugs can be bound yielding so-called prodrugs, which represents a rapidly growing field of pharmaceutical research. The polysaccharide of choice is almost exclusively dextran to give water-soluble prodrugs.

For the introduction of such complex and sensitive functions, the heterogeneous conversion applying carboxylic acid anhydrides or chlorides is not appropriate. A number of new synthesis paths for the defined, homogeneous conversion of dextran to its carboxylic acid esters were therefore developed. Reaction of dextran with carboxylic acids after in situ activation is the most important path among the new synthesis tools for the defined esterification of dextran with complex and sensitive acids. The most popular in situ activating reagents exploited are sulfonic acid chlorides, reagents of the dialkylcarbodiimide type and N,N'-carbonyldiimidazole. Besides basic results for these synthesis paths, a selection of results for the preparation of prodrugs is included in this section to show its usefulness.

4.2.2.1
Activation of Carboxylic Acids with Sulfonic Acid Chlorides

Via the in situ activation, applying sulfonic acid chlorides, covalent binding of bioactive molecules onto dextran was achieved by esterification of the polymer with α-naphthylacetic acid (1), nicotinic acid (2) and naproxen (3) homogeneously in DMF/LiCl using p-toluenesulfonyl chloride (tosyl chloride, TosCl) or methanesulfonic acid chloride (mesyl chloride, MesCl) and pyridine within 22 h at 30–70 °C (Fig. 22).

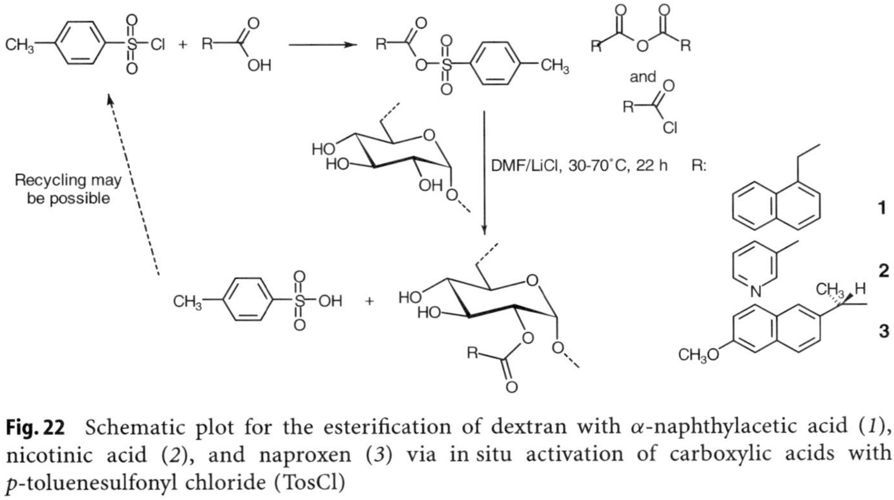

Fig. 22 Schematic plot for the esterification of dextran with α-naphthylacetic acid (1), nicotinic acid (2), and naproxen (3) via in situ activation of carboxylic acids with p-toluenesulfonyl chloride (TosCl)

Both the temperature and the concentration of pyridine as well as the type of sulfonic acid chloride show an influence on the reaction, as displayed in Table 8 [180]. The esterification is even possible without the base.

^{13}C NMR spectra of partially modified dextran with α-naphthylacetate moieties show that the reactivity of the individual hydroxyl groups decreases in the order C2 > C4 > C3. A mechanism for the reaction is suggested, which includes formation of an acylium complex as observed for the reaction with acid chlorides (Fig. 23).

These findings support the NMR results for the acylation with acyl chlorides, i.e. the in situ activation with sulfonic acid chloride succeeds mainly via intermediately formed acyl chlorides of the carboxylic acids [180]. The binding of naproxen is a nice example for the preparation of prodrugs as a valuable approach for the transportation of lipophilic agents in a biological environment [45, 178]. The water solubility of naproxen bound to dextran exceeds that of the acid form of the drug by 500 times. Participation of specific acid–base catalysis in the hydrolysis of the ester in aqueous buffer solutions at 60 °C is observed. An almost identical degradation rate is obtained for the ester in 80% human plasma, excluding catalysis of hydrolysis by plasma

Table 8 Influence of the concentration of pyridine [Py], the reaction temperature, and type of sulfonic acid chloride applied for the esterification of dextran (0.12 mol L^{-1}) in DMF/LiCl with α-naphthylacetic acid (0.37 mol L^{-1}) for 22 h

Conditions [Py] (mol L^{-1})	Temp. (°C)	Sulfonic acid chloride	Product DS
–	50	TosCl	0.13
0.37	50	TosCl	0.15
0.74	50	TosCl	0.19
1.48	50	TosCl	0.23
0.74	30	TosCl	0.13
0.74	60	TosCl	0.22
0.74	70	TosCl	0.23
0.74	50	MesCl	0.18

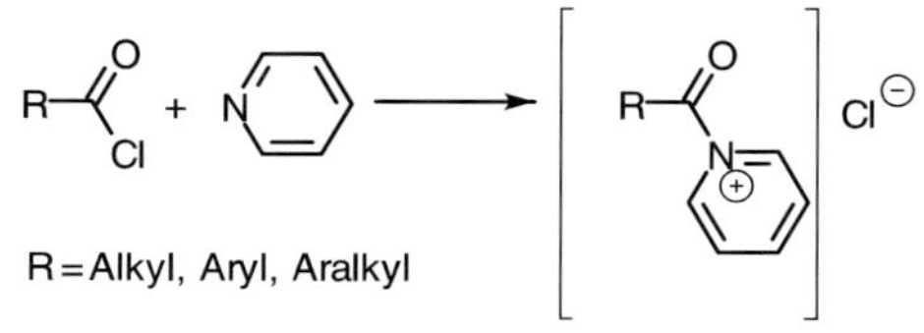

R = Alkyl, Aryl, Aralkyl

Fig. 23 Formation of an acylium complex

enzymes [181, 182]. In the same manner, dextran esters of ketoprofen, di-clofenac, ibuprofen and fenoprofen can be studied showing that the dextran ester prodrug approach provides selective colon delivery systems of drugs possessing a carboxylic acid functional group [183, 184]. The introduction of N-acylamino acid into the dextran backbone is achievable in the same manner [185, 186].

4.2.2.2
Dialkylcarbodiimide-Type Coupling Reagents

Coupling reagents of the dialkylcarbodiimide type are most frequently utilised for the esterification of polysaccharides with complex carboxylic acids [179]. The best known condensation agent of this class, particularly in peptide and protein chemistry, is N,N'-dicyclohexylcarbodiimide (DCC, Fig. 24) [187].

These reagents have a number of drawbacks. First of all, they are toxic especially via contact with skin. The LD_{50} (dermal, rat) of DCC is 71 mg kg^{-1}. This should always be considered if the reaction is used for the preparation of materials for biological applications. Moreover, the N,N'-dialkylurea formed during the reaction is hard to remove from the polymer except for preparation in DMF and DMSO, where it can be filtered off. In case of esterification of polysaccharides in DMSO in the presence of these reagents, oxidation of hydroxyl functions may occur due to a Moffatt type reaction (Fig. 25, [188]). The oxidation products formed can be detected with the aid of 2,4-dinitrophenylhydrazine, e.g. in case of the conversion of dextran with DCC in DMSO [189].

Moreover during the reaction, decomposition of DMSO to dimethylsulfide occurs resulting in a pungent odour. The treatment with DCC may also lead to the formation of isourea ethers according to reaction shown in Fig. 26.

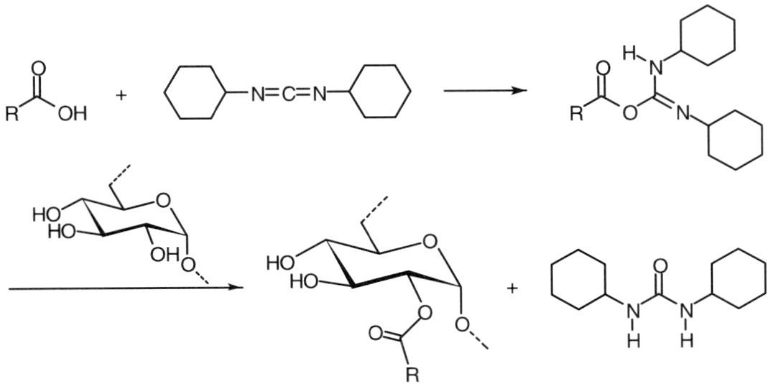

Fig. 24 Esterification of a polysaccharide with carboxylic acid in situ activated with N,N'-dicyclohexylcarbodiimide (DCC)

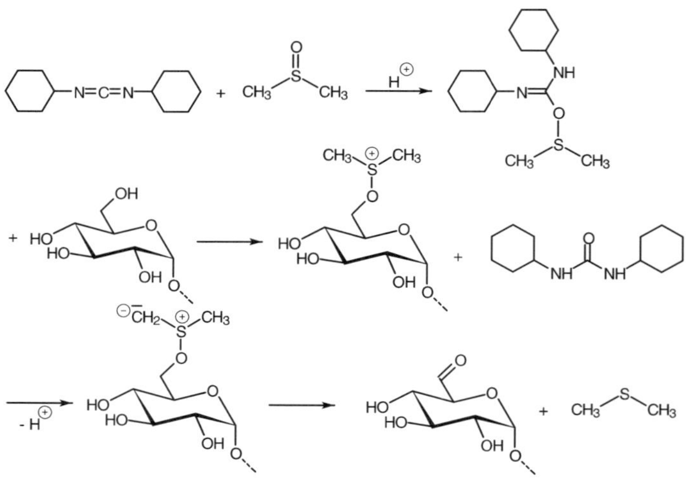

Fig. 25 Moffatt reaction at a non-reducing terminal glucose unit in dextran

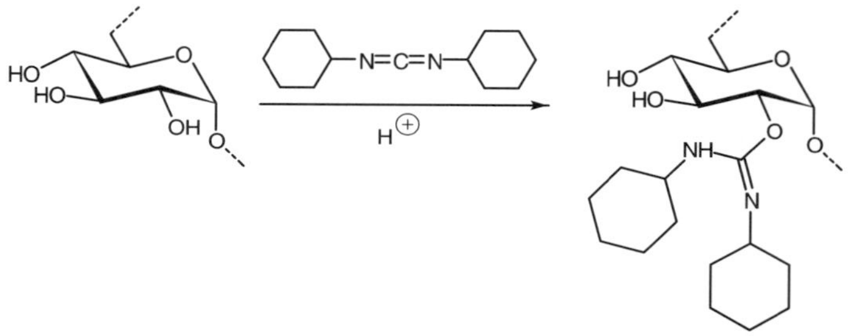

Fig. 26 Formation of isourea ethers as a side reaction during conversion of dextran using DCC as coupling reagent

Nevertheless, the reagent is still used for a large number of esterification reactions of dextran. The synthesis of dextran amino acid esters is achieved by conversion of the polysaccharide in DMSO with the N-benzyloxycarbonyl protected acids for 48 h at 20 °C using DCC and pyridine. O-(N-Benzyloxycarbonylglycyl)dextran with DS 1.1, O-(N-benzyloxycarbonyl-aminoenanthyl) dextran with DS 2.2, and O-(N-acetyl-ʟ-histidinyl)dextran with DS 1.1 are accessible. Deprotection was achieved with oxalic acid and Pd/C [190, 191].

Functionalisation with bulky hydrophobic carboxylic acids/DCC was studied for the synthesis of amphiphilic polymers based on dextran. Bile acid is covalently bound to dextran (Fig. 27) through ester linkage in the presence of DCC/DMAP (added in dichloromethane) as coupling reagent.

A homogeneous reaction occurs if the mixture DMF/formamide is used as solvent. The amount of bound acid (determined by UV/Vis spectroscopy)

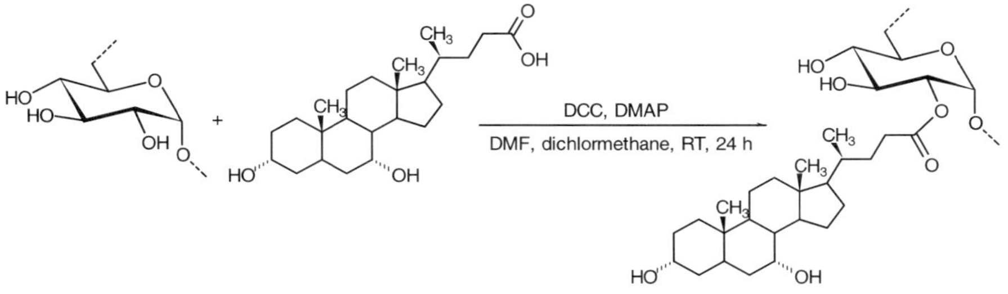

Fig. 27 Synthesis of bile acid esters of dextran using DCC

is in the range from 10.8 to 11.4 mol % [192, 193]. A prodrug accessible via in situ activation of the carboxylic acids with DCC is the metronidazole monosuccinate ester of dextran synthesised in DMSO or DMF using TEA as base (Fig. 28, [194]).

The hydrolysis of the dextran metronidazole succinate over the pH range 7.4–9.2 at 37 °C can be determined with high-performance size exclusion chromatography showing slower release compared to other prodrugs such as the dextran cromoglycate (see below). Interestingly, an intramolecularly catalysed hydrolysis by the neighbouring dextran hydroxyl groups is observed [195]. Metronidazole can also be covalently bound with maleic and glutaric acid. For the dextran metronidazole esters, in which succinic and glutaric acids are incorporated as spacers, the decomposition proceeds through parallel formation of metronidazole and the monoester derivative, as can be demonstrated by reversed-phase HPLC and SEC. Almost identical stability of the individual esters is obtained after incubation in 0.05 M phosphate buffer pH 7.40 and in 80% human plasma, revealing that the hydrolysis in plasma proceeds without enzymic catalysis. The half-lives of the modified dextran derived from maleic, succinic and glutaric acids are 1.5, 32.1 and 50.6 h, respectively [196].

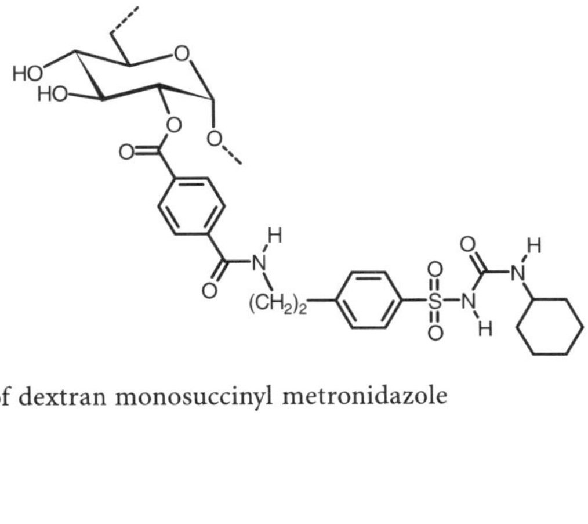

Fig. 28 Structure of dextran monosuccinyl metronidazole

4.2.2.3
N,N′-Carbonyldiimidazole (CDI) as Esterification Reagent

A method with an enormous potential for dextran modification is the homogeneous one-pot synthesis after in situ activation of the carboxylic acids with CDI, which is a rather well known technique in general organic chemistry and was published in 1962 [197]. It is especially suitable for the functionalisation of the biopolymers, because during conversion the reactive imidazolide of the acid is generated and only CO_2 and imidazole are formed as by-products (Fig. 29).

The reagent and by-products are non-toxic. The imidazole is freely soluble in a broad variety of solvents including water, alcohol, ether, chloroform and pyridine and can be easily removed. In addition, the pH is not drastically changed during the conversion, resulting in negligible chain degradation.

In comparison to DCC, the application of CDI is much more efficient, avoids most of the side reactions and allows the use of DMSO, which represents a good solvent for most of the complex carboxylic acids. In case of CDI, no oxidation is observed and no decomposition of the DMSO (no odour of dimethylsulfide).

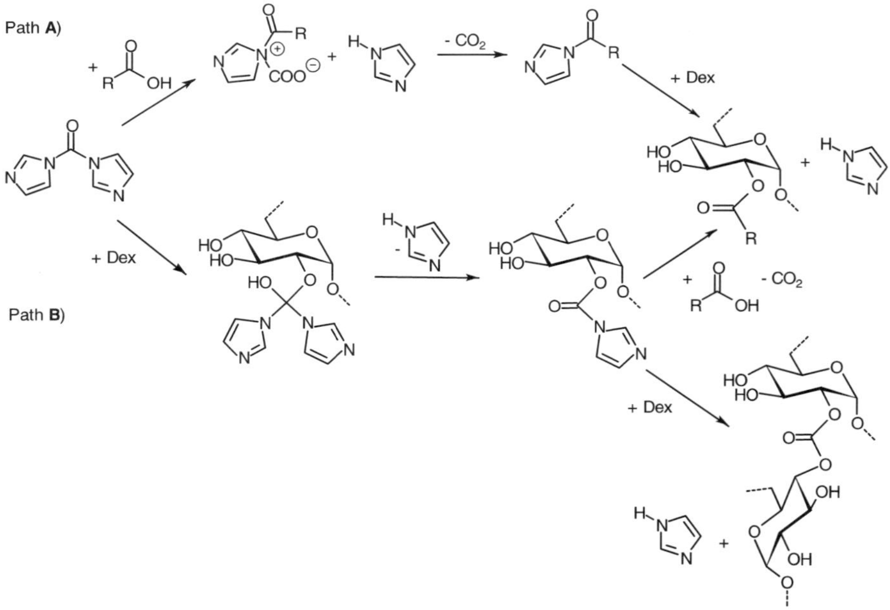

Fig. 29 Reaction paths leading exclusively to esterification (*path A*) or additionally to cross-linking reactions (*path B*) under the action of CDI as reagent for dextran esterification

The conversion is generally carried out as a one-pot reaction in two stages. First, the acid is transformed with the CDI to give the imidazolide. The conversion of the alcohol in the first step is also possible for the esterification but yields undesired cross-linking via carbonate formation in case of a polyol (Fig. 29). The imidazolide of the carboxylic acid should always be firstly synthesised. Model reactions and NMR spectroscopy (Fig. 30) with acetic acid confirm that during a treatment at room temperature CDI is consumed completely within 6 h. Thereby, the tendency of cross-linking initiated by unreacted CDI, which would lead to insoluble products, is avoided.

Basic investigations on conditions for coupling by use of butyric acid and dextran confirm that the imidazolide is formed within 2 h. The reaction at room temperature for 17 h results in butyrate content of 92% of the acid applied. Only 0.25% N is found in the product. The solvent has a pronounced influence; for dextran the solvent of choice is the mixture formamide/DMF/CH$_2$Cl$_2$ [189]. 4-Pyrrolidinopyridine is used as catalyst in this process.

Although CDI was applied as early as 1972 as reagent for the esterification of starch and dextran, it has only scarcely been used up to now. Its renaissance during the last few years may be due to the fact that it became an affordable commercially available product. Among the first attempts for the esterification of polysaccharides via CDI is the binding of amino acids onto dextran. Besides CDI, N,N'-(thiocarbonyl)diimidazole can be utilised to obtain the corresponding imidazolide [198]. The amino acids bound via this path are glycine, L-leucine, L-phenylalanine, L-histidine and L-alanyl-L-histidine. They are protected with N-trifluoroacetyl, N-benzyloxycarbonyl

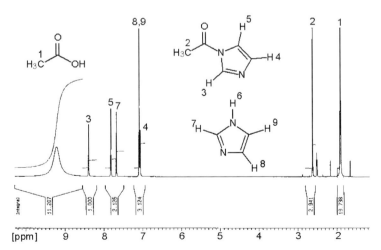

Fig. 30 ^{1}H NMR spectroscopic investigation of the in situ activation of acetic acid with CDI confirming complete consumption of the CDI to the acetyl imidazolide

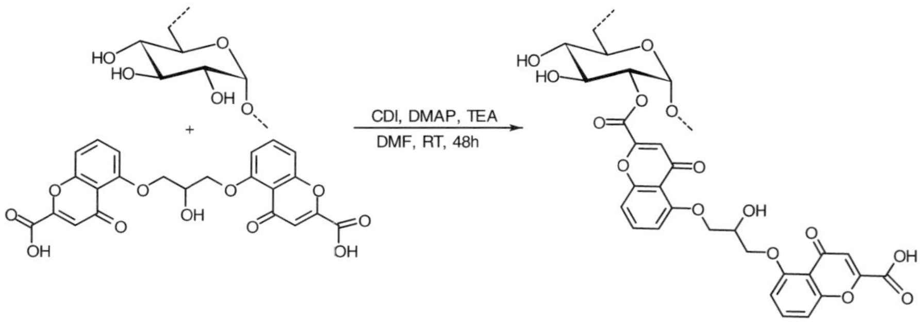

Fig. 31 Esterification of dextran with cromoglycic acid using in situ activation with CDI

and 2,4-dinitrophenyl moieties. The protecting groups can be removed after the esterification of the polysaccharide by hydrolysis or hydrogenation over Pd catalyst [191].

Cromoglycic acid can be covalently bound to dextran (Fig. 31). The acid was transferred into the imidazolide with CDI in DMF in the presence of TEA and 4-N,N-dimethylaminopyridine (DMAP) within 5 h at room temperature. The conversion with dextran dissolved in DMF is achieved within 48 h at room temperature. The procedure gives high yields (up to 50%) with derivatives containing between 0.8 and 40% (w/w) of the acid (DS can not be calculated because there is no structural information excluding the intermolecular esterification of the acid). Comparison with a route involving chlorination of the free acid in a first step, followed by reaction with dextran in formamide, results in low yields (1.5%) of an ester containing only 2.5% (w/w) cromoglycic acid [199].

Studies on the ester of the antiasthmatic drug with dextran indicate that the cromoglycate is released from the ester with a half-life of 10 h, if the acylation is carried out with the chloride of the drug yielding a loading of 2.5% (w/w). The product obtained via the imidazolide releases the cromoglycate (0.8% w/w) with a half-life of 39 min, while another batch containing 40% (w/w) cromoglycate has a release half-life of 290 min in buffer of pH 7.4 at 37 °C [199].

4.2.2.4
Micro- and Nanoparticles Manufactured by Dextran Esterification with CDI

Interestingly, CDI can also be utilised for the introduction of substituents by inter- or intramolecular coupling of OH moieties of the polysaccharide via a carbonate function. This synthesis was used to obtain dextran with 2-hydroxyethyl methacrylate moieties (dex-HEMA) and dex-HEMA with lactate spacer functions (Fig. 32). A new class of dextran derivatives (DS < 0.2) that can be polymerised containing hydrolysable groups is accessible [177].

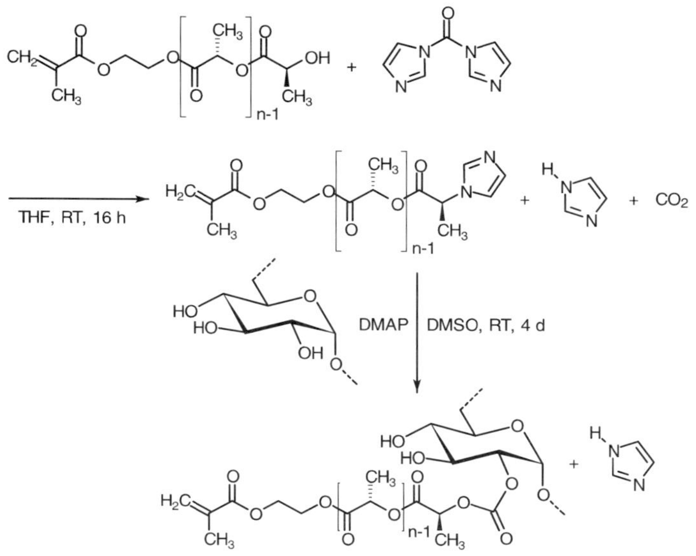

Fig. 32 Conversion of dextran with hydroxyethyl methacrylate lactate using CDI yielding a carbonate-bound ester moiety

These polymerisable dextran derivatives are widely exploited for the preparation of microspheres by water-in-water emulsion polymerisation [200] or for the production of nanogel usable as carriers for intracellular drug delivery with tunable degradation properties [201]. Comparable microgels with a monodisperse size distribution can be obtained using a microfluidic device [202]. Moreover, thermoresponsive hydrogels [203] or self-exploding lipid-coated microgels [204] have been prepared.

The CDI method can be exploited for the adjustment of the hydrophilic–hydrophobic balance, necessary for the formation of polymeric nanoparticles, in a dialysis process. This balance is achieved by esterification of dextran with biocompatible propionate and pyroglutamate moieties, leading to highly functionalised derivatives [34]. The products with DS > 2 can be realised by setting the molar ratio of the two acids to repeating unit in a one-step process or by subsequent propionylation of the remaining hydroxyl groups of the dextran ester. They form spherical nanoparticles by a simple dialysis process [160, 205]. The solvent DMA is slowly exchanged against water, leading to an arrangement of the hydrophobic moieties inside the core. The adjustment of the hydrophilic–hydrophobic balance within the dextran molecule is necessary for the formation of polymeric nanoparticles. The size of the particles can be determined by a particle size distribution analyser (PSDA), dynamic light scattering (DLS), scanning electron microscopy (SEM) and atomic force microscopy. A narrow size distribution of the particles is evidenced by PSDA (Fig. 33).

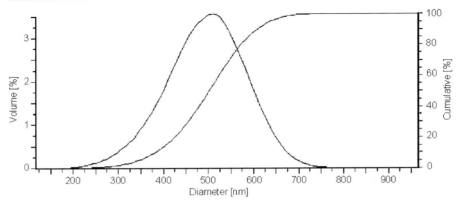

Fig. 33 Particle size distribution of perpropionylated dextran furoate pyroglutamate (DS$_{Fur}$ 0.79; DS$_{Pyr}$ 1.27) nanoparticles

By varying the character and amount of substituents and the M_w of the starting dextran, the design of biopolymer nanoparticles of appropriate size is possible (Table 9). Thus, nanoparticles in the range 90–520 nm are available.

The high DS avoids the collapse of nanoparticles due to the prevention of hydrogen bond formation. The nanospheres in the aqueous suspension do not undergo any morphological changes even after 3 weeks storage. SEM images in Fig. 34 show the uniformity in size and shape and the stability of se-

Table 9 M_w of dextran, DS of dextran furoate pyroglutamate propionate, and mean diameters of nanoparticles determined by DLS or PSDA

Dextran M_w (g mol^{-1})	Dextran ester [a] No.	DS$_{Fur}$	DS$_{Pyr}$	DS$_{Prop}$	Per-acylation [b]	Mean diameter DLS/PSDA (nm)
5400	1	0.79	1.27	–	–	520
5400	2	0.79	1.27	–	+	500
5400	3	0.12	1.13	–	+	460
5400	4	–	0.26	1.70	–	384
18 100	5	–	0.33	2.20	–	413
54 800	6	–	0.26	2.16	–	446
54 800	7	0.22	1.34	–	+	260
54 800	8	0.99	0.33	–	–	214
54 800	9	0.99	0.33	–	+	184
54 800	10	–	1.96	–	+	87

[a] *Fur* furoate, *Pyr* pyroglutamate, *Prop* propionate
[b] Propionylation of all free hydroxyl and amine moieties after CDI esterification

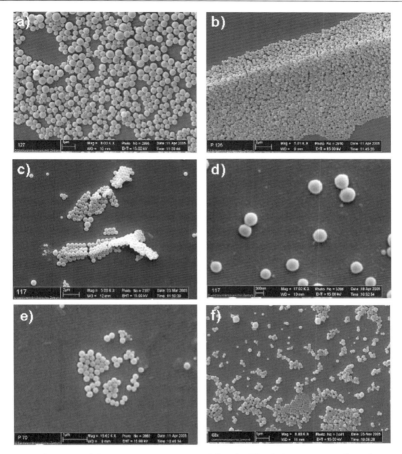

Fig. 34 SEM images of dextran ester nanoparticles (Table 9) from: **a** sample 1, **b** sample 3, **c** sample 4, **d** sample 4 after 3 weeks storage in water, **e** sample 7, and **f** sample 8 (Table 9) on a mica surface

lected dextran nanoparticles. Defined size adjustment, biocompatibility and biodegradability are advantageous for immobilisation of hydrophobic drugs in aqueous systems with subsequent targeting [206].

4.2.2.5
Transesterification

In addition to the conversion of dextran with organic acids after in situ activation, transesterification is an interesting synthesis tool for the introduction of sensitive carboxylic acid moieties, which is illustrated for the acrylation of dextran. Thus, dextran is acylated with vinyl acrylate in the presence of Proleather FG-F and lipase AY, a protease and lipase from *Bacillus sp.* and

Table 10 Ratio of theoretical degree of substitution (DS) versus obtained DS for the conversion of dextran with vinyl acrylate in the presence of the enzyme Proleather FG-F

DS Theoretical	Obtained	Efficiency (%)
0.10	0.072	71.4
0.20	0.151	75.7
0.30	0.224	74.6
0.40	0.315	78.9
0.50	0.370	74.1

Candida rugosa in anhydrous DMSO. Structure analysis by means of NMR spectroscopy indicates functionalisation of positions 2 and 3 of the AGU in equal amounts [207]. The efficiency of the reaction and the DS accessible in the presence of Proleather FG-F is shown in Table 10.

5
Ethers of Dextran

Basically, modification of the dextran backbone by introduction of ether-type moieties leads to comparatively stable dextran derivatives with altered physicochemical properties. A summary of etherification reactions used most frequently for dextran functionalisation is given in Table 11.

The solubility, the hydrophilic–lipophilic balance, the ionic strength and the resistance against hydrolytic or enzymatic degradation can be tailored by etherification. The hydrophilic–lipophilic balance is adjustable by covalent attachment of hydrophobic groups, e.g. long alkyl chains or phenoxy groups and hydrophilic moieties, like hydroxyalkyl, ethyleneglycol or ionic groups. The amphiphilic ethers have emulsifiying properties and can form micelles in water usable as surfactans or for the encapsulation of hydrophobic material such as drugs. Furthermore, the introduction of ether moieties onto the polymer background influences the degradation behaviour. The degradation of ionic dextran derivatives like diethylaminoethyl dextran [253] or carboxymethyl dextran [254] by α-1-glucosidases proceeds slower than for the parent dextran. Nevertheless, cross-linked dextran samples are still degradable by α-1-glucosidases [255, 256]. This approach is applied to the design of biocompatible and biodegradable hydrogels via etherification. Formation of ether bonds is widely exploited for the insertion of spacers usable for subsequent drug fixation. An appropriate length and chemical structure of the spacer can control the drug release [178]. Selective multistep functionalisation including etherification is investigated for the synthesis of heparin mimetica, which represents a rapidly developing field of research.

Table 11 Summary of etherification reactions used most frequently for dextran modification

Type of dextran ether	Functional group R	Refs.
Methyl	$- CH_3$	[22, 208–211]
Ethyl	$- C_2H_5$	[22]
Benzyl	$- CH_2C_6H_5$	[30]
Trityl	$- C(C_6H_5)_3$	[22, 212–214]
Trimethylsilyl	$- Si(CH_3)_3$	[215–218]
Carboxymethyl	$- CH_2COOH$	[108, 219–222]
2-Hydroxyethyl	$- C_2H_4OH$	[223, 224]
2-Mercaptoethyl	$- C_2H_4SH$	[225]
2-Cyanoethyl	$- C_2H_4CN$	[226, 227]
2-Hydroxypropyl	$- CH_2CHOHCH_3$	[224, 228]
2-Hydroxyalkyl	$- CH_2CHOH(CH_2)_nCH_3$	[229–231]
2-Hydroxypropyl-3-phenoxy	$- CH_2CHOHCH_2OC_6H_5$	[229, 232–235]
3-Chloro-2-hydroxypropyl	$- CH_2CHOHCH_2Cl$	[236–238]
2-Diethylaminoethyl	$- C_2H_4N(C_2H_5)_2$	[239–244]
3-Amino-2-hydroxypropyl	$- CH_2CHOHCH_2NH_2$	[245]
3-Dimethylalkylammonium-2-hydroxypropyl	$- CH_2CHOHCH_2N^+(CH_3)_2R$	[246–250]
Polyethyleneglycol cetyl	$- (CH_2CH_2O)_{10}C_{16}H_{33}$	[251, 252]
Polyethyleneglycol stearyl	$- (CH_2CH_2O)_{10}C_{18}H_{37}$	[251, 252]

5.1
Non-ionic Dextran Ethers

5.1.1
Alkyl Dextran

Methylation- or combined methylation–ethylation reactions were used for structure analysis of polysaccharides. The alkylation of dextran can be applied to the investigation of the branching pattern, i.e. the number and length of side chains (Sect. 2.2) [22, 23]. The methylation is carried out in liquid ammonia with sodium iodide and methyl iodide, yielding products that are soluble in chloroform and tetrachloroethane [257].

Partially methylated dextran was synthesised with dimethyl sulfate in 19% sodium hydroxide solution (w/v) [208, 209]. Characterisation of the methyl dextran obtained was carried out by complete hydrolysis and separation of the mixture of glucose, mono-, di- and tri-O-methylated glucoses by paper chromatography. The mono- and difunctionalised glucose can in turn be analysed precisely by GLC. From the portions of 2-*O*-, 3-*O*- and 4-*O*-methyl-glucopyranose in the hydrolysate with a low degree of substitution (DS 0.14, only monomethylation was achieved), the initial rate constants were obtained and the ratios of $k_2 : k_3 : k_4$ were found to be 8 : 1 : 3.5. This pronounced reactivity of the C2-hydroxyl group in etherification procedures is also observed for other polysaccharides [258–260]. As an example for the preparation of an arylether of dextran, the triphenylmethylation (tritylation) should be mentioned. The regioselectivity of the tritylation reaction is used for the characterisation of the non-reducing end groups in the polymers. Tritylation was performed by dissolving dry dextran in formamide, addition of the triphenylmethyl chloride in pyridine and keeping the temperature at 120 °C for 2 h. The degree of tritylation increases as the molecular weight of dextran decreases [212].

5.1.2
Hydroxyalkyl and Hydroxyalkyl Aryl Ethers of Dextran as Emulsifying Agents

The conversion of dextran with 1,2-epoxy-3-phenoxypropane, epoxyoctane or epoxydodecane may be exploited for the preparation of amphiphilic dextran derivatives. Polymeric surfactants prepared by hydrophobic modification of polysaccharides have been widely studied, starting with the pioneering work of Landoll [261]. Neutral water-soluble polymeric surfactants can be obtained by reaction of dextran with 1,2-epoxy-3-phenoxypropane in 1 M aqueous NaOH at ambient temperature (Fig. 35, [229, 233]). The number n of hydrophobic groups per 100 Glc*p* units varies between 7 and 22 depending on the reaction conditions. 2-Hydroxy-3-phenoxy propyl dextran ethers (DexP) behave like classical associative polymers in aqueous solution. In dilute solution, the intrinsic viscosity decreases significantly whereas

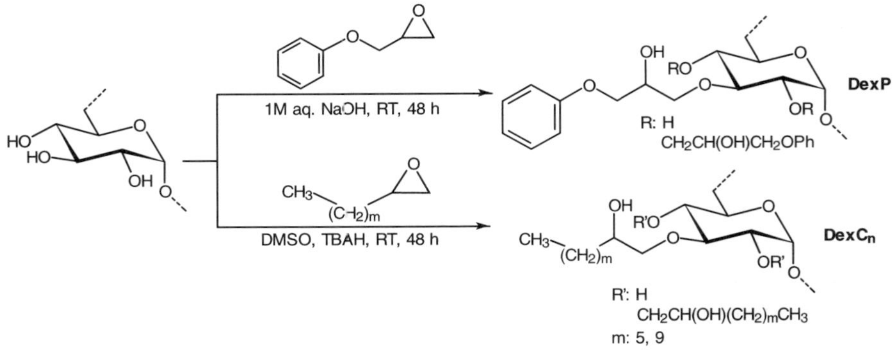

Fig. 35 Synthesis of amphiphilic 2-hydroxy-3-phenoxy propyl (DexP), 2-hydroxyoctyl and 2-hydroxydodecyl dextran (DexC$_n$)

the Huggins coefficient increases with increasing n, interpreted in terms of hydrophobic interaction between the phenoxy groups. Low solvated coils and shrunken conformations are formed. At concentrations above $40\,g\,L^{-1}$ (DexP$_{15}$) and $35\,g\,L^{-1}$ (DexP$_{22}$), the reduced viscosity deviates from linear variation because of intermolecular hydrophobic interactions initiating the formation of aggregates. The surface-active properties were evidenced by surface (air/water) and interfacial (dodecane/water) tension measurements. Both the surface and the interfacial tension decreases with DS and polymer concentration. However, above a critical polymer concentration no further decrease in surface tension is observed, which is assigned to the formation of a dense polymer layer at the liquid–air or liquid–liquid interface. Oil-in-water emulsions can be prepared by sonification in the presence of the amphiphilic copolymer (dissolved in the aqueous phase). The thickness of the adsorbed polymer layer in dodecane/water emulsions was estimated by zeta potential measurements coupled with size measurements. Polymeric surfactants usually show no or only low hemolytic effects.

Amphiphilic 2-hydroxyoctyl and 2-hydroxydodecyl dextran ethers were prepared homogeneously in DMSO at $40\,^{\circ}C$ in the presence of an aqueous solution of tetrabutylammonium hydroxide (TBAH) with aliphatic epoxide (epoxyoctane or epoxydodecane, Fig. 35). The DS, determined by ^{1}H NMR in DMSO-d_6, can be varied by changing the reaction time and the concentration of the epoxide [262]. Viscometric investigations carried out in dilute and semidilute solutions give information about their solution behaviour [230]. The relation between monodisperse dissolved macromolecules and aggregates and their ability to interact with each other depends on the DS, the length of the hydrocarbon chains and the concentration of the polymer. Static and dynamic light scattering measurements allow precise knowledge of the state formed in dilute solutions with varying DS and increasing hydrocarbon chain length [231].

Amphiphilic 2-hydroxy-3-phenoxy propyl dextran ethers (DexP) were also applied as stabilisers of nanoparticles formed by emulsion polymerisation of styrene, leading to a stable hydrophilic surface and showing reduced nonspecific protein adsorption [229, 232, 233]. The styrene-in-water emulsions were prepared by sonification of styrene in the aqueous phase in the presence of DexP, with potassium peroxodisulfate as initiator at $95\,^{\circ}C$. The size of the polysaccharide-coated polystyrene nanoparticles prepared was directly correlated to the styrene concentration and the DS of the DexP. For a concentration of styrene of 10% (v/v), the droplet size is around 160 nm at maximal surface coverage and a coagulate amount below 5% (w/w). With increasing concentration of styrene, the particle size and the amount of coagulate increase, indicating that coalescence processes take place during polymerisation. Furthermore, for dextran ethers of low DS, the coalescence becomes predominant and no latex is observed. Polystyrene particles coated with DexP present markedly higher protein repulsion in contrast to unmodified particles or to dextran-coated particles. The dextran layer formed on the particles is thicker than the layer with DexP, but due to the higher polymer chain flexibility of dextran, the layer is loosely packed and bovine serum albumin (BSA) can penetrate through. In the presence of phenoxy moieties, the stiffness of the adsorbed layer increases due to hydrophobic interactions between the polymer chains and, therefore, the density of the layer increases. The probability of BSA making direct contact to the polystyrene surface is reduced [263].

Comparable experiments were performed with DexP-coated macroporous polystyrene-divinylbenzene (PS-DVB) particles [264] and with DexP, labelled with 4-amino-TEMPO, using EPR spectroscopy to study the conformation of the polymer chains [265]. Low substituted DexP gave thicker layers with lower density than highly substituted derivatives due to the presence of more loops and tails. With increasing DS of DexP, the stiffness of the adsorbed layers and, therefore, the density increases and the non-specific interaction of BSA with the DexP-coated PS-DVB surfaces seems to be restricted to the top of the adsorbed layer.

Moreover, DexP and $DexC_n$ were exploited as stabilisers in an oil-in-water emulsion/evaporation technique for the preparation of biocompatible poly(lactide) (PLA) nanospheres [262]. The influence of the dextran ethers on the particle size, surface density and stability were investigated. DSC studies suggested that the mechanism of interfacial adhesion depends on the nature of the hydrophobic moieties. The dextran ethers produce stable nanoparticles with hydrophilic surfaces and reduced BSA adsorption [266]. Such particles of PLA have been widely applied as carriers for drug delivery [267]. The encapsulation of lidocaine in PLA nanoparticles, adopting DexP and $DexC_n$ as emulsion stabiliser in the preparation of the nanoparticles, was investigated. Using DexP, the encapsulation of lidocaine in PLA nanoparticles is not possible probably due to specific interactions between the drug and DexP. However, lidocaine-loaded nanospheres from about 7 to 18% (w/w) could be obtained in the presence of the 2-hydroxyalkyl dextran ethers. The release of lidocaine of uncoated PLA and of $DexC_n$-coated PLA nanoparticles did not significantly change.

5.1.3
Poly(Ethylene Glycol)-Alkyl Dextran Ether ($DexPEG_{10}C_n$) for Drug Delivery

Amphiphilic poly(ethylene glycol)-alkyl dextran ethers are emerging as vehicles in the oral delivery of poorly water soluble drugs [251, 268, 269]. They form polymer micelles of low critical association concentrations (CAC) and small micelle sizes in aqueous solution. Particulate delivery systems lead to an enhancement of the absorption efficiency and bioavailability of highly lipophilic drugs orally applied, and provide the drug with some level of protection against degradation within the GI tract, prolonged drug transition time, and improved drug absorption [270].

Low molecular weight surfactant micelles are widely used as drug carrier systems due to their good pharmacological characteristics [271, 272]. They are formed above a critical micelle concentration (CMC) and rapidly break apart upon dilution. In contrast to the low molecular weight surfactant micelles, the association of amphiphilic polymers like $DexPEO_{10}C_n$ in water takes place at concentrations (CAC), which are lower by several orders of magnitude than typical surfactant CMC values. The polymeric micelles consist of a hydrophobic core (cetyl or stearyl groups) and the hydrophilic shell (dextran backbone) exposed to the aqueous environment [273, 274]. The hydrophobic cetyl and stearyl groups are attached via short PEG linker to dextran (Fig. 36).

The poly(ethylene glycol)-cetyl and stearyl dextrans are synthesised as follows: the terminal hydroxyl groups of cetyl or stearyl poly(ethylene glycol) are tosylated using p-toluenesulfonyl chloride and either pyridine or a mixture of $Et_3NH^+Cl^-$ and $Me_3NH^+Cl^-$. The tosylated poly(ethylene glycols) are converted with dextran to give the corresponding $DexPEG_{10}C_n$ [251, 252]. The degree of $PEG_{10}C_n$ substitution can be calculated using ^1H NMR spectra. The size of the polymeric micelles ranges from 10 to 100 nm.

CAC values of the copolymers are estimated by fluorescence spectroscopy using pyrene as probe. The excitation spectra of the hydrophobic fluorescence probe, preferentially arranged into the hydrophobic core of the micelle, undergoes a small shift to longer wavelengths (from $\lambda = 333$ nm in a hydrophilic environment to $\lambda = 336$ nm in a hydrophobic environment) [275–277]. An increase in the length of the hydrophobic residue at a given length of the hydrophilic polymer chain causes a decrease in the CAC value and an increase in micelle stability [252, 278]. Cyclosporin A (CsA), a highly effective immunosuppressive agent, was incorporated into $DexPEG_{10}C_n$ micelles by a dialysis method. An aqueous $DexPEG_{10}C_n$ solution was treated with a solution of CsA in ethanol, followed by extensive dialysis against water. The solubility of the lipophilic drug CsA in aqueous solutions of $DexPEG_{10}C_n$ through encapsulation in the hydrophobic core of the micelles can be increased with increasing DS and decreasing molecular weight of the dextran [252]. The cytotoxicity of $DexPEG_{10}C_n$ micelles towards a Caco-2 cell line, deriving from human colon denocarcinoma, is significantly lower than that of unlinked $PEG_{10}C_n$. $DexPEG_{10}C_n$ micelles exhibit high stability in gastric and intestinal

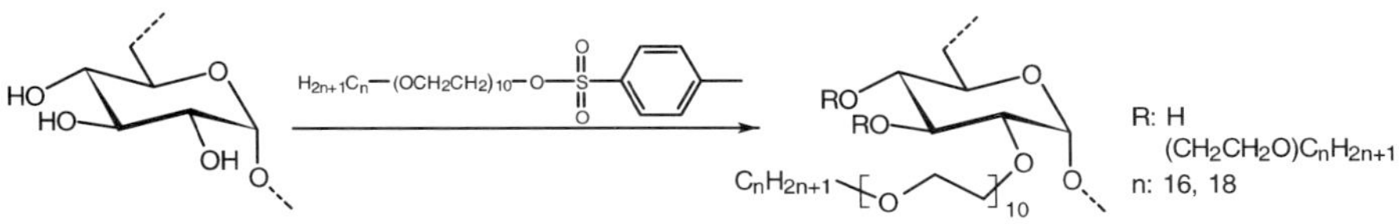

Fig. 36 Synthesis of poly(ethylene glycol)-cetyl and -stearyl dextran ether

fluids and their size is optimal for effective drug delivery. The permeability of CsA encapsulated in $DexPEG_{10}C_n$ micelles across Caco-2 cells is significantly increased compared to free CsA. The application of CsA encapsulated in vitamin B_{12}-modified $DexPEO_{10}C_n$ micelles also enhances the permeability through Caco-2 cell monolayers [279]. The characteristic of the biopolymeric micelle systems indicates that this approach can provide practical opportunities in the oral delivery of hydrophobic drugs.

5.2
Ionic Dextran Ethers

5.2.1
Sulfopropylation

One approach to introduce an anionic moiety onto dextran is sulfopropylation. In combination with a hydrophobic modification, sulfopropylation was utilised for manufacture of anionic amphiphilic dextran ethers. The derivatives were synthesised in two steps. Dextran was reacted with 1,2-epoxy-3-phenoxy propane in aqueous NaOH solution (Fig. 35) or in DMSO using TBAH instead of NaOH. The subsequent step is the conversion with 1,3-propane sultone in DMSO (Fig. 37).

The emulsifying properties of these polymeric surfactants demonstrate that the chemical structure influences the kinetic behaviour of interfacial tension reduction. An increase of sulfopropyl moieties reduces the interfacial tension slower while an increase in 2-hydroxy-3-phenoxy propyl moieties reduces the interfacial tension faster. The ionic strength of the emulsion appears to increase the rate of tension reduction. The average droplet size of oil-in-water emulsions in presence of previously dissolved 2-hydroxy-3-phenoxy propyl sulfopropyl dextran is around 180 nm immediately after preparation and increases with time. The presence of ionic moieties appeared to facilitate emulsification at low polymer concentrations due to electrostatic repulsions between the oil droplets [229].

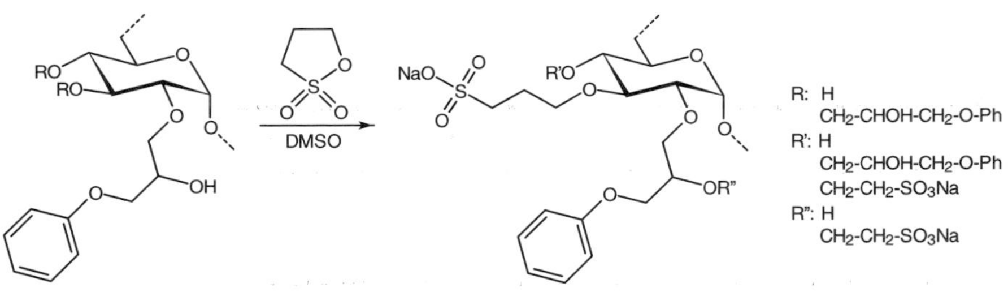

Fig. 37 Sulfopropylation of 2-hydroxy-3-phenoxy propyl applied to the manufacture of anionic amphiphilic 2-hydroxy-3-phenoxy propyl sulfopropyl dextran

5.2.2
Carboxymethyl Dextran (CMD)

Carboxymethylation of dextran in water/organic solvent mixtures using monochloroacetic acid (MCA) under strong alkaline conditions leads to CMD with good reproducibility (Fig. 38).

A DS of up to 1.0 is obtained in one step by applying optimised reaction conditions; 3.8 M aqueous NaOH, reaction temperature of 60 °C for 90 min in tert-butanol/water or isopropanol/water 85 : 15 (v/v) mixtures (Table 12) [219]. The DS of CMD can be increased by repeated carboxymethylations. CMD with DS 1.5 was realised by two-step carboxymethylation. Under optimal conditions, the applied NaOH solution was 3.8 M. The DS value is decreased with lower NaOH concentrations because of incomplete activation of the hydroxyl groups, and also with higher NaOH concentrations due to increasing side reactions of MCA with NaOH forming glycolic acid [280].

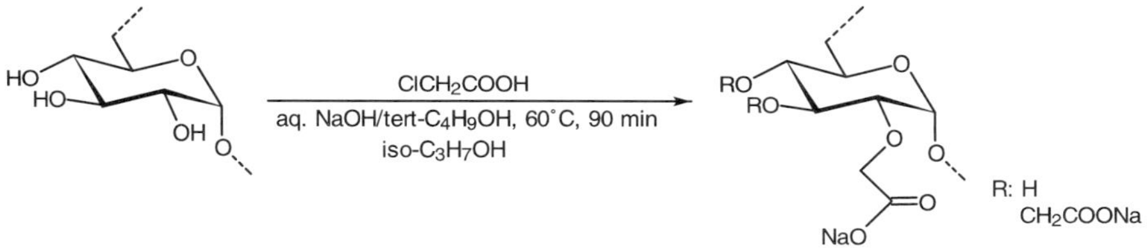

Fig. 38 Synthesis of carboxymethyl dextran (CMD)

Table 12 Influence of solvent mixture and temperature on carboxymethylation of dextran

Solvent mixture (v/v)		Temp.	DS_{CM}
tert-Butanol	Water	(°C)	
100	0	60	0.35
85	15	50	0.82
85	15	60	0.96
85	15	70	0.88
80	20	60	0.74
70	30	60	0.50
60	40	60	0.39
50	50	60	0.31
0	100	60	0.63
85 (isopropanol)	15	60	0.99
85 (M_4U [a])	15	60	0.77

[dextran] = 2 g, [NaOH] = 3.8 M, [MCA] = 3 g, 90 min, total volume 50 mL
[a] 1,1,3,3-Tetramethylurea

Table 13 Molecular weight and polydispersity of carboxymethyl dextran with different degrees of substitution (DS)

Sample	DS_{CM}	M_n	M_w	M_w/M_n
Dextran T40	–	25 500	38 000	1.5
CMD	0.76	49 000	70 400	1.4
CMD	0.83	56 700	77 500	1.4
CMD	1.01	54 700	79 400	1.5
CMD	1.47 [a]	53 500	80 600	1.5

[a] After second carboxymethylation

The isopropanol/water or tert-butanol/water mixtures (85 : 15, v/v) are favourable regarding the miscibility and solubility of the reactants. Carboxymethylation proceeds very fast. After 15 min DS 0.8 and 90 min DS 1.0 is obtained. Further prolongation of the reaction time does not yield higher values. The conversion of dextran T10 ($10 000 \, \mathrm{g \, mol^{-1}}$) and T40 ($40 000 \, \mathrm{g \, mol^{-1}}$) leads to comparable DS values by using optimal reaction conditions. By carboxymethylation, no degradation of dextran was observed although the molecular weight of the CMD samples compared to the starting dextran (T40) is rather high, obviously due to the problems of SEC measurements of polyelectrolytes (Table 13).

The NMR data of ^{1}H-COSY and HMQC spectra reveal that at the beginning of the reaction (below a DS of about 0.3) only three possible monosubstituted glucoses prevail as building blocks of the polymer chain. The values for the rate coefficients are $k_2 = 1.2$, $k_3 = 0.5$ and $k_4 = 0.8$. Beyond this DS (up to a value of about 0.8), disubstituted units are also observed but no trisubstitution occurs [220, 221].

5.2.2.1
Bioactivity of CMD-Based Derivatives (Heparan Sulfate Mimetics)

CMD exhibits no anticoagulant and anticomplementary activities (Table 14). Nevertheless, carboxymethylation of dextran is a crucial step in the preparation of heparin-like polymers. Heparin is a natural polyanionic polysaccharide with repeating uronic acid-N-acetyl-D-glucosamine units, which is able to interfere with blood coagulation. It accelerates the inactivation of coagulation enzymes by natural inhibitors like antithrombin (AT) and heparin cofactor II (HCII). Commercial heparin contains essentially sulfate, sulfamate and carboxylic groups with different size, charge and chemical composition. Carboxymethyl dextran benzylamidesulfonate sulfate (CMDBSSu) was designed to mimic heparin-like properties offering the benefit of no potential risk of virus contamination and with a relatively simple synthesis [281]. It

Table 14 Composition, anticoagulant (antithrombic), and anticomplementary activities of dextran derivatives with different DS of carboxymethyl (CM), benzylamide (B) and sulfonate/sulfate (S) functions

Samples	DS CM ±0.01	B ±0.01	S ±0.01	NIH units mg^{-1} [a]	µg/ 10^7 EAC4b,3b [b]
CMD	0.69	–	–	–	120
CMDB	0.51	0.14	–	–	120
CMDSu	0.45	–	0.04	15	13
CMDSu	0.48	–	0.05	25	9
CMDSu	0.64	–	0.05	18	11
CMDSu	0.72	–	0.10	40	3
CMDSu	0.75	–	0.14	65	1
CMDBSSu	0.40	0.12	0.03	3	60
CMDBSSu	0.60	0.14	0.05	16	10
CMDBSSu	0.43	0.04	0.11	23	7.5
CMDBSSu	0.37	0.15	0.45	1	60
CMDBSSu	0.37	0.11	0.90	1	45
CMDBSSu	0.50	0.03	0.10	22	4
CMDBSSu	0.68	0.04	0.12	31	4

[a] Specific anticoagulant activity is expressed as the number of NIH units of human thrombin (1094 NIH units mg^{-1}) inactivated by 1 mg of dextran derivative
[b] The anticomplementary activity was expressed as the input of dextran derivative that inhibits 50% convertase formation on 10^7 EAC4b,3b

can delay the coagulation of plasma by catalysing the inactivation of thrombin [107, 282–284]. Thrombin is a multifunctional serine proteinase that activates platelets, converts fibrinogen into clottable fibrin, and amplifies blood coagulation by activating factors V, VIII, XI. The most potent inhibitor of thrombin in plasma is AT, forming an inactive covalent complex with thrombin and other serine proteinases. Thrombin is the only clotting proteinase that is inhibited by HCII [283]. CMDBSSu also displays anticomplementary activity both in vitro [285] and in vivo [286] and can modulate the proliferation of vascular cells [287–289].

The synthesis of CMDBSSu involves the random carboxymethylation of hydroxyl groups on the glucose units, the benzylamidation of some of the carboxylic groups with benzylamine in the presence of N-ethoxycarbonyl-2-ethoxy-1,2-dihydroquinoline (EEDQ) to carboxymethyl dextran benzylamide (CMDB), the sulfonation of phenyl rings and sulfation of remaining hydroxyl groups at the polymer backbone (Fig. 39, [108, 222]).

The synthesis steps described may also lead to carboxymethyl dextran sulfate (CMDSu) and carboxymethyl dextran benzylamide sulfate (CMDBSu, see Fig. 39). The anticoagulant activities depend on the proportion and distribution of the substituents. CMDBSSu represents a family of soluble

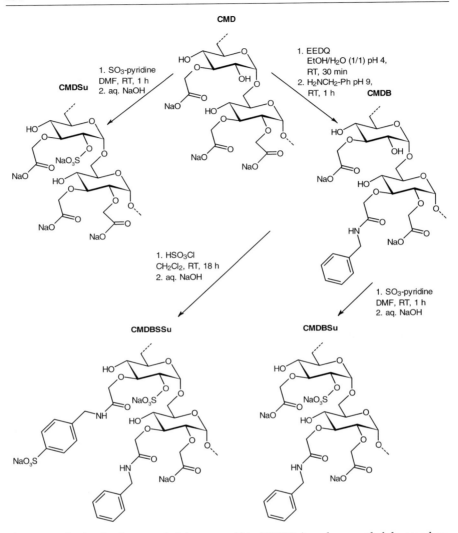

Fig. 39 Synthesis of carboxymethyl dextran sulfate (CMDSu), carboxymethyl dextran benzylamide sulfate (CMDBSu) and carboxymethyl dextran benzylamide sulfonate sulfate (CMDBSSu), respectively

macromolecular compounds, designed for interaction with specific protein domains. The results in Table 14 show that CMD and CMDB have no anticoagulant ability, whereas the CMDBSSu and CMDSu samples reveal a significant activity. A pronounced effect is observed when the carboxymethyl group content is beyond a DS of 0.40 and activity can be increased both by higher degree of sulfation and benzylsulfonation, expressed as S content (Fig. 40, [107, 281]).

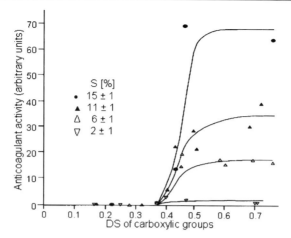

Fig. 40 Anticoagulant activity of CMDBSSu in relation to the carboxymethyl group and sulfonate/sulfate content (adapted from [107, 281])

The anticoagulant activity of dextran derivatives were assessed by measuring the thrombin clotting time (ThNIH units) of freshly prepared platelets from plasma in the presence of the CMD, CMDB, CMDBSSu and CMDSu polymers and of human thrombin (1094 NIH units mL^{-1}). The anticomplementary activity was expressed as the amount of polymer that inhibits 50% formation of the alternative and classical pathway C3 convertase [220, 290, 291].

A prerequisite for CMDBSSu activity is the availability of active sites (OH −, $CH_2COOH(Na)$, SO_3Na and $CH_2CONHCH_2C_6H_5SO_3Na$), capable of binding thrombin and catalysing interactions with its inhibitors. Highly substituted CMDBSu with a DS of carboxymethyl moieties up to 1.1, of benzylamide up to 0.35, and of sulfate up to 1.5 was synthesised by benzylamidation of carboxymethyl groups with benzylamine in the presence of water-soluble 1-cyclohexyl-3-(2-morpholino-ethyl)carbodiimide and subsequent partial sulfation of remaining hydroxyl groups with SO_3-pyridine in DMF [292]. The anticoagulant activity of CMDBSu increases with the sulfate content and reaches a heparin activity of 20% at a DS of sulfate of 1.3. The presence of O-sulfate groups is essential for the biological activity (compare Sect. 4.1.2). Furthermore, the presence of benzylamide groups improves the activity.

For CMDBSSu, the anticoagulant and anticomplementary activities increase with increasing molecular weight and reach a plateau at about $40\,000\,g\,mol^{-1}$ [285]. Heparin [293, 294] and synthetic heparinoids derived from cis-1,4-polyisoprene containing carboxylic and sulfonate groups [295] show a comparable plateau. For CMDBSSu with a high degree of substitution (DS of CM 0.72–1.09, DS of benzylamide 0.14–0.66, DS of sulfonate ≤ 0.09, DS of sulfate 0.03–0.76), the anticoagulant activity increases in the range of molecular weights $10\,000$–$80\,000\,g\,mol^{-1}$ [283].

In addition to anticoagulant activities, CMDBSSu has the capacity to inhibit smooth muscle cells (SMC) and stimulate endothelial cells (EC) proliferation. The proliferative nature is not related to the anticoagulant properties [109, 288]. Some CMDB derivates exhibit an inhibitory effect on human breast cancer lines [296–298]. The inhibitory effect on cell growth appears to be independent of the sulfate and sulfonate groups but depends rather on the benzylamide content.

A CMDBSSu (DS of CM 1.1, DS of benzylamide 0.025, DS of sulfonate/sulfate 0.37) can be applied as tissue repair agent, efficient at protecting the heparin-binding growth factors FGF2 and TGFβ1 against pH, thermal and trypsin degradation [299, 300]. A test with an in vivo wound-healing model of colonic anastomosis indicated that the polymer effected after 48 h a twofold increase of the anastomosis resistance to leakage compared to untreated controls.

CMD and mixed derivatives based on CMD may be used for the fixation of antibiotics and enzymes on a polysaccharide matrix. A procedure was developed for amidation of CMD and its ethyl ester (CMDEE) with various aromatic amines in dioxane (100 °C), ethanol (78 °C), n-propanol (97 °C) and 2-propanol (82 °C) in the presence of a small amount of water [301]. CMDEE reacts with aliphatic and aromatic amines but in the case of amino acids the degree of amidation depends on the position of the amino moieties in the acids [302, 303]. The DS decreases from δ-, γ-, β, to α-amino acids. For the reaction with α-amino acids, carboxymethyl and carboxyethyl dextran azides are used [304].

5.2.2.2
CMD as Coating Material

The cytocompatible CMDs are able to stabilise iron oxide-based nanoparticles by coating. The commonly used synthesis route to obtain dextran-coated iron oxide-based nanoparticles is the alkaline precipitation of magnetite-like compounds (Fe^{2+} and Fe^{3+} salt solutions) in water in the presence of the colloid stabilising CDM [305, 306]. The separation from unbound polysaccharide is carried out by gel filtration chromatography. The coated superparamagnetic iron oxide particles are more stable in biological media because the CMD coating separates the iron oxide core from solutes and minimises protein adsorption on the nanoparticle surface. According to the synthesis conditions, the structure of the CMD coating and the persistence time in the blood can be influenced.

These nanoparticles can interact cell-type-specifically with viable human cells [307]. The functional groups are accessible to subsequently covalent fixation of bioactive molecules such as oligonucleotide (Fig. 41). Such oligonucleotide-functionalised magnetic nanoparticles can be taken up intracellularely by endocytosis [308].

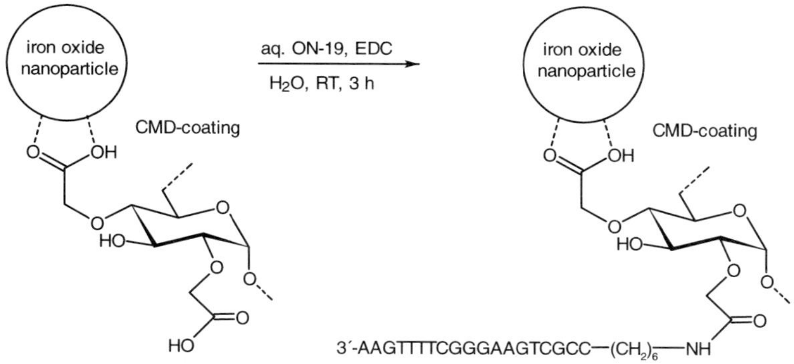

Fig. 41 Coupling of the oligonucleotide ON-19 with carboxymethyl dextran-coated iron oxide nanoparticles

Nanoparticles may be applied in tumour diagnosis and therapy [309]. The nanoparticles have several advantages such as high drug encapsulation, efficient drug protection against chemical or enzymatic degradation, unique ability to create a controlled release of drugs, cell internalisation and the ability to reverse the multidrug resistance of tumour cells [310]. The dextran-coated nanoparticles show prolonged blood flow time and allow targeting to specific tissue, such as lymph nodes or brain tumour [88].

5.2.2.3
CMD Complexes

CMD and dicarboxymethyl dextran (DCMD) as biocompatible and water-soluble polymeric carriers can react with *cis*-dihydroxo(cyclohexane-*trans*-L-1,2-diamine)platinum II (Dach-Pt) via a complex-type bond (Fig. 42). Low molecular weight platinum complexes are poorly water soluble. Their cytotoxic activity decreases in the bloodstream because of ligand exchange reactions with proteins, amino acids and other amino compounds. The immobilisation of Dach-Pt to DCMD by a chelate-type coordination bond is stable to maintain the cytotoxic activity in contrast to a single coordination of Dach-Pt to CMD, where the cytotoxic activity decreases gradually with the preincubation time. DCMD/Dach-Pt should also show a longer half-life in the body and a larger accumulation at the inflammatory tumour site than low molecular weight platinum complexes [311].

5.2.2.4
CMD for Protein Immobilisation

Multiple funtional improvements in β-lactoglobulin (β-LG) could be achieved by covalent binding to CMD ($M_n \sim 10\,000\,\mathrm{g\,mol^{-1}}$) using the water-soluble

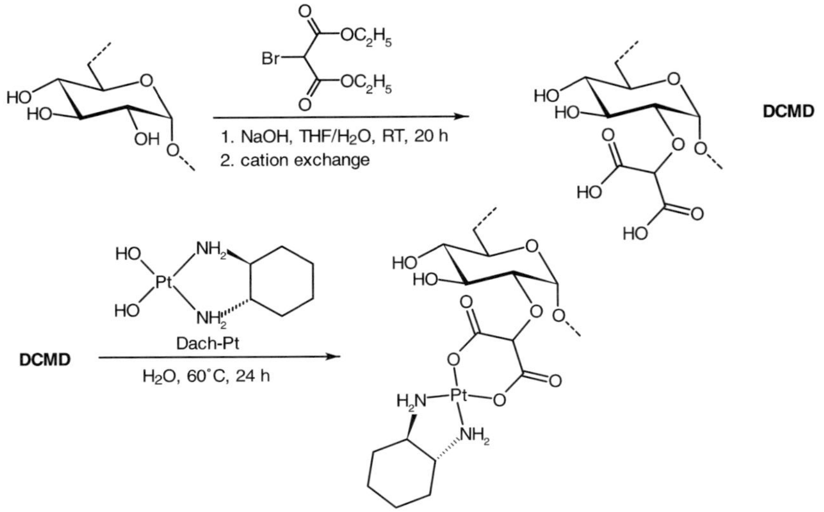

Fig. 42 Synthesis of dicarboxymethyl dextran (DCMD)/*cis*-dihydroxo(cyclohexane-*trans*-L-1,2-diamine)platinum II (Dach-Pt) complex

1-ethyl-3-[3-(dimethylamino)propyl]carbodiimide (EDC) as coupling reagent [312, 313]. β-LG is a predominant whey protein and known as a potent allergen of milk allergy (82% of milk allergy patients are sensitive to β-LG). The modification of the protein by covalent binding to CMD leads to reduced allergenicity, improved heat stability and emulsifying properties of the protein by maintaining the retinol binding activity [314]. The emulsifying properties of β-LG can be markedly improved with the molecular weight of the CMD used ($M_n \sim 40\,000$ g mol^{-1}, $\sim 70\,000$ g mol^{-1}, $\sim 162\,000$ g mol^{-1}), especially in the pH region lower than 7, in the presence of salt, and after heating. It is assumed that the physical properties of CMD (such as hydrophilicity, negative charge and viscosity) influence the emulsifying activities rather than the effect of the conformational changes [315]. The immunogenicity of β-LG also depends on the molecular weight of the CMD. The higher the molecular weight of CMD ($M_n \sim 40\,000$ g mol^{-1}, molar ratio of β-LG to CMD 8 : 1 and $M_n \sim 162\,000$ g mol^{-1}, molar ratio of β-LG to CMD 7 : 1) the lower is the immunogenicity of β-LG [316, 317]. By means of intrinsic fluorescence spectra, circular dichroism spectra and enzyme-linked immunosorbent assay (ELISA) with monoclonal antibodies, it can be shown that the surface of β-LG is covered by CMD with retention of the native conformation. Local conformational changes of β-LG by covalent binding of CMD were evaluated by competitive ELISA. Whereas the conformation around ^{125}Thr–^{135}Lys (α-helix) and the regions near the epitope 61B4 maintain their native form, the conformations around ^{15}Val – ^{29}Ile (β-sheet) and ^{8}Lys – ^{19}Trp (random coil, β-sheet and short helix) are changed. The suppression of the generation of T cell epitopes by covalent binding of CMD is important for the explanation of the reduced immunogenicity of β-LG. The substances could be applied for hypoallergenic formulations and hypoallergenic emulsifying agents also suitable for novel foods with low allergenicity.

5.2.2.5
Drug Support

One of the serious problems in chemotherapy is the limited selectivity of most of the common anticancer drugs. Because of the low cell selectivity, they do not only interfere with tumour cells but also with healthy cells, and lead therefore to serious side effects. New macromolecular prodrugs were synthesised that are composed of CMD, Gly – Gly – Gly linker, and camptothens (CPTs), one of the most promising classes of anticancer drugs [318].

T-2513 (7-ethyl-10-(3-aminopropyloxy)-camptothecin, Fig. 43), which exhibits a good hydrophobic/hydrophilic balance, can be bound to the polymer via an amide bond to the peptidyl linker. The macromolecular prodrug is stable and pharmacologically inactive during circulation in the bloodstream but becomes active after reaching the target site. The entry of macromolecules into the cells is restricted to endocytosis and pinocytosis that can be highly cell specific. A comparable macromolecular chemotherapeutic with tumour targeting capability is methotrexate, covalently bound via peptide linker (Pro – Val – Gly – Leu – Ile – Gly) onto CMD [319].

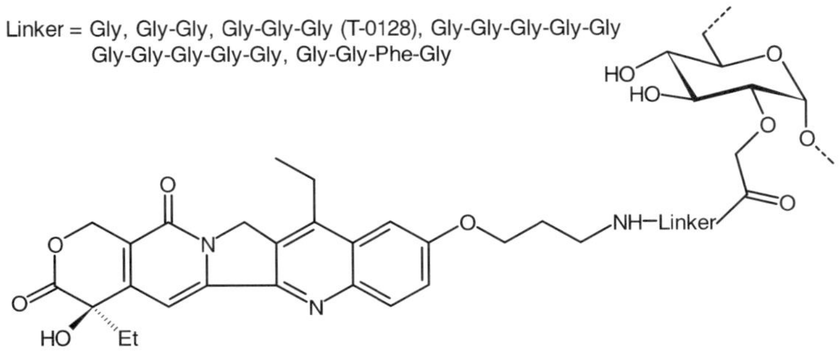

Linker = Gly, Gly-Gly, Gly-Gly-Gly (T-0128), Gly-Gly-Gly-Gly-Gly
Gly-Gly-Gly-Gly-Gly, Gly-Gly-Phe-Gly

Fig. 43 Chemical structure of T-2513 covalently bonded to CMD via linker [318]

5.2.2.6
Stimuli-Responsive Hydrogels Based on CMD

Stimuli-responsive homo- and copolymeric hydrogels with pH, temperature, ionic strength, solvation, electrical field, or magnetic field-sensitive pendent groups have been produced for drug, peptide and protein delivery devices, for biosensors and for tissue engineering. Their swelling behaviour, net-

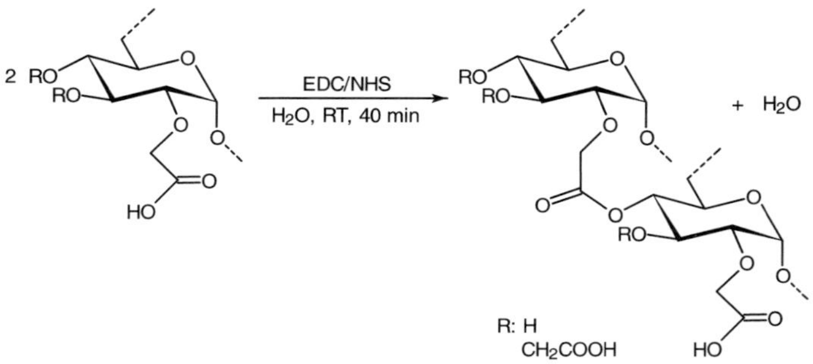

Fig. 44 pH- and ionic strength-sensitive hydrogel synthesised by inter- and intramolecular cross-linking of carboxymethyl dextran

work structure, permeability and mechanical strength changes in response to external stimuli [320–324]. The pH and ionic-strength sensitive hydrogel membranes for drug delivery and tissue engineering applications can be synthesised by intermolecular cross-linking of CMD in the presence of EDC and N-hydroxysuccinimide (NHS) (Fig. 44) [325]. The synthesis method is similar to the procedure described in [326]. EDC/NHS is not incorporated into the hydrogel in contrast to conventional cross-linking agents because the water-soluble urea derivatives formed can be simply washed out [327].

Both the DS of carboxymethyl functions and the concentration of the cross-linking reagent control the degree of cross-linking and the charge density of the hydrogel. Hence, the sensitivity of this hydrogel is tailored for specific applications. CMD of low DS shows no pH sensitivity. With increasing DS, the sensitivity arises and, therefore, the porosity of the hydrogel can be reversibly changed in response to changes in environmental pH. This response can control the transport of proteins through the membrane. At high pH, the COOH groups are dissociated inducing electrostatic repulsion, expanding the conformation and increasing the porosity of the hydrogel. Similar behaviour has been observed with pH-sensitive methacrylated dextran hydrogels, polypeptide hydrogels and chitosan-polyvinyl pyrrolidone hydrogels [324, 328, 329]. The structure of the hydrogel membrane is also sensitive to changes in ionic strength and the protein diffusion rates. At low ionic strengths, the protein diffusion rate increases with the ionic strengths; however, above a certain value the expanded matrix changes to a more compact one and the diffusion rate drops with further increase of ionic strengths.

5.2.3
2-(Diethylamino)ethyl (DEAE) Dextran

2-(Diethylamino)ethyl dextran (DEAE dextran) is synthesised by reaction of dextran with (2-chloroethyl)diethylammonium chloride in the presence of NaBH₄ in an alkaline solution at 85–90 °C (Fig. 45).

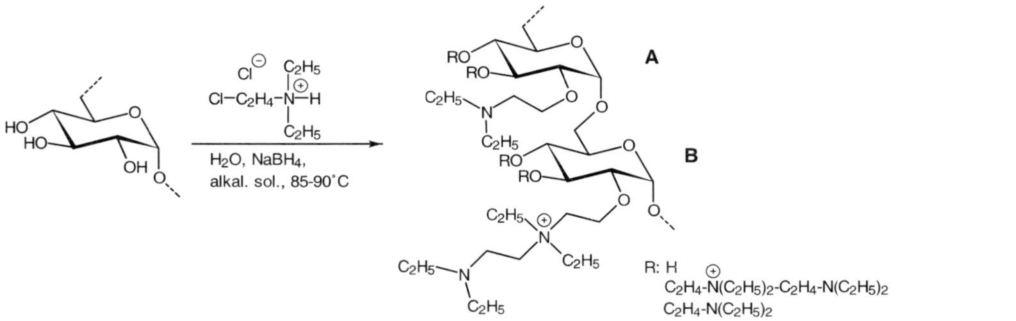

Fig. 45 Synthesis and structure of DEAE dextran containing 2-(diethylamino)ethyl (A) and 2-[[2-(diethylamino)ethyl]diethylammonium]ethyl (B) groups

DEAE dextran contains three basic groups with different pK_a values (Fig. 45). One (2-(diethylamino)ethyl, A) consists of a tertiary amino group with a pK_a of 9.5. The other two moieties are (2-[[2-(diethylamino)ethyl]-diethylammonium]ethyl, B) a tertiary amino group with a pK_a of 5.7, and the strong basic quaternary ammonium group with a pK_a of 14 [242, 243]. The pH value and ionic strength influences the dissociation behaviour and the conformational change of DEAE dextran.

DEAE dextran is a biocompatible derivative with pharmacological and therapeutic properties [330]. In particular, DEAE dextran is suitable for binding bile acids and reducing the absorption of dietary cholesterol and fatty acids in the gastrointestinal lumen. A hypocholesterolaemic effect [331, 332] and reduction of triglycerides in presence of DEAE dextran have been proved [333]. Furthermore, DEAE dextran has the ability to enhance viral infectivity in cell cultures and is suitable as an agent for transfection because it enhances the uptake of proteins and nucleic acids by cells. Cultured human lymphocytes can be utilised for the synthesis of DNA in the presence of DEAE dextran deoxynucleotides [334]. The DNA synthesis in cells, treated with DEAE dextran, resembles DNA synthesis in vivo [244]. Plasmid DNA mixed with DEAE dextran is taken up by colon epithelial cells. This transfection technique is useful in gene therapy, e.g. for colon disease [335–338]. Vacuum-dried enzymes and enzymes in solution can be stabilised by DEAE dextran. Thus, the activity of vacuum-dried glycerol kinase is preserved in the presence of DEAE dextran and lactitol [339].

Water-insoluble PEC were prepared by mixing oppositely charged polyelectrolytes, e.g. DEAE dextran with CMD [340, 341], sodium dextran sulfate, poly(styrene sulfonate) (NaSS) [342], poly(sodium L-glutamate) (PSLG), poly(vinyl alcohol)sulfate [343], or potassium metaphosphate (MPK) [240]. They are useful as membranes or in biomedical applications [343, 344].

5.2.3.1
Hydrophobically Modified DEAE Dextran

In contrast to the hydrophilic DEAE dextran, new hydrophobically modified polycationic dextran (HMPC) was synthesised, carrying pendant N-alkyl-N,N-dimethyl-N-(2-hydroxypropyl)ammonium chloride groups, with alkyl being an octyl, dodecyl or cetyl moiety (Fig. 46, [193]).

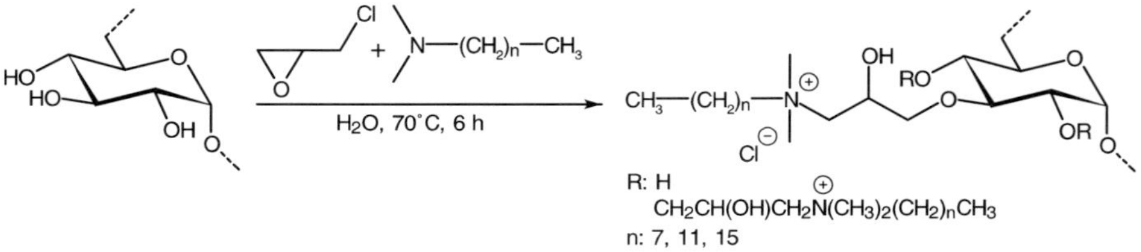

Fig. 46 Synthesis of amphiphilic polycationic dextran derivatives

These polymers show self-assembly into intramolecular (hydrophobic interactions in the same polymer chain) or intermolecular (between different chains) micelle-like clusters. Below the CAC, the polymer is extended because of electrostatic repulsion that does not allow a significant association between the hydrophobic pendant groups. Above the CAC, the higher ionic strength changes the polymer conformation to a less extended form, resulting in association of the hydrophobic side chains. The properties of the micelles depend significantly on the length of the hydrophobic side chains and the degree of substitution. The micelles with dodecyl- and cetyl alkyl side groups are formed mainly intramolecularly, whereas octyl side groups lead to intra- and intermolecular interactions. The association of hydrophobically modified polycationic dextran derivatives with oppositely charged surfactants like sodium octyl, dodecyl and tetradecyl sulfate leads to aqueous solutions with complex phase behaviour and self-assembling morphologies in solution [247–250]. These properties, e.g. the spontaneous formation of hydrophobic cores in aqueous solution, exhibit a great potential in drug/gene delivery research and for other biomedical applications. The mechanism and strength of the interaction between HMPC and oppositely charged surfactants depend on the structure of the surfactants, their hydrophobic part, the charge density, DS, chain conformation, hydrophobicity and molecular weight of the HMPC. In particular, the effect depends on the molar ratio of surfactant and HMPC.

6
Miscellaneous Dextran Derivatives

6.1
Dextran Conjugates

An enormous amount of publications have appeared over the last three decades dealing with the preparation of so-called dextran conjugates. These are either dextrans carrying bioactive molecules such as pharmaca, enzymes, proteins and hormones or dextrans modified with fluorescent dyes. These conjugates usually have low DS values and structural data are not accessible. A number of different paths for the covalent coupling are known. The preparation of ester-type conjugates mainly prodrugs is described in Sect. 4.2. In addition, binding via carbonate and carbamate moieties, using cyanogen bromide activation of the dextran, periodate oxidised dextran and binding as Schiff base, and the application of spacer molecules are all broadly exploited techniques. In this section, the chemistry of dextran conjugates is briefly discussed for representative examples. For more detailed information on the application of conjugates, the excellent reviews by DeBelder [45], Larsen [178] and Mehvar [179] are recommended. In addition, the importance and the influence of the linker between the polymer and the drug on the biological interactions and the release mechanism was reviewed [345].

6.1.1
Coupling via Formation of a Schiff Base

The most versatile method for the introduction of bioactive molecules with terminal NH_2 functions such as proteins and enzymes is the coupling via formation of a Schiff base (Fig. 47). It is necessary to oxidise the dextran backbone in the first step to obtain reactive aldehyde moieties. The Schiff base formed between the amine group of the bioactive molecule and the oxidised dextran is usually reduced in a separate step to obtain a stable conjugate.

The activity of proteins is not significantly diminished. Complex structures such as monoclonal and polyclonal antibodies [346] and bacterial proteins usable in the tumour-specific delivery of oncolytic drugs [347] can be covalently attached. Recently, such a system was used for immunodetection. Thus, proteins covered with dialdehyde dextran have been exploited as carrier molecules to detect the immunoresponse against an aminated hapten [348]. In addition, dialdehyde dextran has been exploited to detect protein–protein interactions. A very simple strategy, based on the intermolecular cross-linking of associated proteins by using dialdehyde dextran, has been proposed [349]. Dextran dialdehyde is applied as coating material for magnetic nanoparticles and immobilisation of enzymes such as glucose oxidase on the particle surface [350].

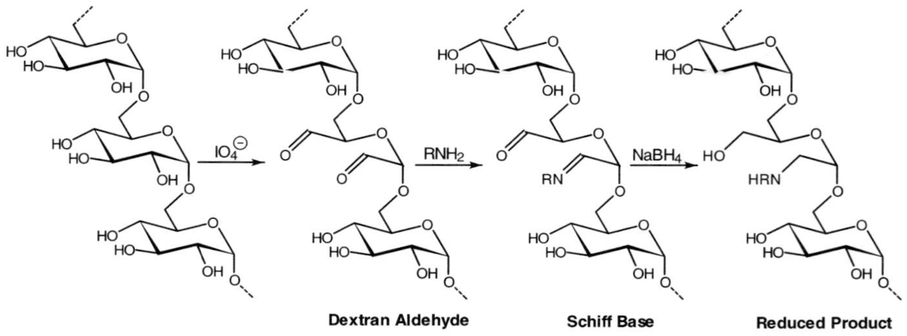

Fig. 47 Covalent binding of bioactive molecules with terminal NH$_2$ functions to dextran via oxidation of the dextran, formation of a Schiff base, and reduction

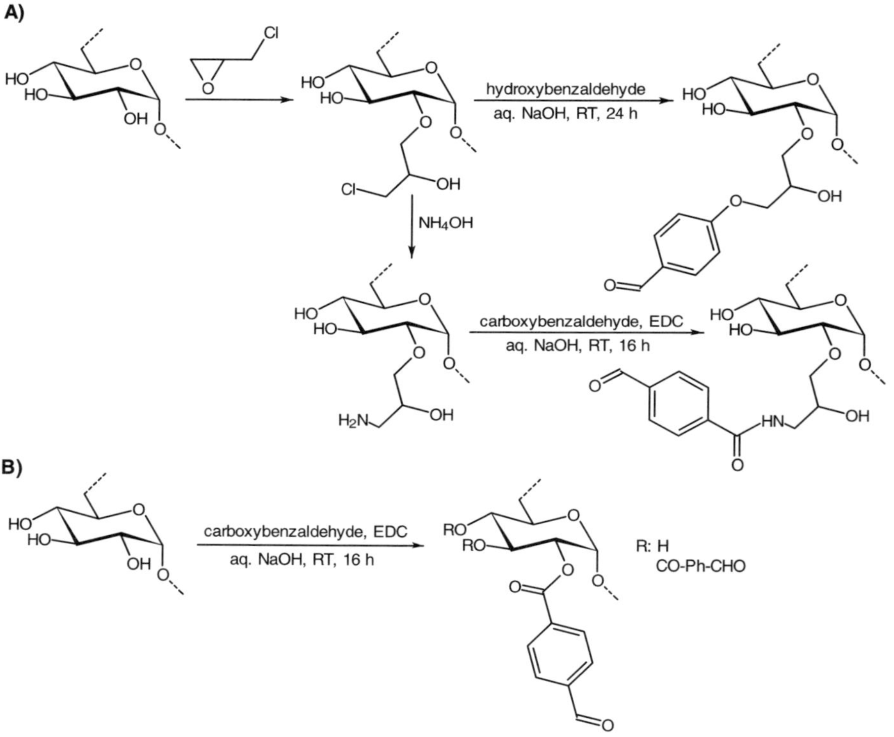

Fig. 48 Introduction of aldehyde functions by conversion of 3-chloro-2-hydroxypropyl dextran (activated dextran) with hydroxybenzaldehyde *A* or carboxybenzaldehyde (*A, lower part, B*)

It should be mentioned that introduction of a reactive aldehyde function can also be achieved by activation of dextran with epichlorhydrin and subsequent conversion with carboxybenzaldehyde or hydroxybenzaldehyde

(Fig. 48) [351]. The reaction via the aminated dextran (path A lower part) exhibits almost complete conversion of the amine functions, resulting in a fairly high amount of aldehyde moieties. Nevertheless, the amount of aldehyde is in the range of one to seven aldehyde groups per 100 mol anhydroglucose units.

6.1.2
Cyanogen Bromide as Coupling Agent

A method with the same focus, i.e. conjugation of amine-containing drugs and proteins to dextran, is the activation of dextran with cyanogen halides (Fig. 49). This procedure was among the first synthesis tools exploited for the covalent binding of peptides and proteins to polysaccharides [352]. Cyanogen bromide is the most efficient coupling agent. Despite its popularity, the precise total structure of the cyanogen bromide-activated polysaccharides and the endogenous functional groups are obscure. The most reasonable intermediate is shown in Fig. 49 [353].

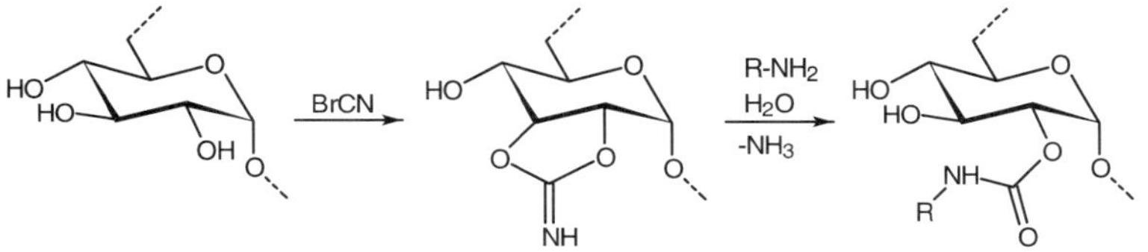

Fig. 49 Use of cyanogen bromide for the conjugation of peptides and proteins to polysaccharides

The technique is also used to introduce diene moieties, which enable the polymer for a highly efficient modification with dienophile components via Diels–Alder reaction [354]. Therefore, methyl octa-4,6-dienoate is converted to octa-4,6-dienoic acid hydrazide, which is bound to dextran via reaction with BrCN. The maleimide-modified protein (albumin) is easily attached to the polymer backbone by Diels–Alder reaction (Fig. 50).

6.1.3
Carbonates and Carbamates

First attempts to use the carbonate or the carbamate function for the covalent binding of amino acids and proteins applying the activation of dextran with phosgene showed that this approach is limited due to the fact that it is combined with a number of side reactions (Fig. 51) [355].

More useful is the conversion of dextran with ethyl chloroformate or 4-nitrophenyl chloroformate carried out in DMF/LiCl to give the corresponding carbonates [356]. Analysis of the total carbonate content and the content of 4-nitrophenyl carbonate moieties during the course of the reac-

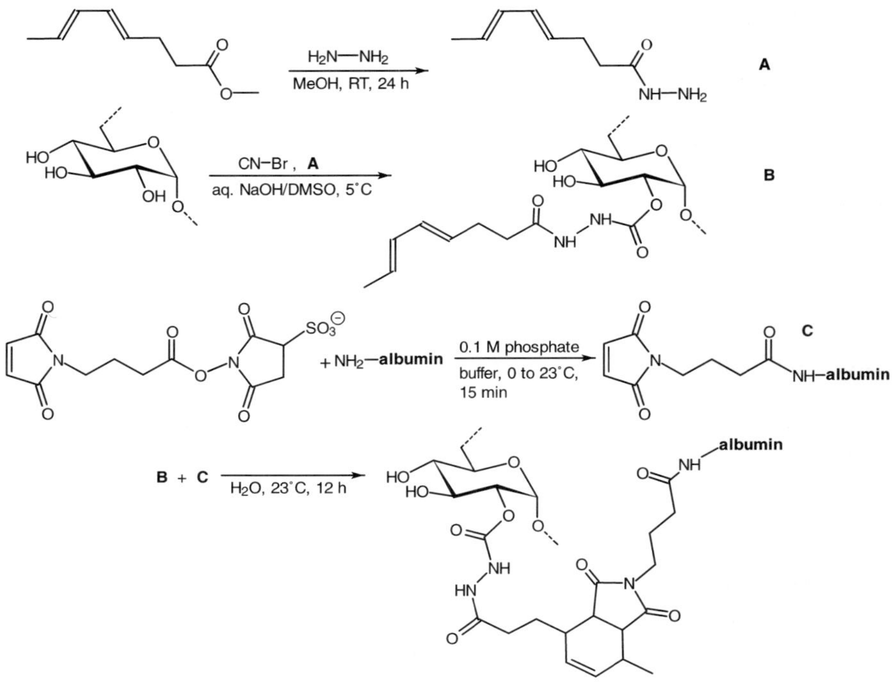

Fig. 50 Synthesis path applied for the binding of a maleimide-modified protein, e.g. albumin, to dextran via Diels–Alder reaction after introduction of a dienophil component using BrCN

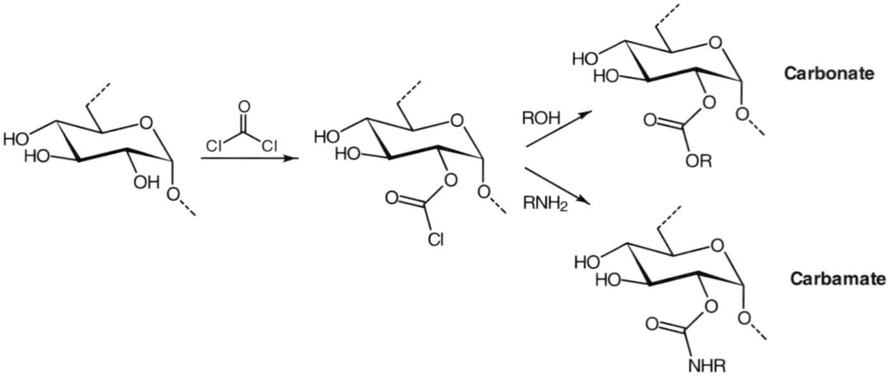

Fig. 51 Preparation of dextran carbonates and carbamantes after phosgene treatment

tion demonstrates the formation of different types of carbonate moieties. The 4-nitrophenyl carbonate moieties are transformed into other carbonate structures, most likely by reaction with neighbouring polymeric hydroxyl groups as revealed by NMR spectroscopic studies. The formation of cyclic carbonates is discussed, which could be inter- or intramolecular (Fig. 52). The data suggest that the process occurs predominantly within the polymer chains. It is strongly enhanced by addition of a strong base. The maximum total DS reached is in the region of 0.8 after 40 min. If longer reaction times are applied, the DS values decrease because of the attack of a neighbouring hydroxyl group of the dextran [357]. In model experiments with phenylamine and tyramine it was shown that conversion to carbamate moieties can be achieved (Fig. 52).

A useful approach for the introduction of a carbonate function is the reaction with N,N'-carbonyldiimidazole as discussed in Sect. 4.2.2.3 (Fig. 29), which was applied for the preparation of a new class of polymerisable dextrans [177].

More often, carbamates of dextran are prepared by one-step conversion of the polysaccharide with isocyanates or isothiocyanates. This path is especially useful for the conjugation of fluorescent dyes. Fluorescent-marked dextrans are commercial products today. Fluorescent dextran derivatives with different molecular weights and substituents are available as invaluable tracers for studies on microcirculation and vascular permeability in health and disease [8]. Fluorescein isothiocyanate (FITC) dextran (actually the thiocarbamate) has been well established in this area of research since the 1970s [358]. The isothiocyanate of fluorescein is covalently bound to dextran leading to

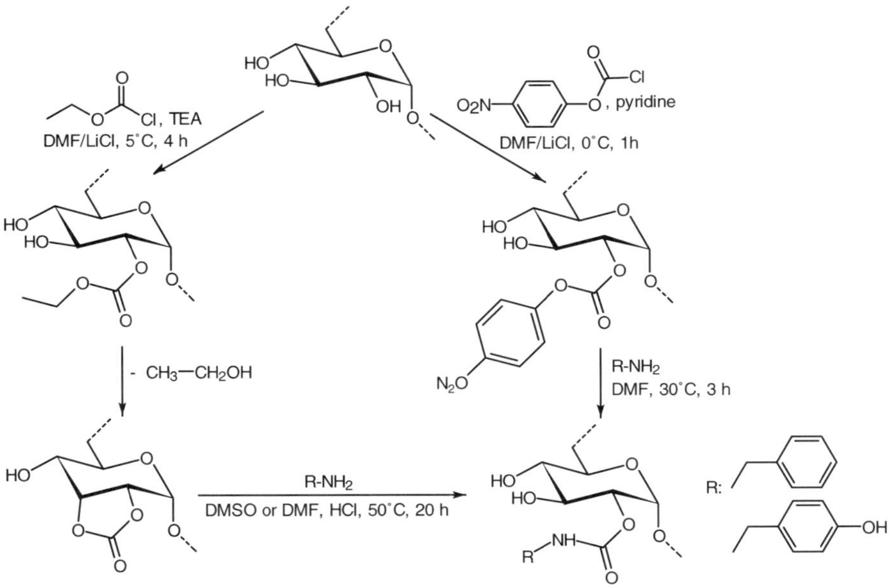

Fig. 52 Binding of phenylamine and tyramine to dextran via functionalisation with 4-nitrophenyl carbonate and subsequent formation of a carbamate

a low substituted fluorescein-labelled dextran (DS 0.004–0.01), as schematically shown in Fig. 53 [359].

An alternative method for the synthesis of different dextran conjugates with fluorescent dyes is the reaction of a water-soluble amino dextran with the succinimidyl ester of the corresponding dye [360]. The amide linkage is more stable than the corresponding thiocarbamate formed with isothiocyanates. This method is displayed schematically in Fig. 54 for the preparation of 5- (and 6-) carboxynaphthofluorescein-labelled dextran from amino dextran (Fig. 47, [361, 362]).

Fluorophores (including coumarin, cascade blue, tetramethylrhodamine and texas red), pH and Ca^{2+} indicators (fura, indo, calcium green-1) and coenzymes (biotine) are coupled to dextran by using the appropriate technique [360]. The investigation of hydrodynamic properties of the cytoplasmic matrix [363], the observation of uptake and internal processing of exogenous materials by fluid-phase endocytis [364], and the in vivo tracing of cell lineage [365] are examples of the importance of dextran conjugates

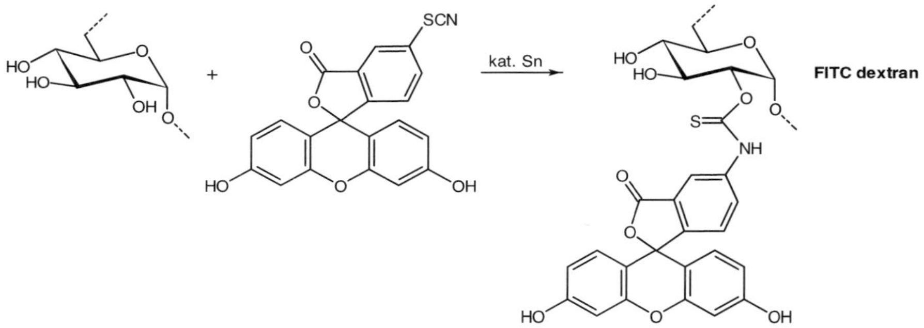

Fig. 53 Preparation of fluorescein-labelled dextran (FITC dextran) by the isothiocyanate technique [359]

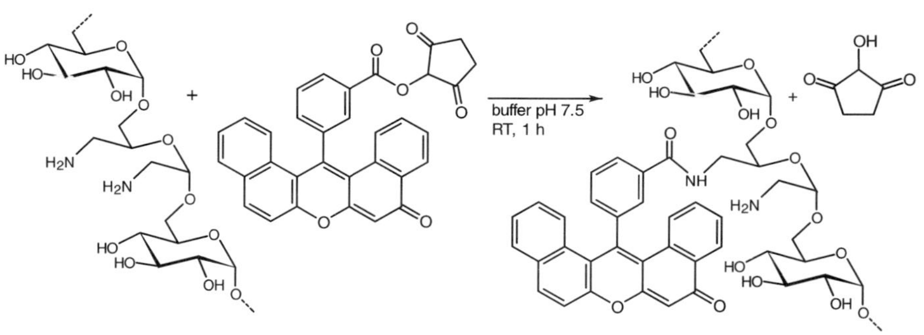

Fig. 54 Preparation of 5-carboxynaphthofluorescein-labelled dextran by the succinimide technique [361]

in medical research. One further application is the use as a size exclusion probe for determining relative pore sizes in both synthetic and natural membranes [366].

6.2
p-Toluenesulfonic Acid Ester of Dextran

Chemical functionalisation of polysaccharides directly at the C-atoms of the anhydrosugar unit may be carried out by nucleophilic displacement reactions (S_N). An indispensable prerequisite for S_N is the transformation of the hydroxyl moieties to a leaving group, which can be achieved by sulfonic acid ester formation, especially by introduction of a tosyl group. It has been shown that tosylation of dextran under homogeneous conditions is an efficient modification yielding polymers with a broad spectrum of solubility (Table 15). Preliminary studies on the S_N reaction of tosyl dextran (DS 1.39) with NaN_3 revealed that a polymer with DS_{Azide} 0.62 and DS_{Tosyl} 0.66 can be synthesised, i.e. almost 45% of the tosyl moieties were substituted [367].

Table 15 Conditions and results for the reaction of dextran dissolved in DMAc/LiCl with p-toluenesulfonic acid chloride in the presence of triethylamine for 24 h at 8 °C

Molar ratio [a]	Elemental analysis (%)		DS [b]		Solubility
	S	Cl	DS_{Tosyl}	DS_{Cl}	
1 : 0.5	1.54	0.32	0.08	0.03	DMSO, DMA, DMF
1 : 1.5	8.63	0.21	0.75	0.02	DMSO, DMA, DMF, THF
1 : 5.0	11.83	0.22	1.39	0.02	DMSO, DMA, DMF, THF, acetone, $CHCl_3$

[a] Glucopyranosyl unit: p-toluenesulfonic acid chloride, 2 mol triethylamine per mol p-toluenesulfonic acid chloride were used as base
[b] Degree of substitution of p-toluenesulfonic acid ester (index Tosyl) and chlorodeoxy (index Cl) groups

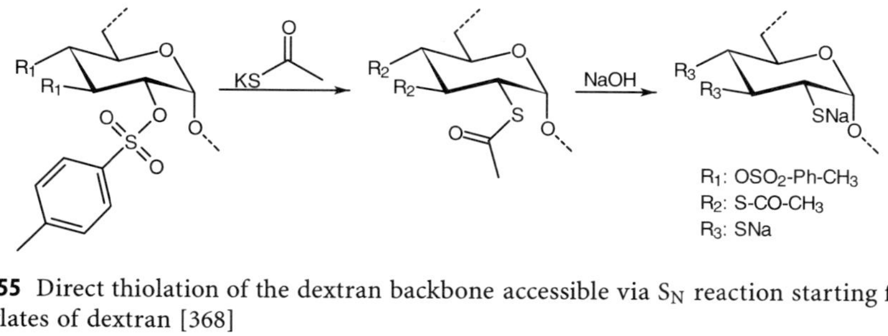

R1: OSO_2-Ph-CH_3
R2: S-CO-CH_3
R3: SNa

Fig. 55 Direct thiolation of the dextran backbone accessible via S_N reaction starting from tosylates of dextran [368]

In addition, dextran tosylates may be used for the preparation of thiol-containing derivatives usable for self-assembly structures (Fig. 55) [368].

6.3
Thiolated Dextran

Besides the direct thiolation of the polysaccharide backbone accessible via S_N reaction starting from tosylates of dextran, as discussed above, a number of paths for the introduction of spacer-bound thiol functions are described. Early attempts towards thiolation of dextran applied acetylthiosuccinic anhydride yielding the mercaptosuccinyl derivative of dextran (Fig. 56). The reaction was carried out by addition of acetylthiosuccinic anhydride to a solution of the macromolecule while the pH was maintained at 8 by addition of NaOH and subsequent treatment of the product with Amberlite [369].

More recently, thiolated dextran was synthesised by modification of dextran with 4-nitrophenyl chloroformate in DMSO/pyridine in the presence of DMAP at 0 °C yielding a carbonate (content 6%), which can be substituted with cystamine. Subsequent reduction yields the thiolated dextran containing between 1 and 4% thiol moieties (Fig. 57) [370].

This thiolated dextran may be immobilised by chemisorption on a metal surface such as silver and can thus be used as an inhibitor of non-specific protein adsorption. Surface plasmon resonance (SPR) was employed to monitor BSA adsorption from a flowing buffered solution, revealing a significant reduction of BSA adsorption for coated silver compared to the uncoated surface. The results obtained by SPR have been complemented and validated by ellipsometry, atomic force microscopy (AFM) and X-ray photoelectron spectroscopy (XPS) analysis. The amount of thiol functions and the molecular weight of the polymer affect the protein resistant performance of the dextran layer. By increasing the proportion of thiol substituents, a greater coverage of the silver surface is observed, conferring improved protein resistance. Higher molecular weight dextran derivatives (M_w ranging from 5000 to 500 000 g mol^{-1}) are shown to pack less efficiently to produce layers with less effective protein resistance [370, 371]. In addition, SPR has been employed to

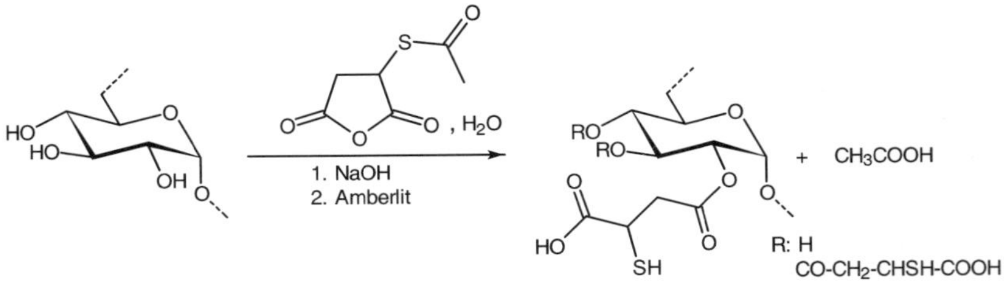

Fig. 56 Synthesis of thiol-functionalised dextran by conversion of dextran with acetylthiosuccinic anhydride

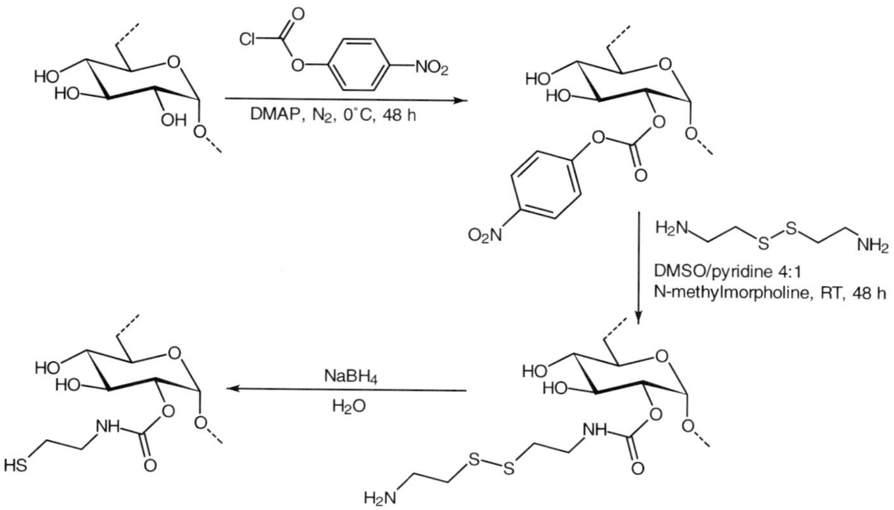

Fig. 57 Synthesis of thiolated dextran via formation of a nitrophenyl carbonate [370]

investigate the hydrolytic degradation of thiolated dextran monolayers on silver by the enzyme dextranase. It was demonstrated that dextranase does not completely remove the thiolated dextran monolayers, even at the enzyme's most active pH. A significant protein resistance is still found after degradation [372]. AFM measurements in liquid environment have demonstrated the ability to measure corresponding changes in both monolayer morphology and elasticity due to the hydration state of the dextran derivative [373].

Two methods for the derivatisation of dextran with boron-containing substituents via sulfur bridges were developed for substances useful in boron neutron capture therapy (BNCT). One method comprises activation of dextran with 1-cyano-4-(dimethylamino)pyridine (CDAP) with subsequent coupling of 2-aminoethylpyridyl disulfide (Fig. 58, method A). The thiolated dextran could couple the boron-containing molecule ($Na_2B_{12}H_{11}SH$, BSH) with a terminal thiol function in a disulfide exchange reaction. In another procedure, dextran was derivatised to an allyl dextran (Fig. 58, method B), which reacted with BSH in a free-radical-initiated addition reaction. The amount of functionalisation is rather low. For method A, 10–20 boron-containing functions were determined per dextran molecule (starting dextran was T70) and 100–125 units per dextran chain in the case of method B. Nevertheless, the derivatives are useful for tumour targeting applications [374].

The binding of proteins such as tetanus toxoid (TTd) onto dextran can also be achieved via thiolation. Therefore, dextran is converted selectively at the terminal reducing ends. This is possible either after amination and modification of the terminal amino function with 2-iminothiolan hydrochlo-

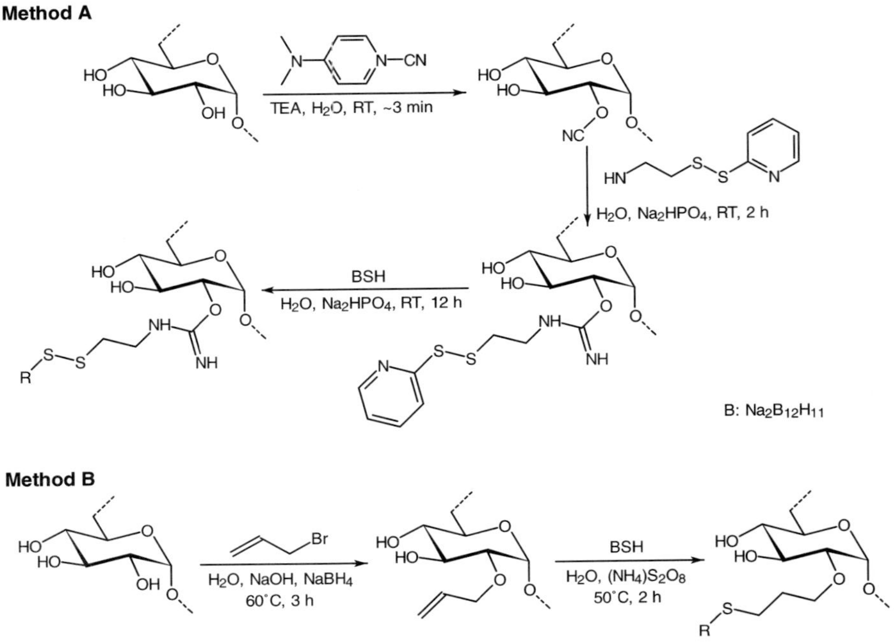

Method A

B: $Na_2B_{12}H_{11}$

Method B

Fig. 58 Methods for the derivatisation of dextran with boron-containing substituents via sulfur bridges developed for boron neutron capture therapy (BNCT)

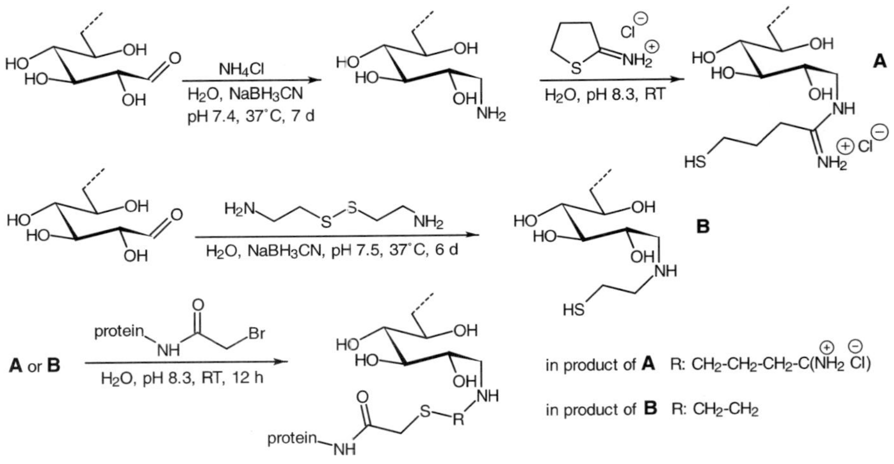

in product of **A** R: $CH_2-CH_2-CH_2-C(NH_2^+ Cl^-)$

in product of **B** R: CH_2-CH_2

Fig. 59 Selective binding of proteins onto dextran by activation of the terminal reducing ends

ride (Fig. 59, path A) or by direct treatment of dextran with cystamine (Fig. 59, path B). The molecular weight of the dextran investigated for modification was in the range of 500 to 150000 g mol^{-1}. The conjugation of the protein (e.g. TTd) onto the polysaccharide was performed after activation with *N*-hydroxysuccinimide ester of bromoacetic acid (Fig. 59). In contrast to random activation of the polysaccharide, this method avoids cross-linking leading to conjugates that induce strong antipolysaccharide T-helper-cell dependent immune responses. Unlike direct reductive amination, the 2-iminothiolan based conjugation technique is fast and made it possible to couple fairly large polysaccharides to TTd [375].

6.4
Silyl Dextran

Silylation reactions on polysaccharides with chlorosilanes and silazanes were attempted more than 50 years ago resulting in hydrophobic silyl ethers with both increased thermal stability and solubility in organic solvents [376]. The silylation reaction for the protection of hydroxyl groups in mono- and polysaccharides exhibits many advantages, e.g. fast silylation, solubility of silylether in organic solvents suitable for subsequent derivatisation, stability of the resulting silylether under basic conditions but easy deprotection of the silyl moieties by acid hydrolysis or nucleophilic agents like fluoride and cyanide ions [377]. The partial and complete silylation of dextran was studied in detail by Ydens and Nouvel [215–217].

The partial silylation was carried out with 1,1,1,3,3,3,-hexamethyldisilazane (HMDS) in DMSO at 50 °C (Fig. 60). The hydroxyl groups react with HMDS resulting in trimethylsilyl dextran and $(CH_3)_3SiNH_2$, which further reacts with hydroxyl groups to silyl ether and NH_3. The silylation reaction starts ho-

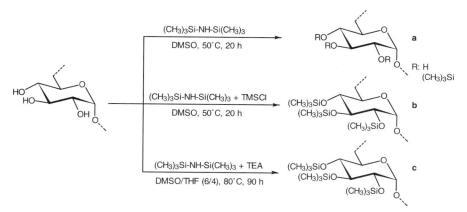

Fig. 60 Partial **a** and complete silylation (**b, c**) of dextran

mogeneously, but with increasing degree of silylation the solubility in DMSO decreases. Highly substituted products are soluble in THF, chloroform or toluene. The DS of silylation increases with reaction time, temperature and HMDS/OH ratio and decreases with the molecular weight of dextran; a complete silylation is not reached under these conditions (samples 1–5 and 13 in Table 16). Although acidic catalysts like HCl, p-toluenesulfonic acid or trifluoroacetic acid can increase the reactivity of HMDS, they are excluded due to their degrading potential.

Table 16 shows the influence of reaction conditions and of activation reagents such as TMSCl, hexamethylphosphoric triamide (HMPA), N-methylimidazole (NMI), saccharin and tetrabutylammonium fluoride (TBAF) on the DS of silyl dextran. For complete silylation, a mixture of 1.8 mol HMDS and 0.2 mol chlorotrimethylsilane (TMSCl) per mol OH groups have to be used. The more reactive TMSCl reacts in a first step with hydroxyl groups to trimethylsilyl dextran and HCl. Subsequently, the HCl cleaves HMDS in TMSCl and NH_3 [218]. Complete silylation of dextran could also be carried

Table 16 Influence of HMDS/OH molar ratio, co-solvent and activation reagent on the degree of substitution (DS_{Silyl})

Dextran	Sample	Medium	Molar ratio [HMDS]/ [OH]	Activation reagent [a] (mol)	Temp. (°C)	Time (h)	DS_{Silyl}
T40	1	DMSO	0.25	–	50	20	1.1
	2	DMSO	0.50	–	50	20	2.0
	3	DMSO	2.0	–	50	20	2.3
	4	DMSO	2.0	–	50	16	2.1
	5	DMSO	5.0	–	50	20	2.3
	6	DMSO/THF (6/4)	4.0	0.2 NEt3	60	48	3.0
	7	DMSO/toluene (1/1)	2.0	–	50	20	2.5
	8	DMSO	1.8	0.2 TMSCl[b]	50	20	3.0
	9	DMSO	2.0	0.1 HMPA[c]	50	20	2.5
	10	DMSO	2.0	0.1 NMI[d]	50	20	2.3
	11	DMSO	2.0	0.1 Saccharine	50	20	2.7
	12	DMSO	2.0	0.1 TBAF[e]	50	20	2.6
T10	13	DMSO	2.0	–	50	16	2.3
	14	DMSO/THF (5/2)	4.0	0.2 NEt3	60	48	2.8
	15	DMSO/THF (6/4)	4.0	0.2 NEt3	80	90	3.0

[a] With respect to mol HMDS
[b] Chlorotrimethylsilane
[c] Hexamethylphosphoric triamide
[d] N-Methylimidazole
[e] Tetrabutylammonium fluoride

out in DMSO using THF as co-solvent in a ratio of 6 : 4 and 0.2 mol triethyl-amine as activation agent with respect to mol HMDS and a HMDS to OH molar ratio of 4 : 1. Co-solvents such as THF and toluene guarantee homogeneous conditions for the whole conversion. At 50 °C, silylation of dextran with HMDS proceeds without chain degradation independent of the reaction times, co-solvents, or activation reagents. Nevertheless, degradation of dextran was observed at 80 °C. More reactive silylation reagents such as *N,O*-*bis*(trimethylsilyl)-acetamide lead to significant chain degradation even at 50 °C.

The degree of silylation can be determined by ^{1}H NMR in CDCl$_3$ or in DMSO-d_6. In the case of slightly silylated dextran, the analysis of the ^{1}H NMR spectra in D$_2$O and DMSO-d_6 give similar results. A precise assignment of the NMR signals was achieved by 2D ^{1}H – ^{1}H – and ^{1}H – ^{13}C techniques (COSY 45° and HMQC, Fig. 61).

The silylation yield can be calculated using the equation:

$$\text{Yield}(\%) = \frac{A(\text{OSiMe}_3)}{A(\text{anomeric H})} \times \frac{100}{27} .$$

$A(\text{OSiMe}_3)$ is the area of the trimethylsilyl group at 0.15 ppm in CDCl$_3$ or 0.18 ppm in DMSO-d_6. The area of the anomeric protons $A(\text{anomeric H})$ is centred at 4.7 ppm and the area of the glucosidic protons $A(\text{glycosidic H})$ is

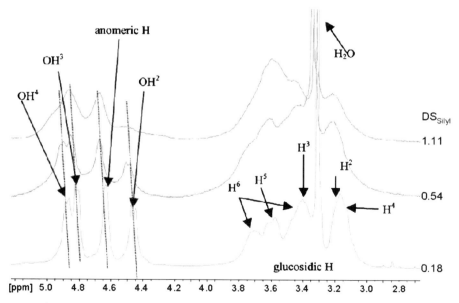

Fig. 61 ^{1}H-NMR spectra of silylated dextrans in DMSO-d_6 with different DS$_{\text{Silyl}}$ (adapted from [217])

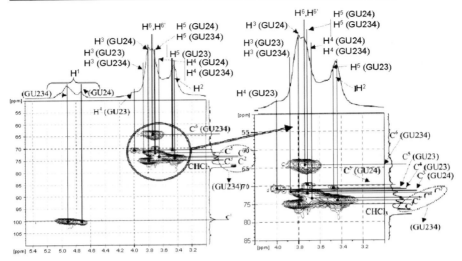

Fig. 62 HMQC spectrum of silylated dextran (DS_{Silyl} = 2.85, adapted from [217])

situated from 3.2 to 4.2 ppm in CDCl$_3$, and from 3.0 to 4.0 ppm in DMSO-d_6 (Fig. 61).

The DS can also be determined by subsequent reaction of the remaining OH groups with phenylisocyanate [378] or with trichloroacetylisocyanate [379] to the carbamate. The calculation from NMR data can be carried out either before or after hydrolysis of the TMS groups by using the ratio of the A(NH) of the urethane protons at 8.5–9.5 ppm to the total of A(anomeric H) and A(glycosidic H) between 3 and 5 ppm. Both methods give comparable values. The analyses of the NMR spectra in Fig. 62 allow determination of the different reactivity of the OH groups. The rapid decrease of the hydroxyl proton signal at position 2 means that OH2 is the most reactive group. Nouvel found the following order of reactivity for the silylation of dextran: OH2 > OH4 > OH3 [217]. The HMQC spectrum of silyl dextran with a DS of 2.85 (Fig. 62) confirms the presence of 2,3,4-tri-O-; 2,3-di-O-; and 2,4-di-O-silylated glucose units.

6.4.1
Poly(ε-Caprolactone)-grafted (PCL) Dextran Copolymers

The combination of biodegradable hydrophobic polyesters with biodegradable hydrophilic dextran leads to completely biodegradable amphiphilic polymer architectures with promising properties, useful as environmentally friendly surfactants or as a matrix in drug delivery systems. For instance, the synthesis of PLA-grafted dextran copolymers obtained by ring-opening polymerisation (ROP) of lactide with dextran in presence of stannous octoate was described [380–384]. A controlled synthesis of the amphiphilic PLA-grafted dextran copolymers was achieved by partial silylation of the hydroxyl functions in dextran, with subsequent ROP of the lactide with the remaining hydroxyl groups and, finally, the deprotection of the silyl ether [216, 217, 385, 386]. A three-step procedure was also suggested by Ydens [215] and Nouvel [216, 217] for the controlled synthesis of amphiphilic PCL-grafted copolymers of dextran (Fig. 63).

The remaining hydroxyl groups of partially silylated dextran may initiate the ring-opening polymerisation of ε-caprolactone (Cl). Through a controlled silylation reaction, the number of hydroxyl groups available for further ROP of ε-caprolactone is adjusted. The hydroxyl groups along the silylated dextran backbone are catalytically activated by AlEt$_3$ or Al(O-isoPr)$_3$ at 60 °C in toluene or Sn(Oct)$_2$ at 100 °C (Table 17).

Assuming that each hydroxyl group initiates the ring-opening polymerisation of ε-caprolactone (well-known as "living" character of the "coordination–insertion" mechanism), high efficiency of grafting is observed, including the control of the average length of every PCL branch (DP$_{PCL}$) and the average number of grafts per glucosidic unit (N_g) (Table 17).

The third step is the deprotection, i.e. splitting off the trimethylsilyl groups under mild conditions (THF, aqueous HCl). Both the silylated and the deprotected PCL-grafted dextran copolymers are semicrystalline with a melting temperature of 57 °C, attributed to the PCL grafts. The PCL-grafted dextran copolymers with low PCL content are soluble in water but insoluble in organic solvents such as toluene or THF. For high PCL content, the solubility in water is limited. The amphiphilic copolymers show a decreased surface tension compared to dextran. An aqueous solution of dextran T10 (M_n 6600 g mol^{-1}) has a surface tension in the range of water (73 mN m^{-1}). PCL-grafted dextran copolymer with a PCl weight fraction (F_{PCl}) of 0.20 and a DP$_{PCL}$ of 6.6 displays a surface tension of 54 mN m^{-1} in aqueous solution (5 g L^{-1}).

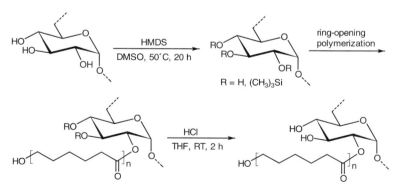

Fig. 63 Controlled synthesis of poly(ε-caprolactone) grafted onto dextran via partially silylated dextran

Table 17 Dependence of the weight fraction of poly(ε-caprolactone) (PCL) grafted on silyldextran on the catalysator and the initial weight fraction of ε-caprolactone by ROP in toluene

Silyldextran DS	M_n	[Cl]/[OH]	Catalyst [a]	f_{Cl} [b]	F_{PCl} [c]	AL_{PCL} [d]	M_n	M_w/M_n
2.8 [e]	10 100	20	Al(Et)$_3$	0.64	0.41	13	34 700	1.4
2.8 [e]	10 100	20	Al(OiPr)$_3$ [g]	0.64	0.20	6.6	11 600	1.2
2.6 [e]	12 600	20	Al(OiPr)$_3$	0.77	0.73	18	31 300	1.4
2.6 [e]	12 600	30	Al(OiPr)$_3$	0.83	0.77	30	28 300	1.3
2.8 [e]	10 100	20	Sn(Oct)$_2$	0.64	0.50	17	36 500	1.5
2.7 [f]	32 900	5	Sn(Oct)$_2$	0.32	0.26	5	34 300	1.6
2.7 [f]	32 900	10	Sn(Oct)$_2$	0.50	–	–	43 100	1.7

[a] Molar ratio of [catalyst]/[OH] = 0.05
[b] Initial weight fraction of ε-caprolactone
[c] Weight fraction of ε-caprolactone in the copolymers determined by gravimetry
[d] Average length of every PCL branch
[e] 60 °C, 68 h, dextran T10
[f] 100 °C, 44 h, dextran T40
[g] Molar ratio = 0.02

Well defined amphiphilic PCL-grafted dextran copolymers have been used to prepare nanoparticles by the nanoprecipitation technique in water. The copolymers were first dissolved in DMSO at 50 °C and then added drop-wise under vigorous stirring into a large volume excess of water thermostated at 50 °C. The suspension turned opalescent as a result of colloidal nanoparticle formation and was dialysed against distilled water to separate DMSO. The nanoparticle suspension was finally concentrated and is stable over a period of 3 weeks. The apparent mean diameter of the nanoparticles is close to 200 nm with a narrow size distribution. Such behaviour suggests a core-shell structure consisting of a hydrophobic PCL core surrounded by a hydrophilic dextran shell, which prevents rapid agglomeration [387, 388].

7
Outlook

Dextran is a unique polysaccharide because of its structure (only glucose units), purity, defined branching pattern depending on the microbial sources and defined molecular weight. Today, it is produced on a commercial scale resulting from optimised biotechnological processes for the biosynthesis of dextran using preferably *Leuconostoc mesenteroides*.

The commercial availability and the outstanding features of this important biopolymer make it a desired starting material for products in biological and medical applications where reproducible properties and biocompatibility are essential prerequisites. In addition, dextran is of growing interest for the construction of supramolecular architecture because of the flexibility of the $1 \rightarrow 6$ linked polymer backbone. The use of dextran and its derivatives is focused on the synthesis of advanced materials and bioactive compounds. An additional aspect of dextran is its conformity with the principles of green chemistry because it calls for the design of biodegradable end products, principally by employing starting materials from renewable resources. Together with other polysaccharides, dextran will become an important starting material; however, it requires detailed understanding of various aspects including:

- Biosynthesis to obtain dextran with controlled and reproducible molecular structure and molecular weight
- Controlled biosynthesis to design the material with controlled reactivity and in a special shape, e.g. as nanoparticles
- Solubility in simple, non-toxic media and understanding of activation prior to a modification reaction by methods that meet the principles of green chemistry
- Details of the reaction steps both by chemical and biochemical (enzyme-catalysed) conversions
- Functionalisation of the polymer backbone through reaction of the hydroxyl groups (esterification and etherification), nucleophilic displacement reactions, oxidation and selective reactions at the terminal reducing ends
- Modification of the biopolymer with naturally occurring molecules in order to obtain completely bio-based materials.

A huge variety of functionalisation reactions on dextran have been carried out over the last three decades. In comparison to other polysaccharides such as cellulose and starch, the esters of simple carboxylic acids and short chain alkyl ethers did not find significant interest. In contrast, the sulfuric acid halfesters of dextran (dextran sulfates) are among the most promising polysaccharide derivatives in the pharmaceutical field. A variety of highly functionalised dextran sulfates are commercial products. They exhibit pronounced anticoagulating, cancerostatic, antiviral and even antiprion activity. Nevertheless, the chemical structure of the polysaccharide derivative was not specifically modified for most of the investigations towards the bioactivity of dextran sulfates, i.e. it was not considered as a parameter of influence for their effectiveness. From the authors' point of view, this is one of the weaknesses of the majority of the studies in the field. The comprehensive analysis of the derivatives and a defined variation of structural features, such as the amount of sulfate groups and their distribution within the basic units and along the polymer chain, should be used for the establishment of the structure–

property relationship. The need for selective synthesis and analysis is even more pronounced in the case of mixed derivatives such as carboxymethylated dextran sulfates, which are investigated as heparin mimics. Only the exploitation of structure–property relationships can lead to products with tailored biological activity.

The search for new synthesis tools for the binding of different chemical functionalities onto dextran is stimulated by the growing number of poorly water soluble pharmaceuticals, which can be bound to the polymer to increase their bioavailability and to protect them from fast metabolism. There is still a need for efficient reagents that can realise the conjugation of the polymer with the bioactive molecule without pronounced side reactions and side products. For most of the modification methods, the detailed mechanisms and structures are poorly understood, as in case of the broadly used conjugation reactions with cyanogen halides. Esterification reactions of dextran after in situ activation of the corresponding carboxylic acids with efficient, non-toxic reagents such as CDI have demonstrated its usefulness. These approaches guarantee fast and complete conversion to structurally homogeneous derivatives giving only non-toxic and easily removable by-products. In addition to the binding of drugs, they are also applied to the introduction of photo cross-linkable substituents. The preparation of hydrogels and nanostructured materials is achieved. The fine-tuning of the chemical structure results in compounds with an adjusted hydrophilic–hydrophobic balance or with environmental sensitivity. Moreover, modified dextran is of growing interest in the field of surface modification. In this case, ionic substituents are introduced to yield self assembly via electrostatic interactions or specific binding on a surface, initiated by using substituents capable of recognition effects such as thiol moieties.

All in all, it is obvious that dextran will gain increasing importance as a carrier material in pharmaceutical applications, as a basis for bioactive derivatives and as a nanostructured device. Dextran and modified dextrans should always be considered as a biocompatible material with a high structure-forming potential.

References

1. Klemm D, Schmauder HP, Heinze T (2002) Cellulose. In: DeBaets S, Vandamme EJ, Steinbüchel A (eds) Biopolymers: polysaccharides 2, vol 6. Wiley, Weinheim, p 275
2. Shogren RL (1998) Starch: properties and material applications. In: Kaplan DL (ed) Biopolymers from renewable resources. Springer, Berlin Heidelberg New York, p 30
3. Taylor C, Cheetham NWH, Walker GJ (1985) Carbohydr Res 137:1
4. Pasteur L (1861) Bull Soc Chim Paris 30
5. VanTieghem P (1878) Ann Sci Nature Bot Biol Veg 7:180
6. Jeanes A, Haynes WC, Wilham CA, Rankin JC, Melvin EH, Austin MJ, Cluskey JE, Fisher BE, Tsuchiya HM, Rist CE (1954) J Am Chem Soc 76:5041

7. Vandamme EJ, Bruggeman G, DeBaets S, Vanhooren PT (1996) Agro-Food-Ind Hi-Tech 7:21
8. DeBelder AN (1996) Medical applications of dextran and its derivatives. In: Dimitriu S (ed) Polysaccharides in medicinal applications. Marcel Dekker, New York, p 505
9. ElSeoud O, Heinze T (2005) Organic esters of cellulose: New perspectives for old polymers. In: Heinze T (ed) Advances in polymer science: polysaccharide I, structure, characterization and use, vol 186. Springer, Berlin Heidelberg New York, p 103
10. Heinze T (2004) Chemical functionalization of cellulose. In: Dumitriu S (ed) Polysaccharide: structural diversity and functional versatility, 2nd edn. Marcel Dekker, New York, p 551
11. Holzapfel WH, Schillinger U (1992) The genus leuconostoc. In: Ballows A, Truper HG, Dworkin M, Harder W, Schleifer KH (eds) The prokaryotes, 2nd edn. Springer, Berlin Heidelberg New-York, p 1508
12. Naessens M, Cerdobbel A, Soetaert W, Vandamme EJ (2005) J Chem Technol Biotechnol 80:845
13. Leathers TD (2002) Dextran. In: Vandamme EJ, DeBaets S, Steinbüchel A (eds) Biopolymers: polysaccharides 1, polysaccharides from prokaryotes. Wiley, Weinheim, p 300
14. Hamada S, Slade DH (1980) Microbiol Rev 44:3319
15. Leach SA, Hayes ML (1968) Caries res 2:38
16. Hare MD, Svensson S, Walker GJ (1978) Carbohydr Res 66:245
17. Marotta M, Martino A, DeRosa A, Farina E, Carteni M, DeRosa M (2002) Process Biochem 38:101
18. Khalikova E, Susi P, Korpela T (2005) Microbiol Mol Biol Rev 69:306
19. Harris PJ, Henry RJ, Blakeney AB, Stone BA (1984) Carbohydr Res 127:59
20. Seymour FR, Slodki ME, Plattner RD, Jeanes A (1977) Carbohydr Res 53:153
21. Slodki ME, England RE, Plattner RD, Dick WE (1986) Carbohydr Res 156:199
22. Larm O, Lindberg B, Svensson S (1971) Carbohydr Res 20:39
23. VanCleve JW, Schaefer WC, Rist EC (1956) J Am Chem Soc 78:4435
24. Seymour FR, Chen ECM, Bishop SH (1979) Carbohydr Res 68:113
25. Dols M, Remaud-Simeon M, Willemot RM, Vignon M, Monsan PF (1997) Appl Biochem Biotechnol 62:47
26. Shimamura A, Tsumori H, Mukasa H (1982) Biochim Biophys Acta 702:72
27. Honda O, Kato C, Kuramitsu HK (1990) J Gen Microbiol 136:2099
28. Gilmore KS, Russell RR, Ferretti JJ (1990) Infect Immun 58:2452
29. Usui T, Kobayashi M, Yamaoka N, Matsuda K, Tuzimura K (1973) Tetrahedron Lett 36:3397
30. Gagnaire D, Vignon M (1977) Makromol Chem 178:2321
31. Cheetham NWH, Fiala-Beer E (1991) Carbohydr Polym 14:149
32. Seymour FR, Knapp RD, Bishop SH (1976) Carbohydr Res 51:179
33. Heinze T, Liebert T (2004) Macromol Symp 208:167
34. Hornig S (2005) Diploma Thesis, Friedrich Schiller University of Jena, Germany
35. Alsop RM, Byrne GA, Done JN, Earl IE, Gibbs R (1977) Process Biochem 12:15
36. Bovey FA (1959) J Polym Sci 35:167
37. Senti FR, Hellmann NN, Ludwig NH, Babcock GE, Tobin R, Glass CA, Lamberts BL (1955) J Polym Sci 17:527
38. Antonini E, Bellelli L, Bruzzesi MR, Caputo A, Chiancone E, Rossi-Fanelli A (1964) Biopolymers 2:27
39. Ioan CE, Aberle T, Burchard W (2000) Macromolecules 33:5

40. Gekko K (1971) Makromol Chem 148:229
41. Hirata Y, Sano Y, Aoki M, Shohji H, Katoh S, Abe J, Hitsukuri S, Yamamoto H (2003) Carbohydr Polym 53:331
42. Hirata Y, Sano Y, Aoki M, Kobatake H, Kato S, Yamamoto H (1999) J Colloid Interface Sci 212:530
43. McCurdy RD, Goff HD, Stanley DW, Stone AP (1994) Food Hydrocolloids 8:609
44. Ioan CE, Aberle T, Burchard W (2001) Macromolecules 34:326
45. DeBelder AN (2003) Dextran. In: Amersham bioscience. Article no 18-1166-12
46. Granath KA (1958) J Colloid Sci 13:308
47. Chanzy H, Excoffier G, Guizard C (1981) Carbohydr Polym 1:67
48. Guizard C, Chanzy H, Sarko A (1984) Macromolecules 17:100
49. Shingel KI (2002) Carbohydr Res 337:1445
50. Hirata Y, Aoki M, Kobatake H, Yamamoto H (1999) Biomaterials 20:303
51. Stenekes RJH, Talsma H, Hennink WE (2001) Biomaterials 22:1891
52. Amersham bioscience, data file dextran. http://www.dextran.nu/media/18-1153-41AA.pdf
53. Carrasco F, Chornet E, Overend RP, Costa J (1989) J Appl Polym Sci 37:2087
54. DeGroot CJ, VanLuyn MJA, VanDiek-Wolthuis WNE, Cadée JA, Plantinga JA, Otter WD, Hennink WE (2001) Biomaterials 22:1197
55. Cadee JA, VanLuyn MJA, Brouwer LA, Plantinga JA, VanWachem PB, DeGroot CJ, DenOtter W, Hennink WE (2000) J Biomed Mater Res 50:397
56. Hehre EJ, Hamilton DM (1951) J Biol Chem 192:161
57. Vedyashkina TA, Revin VV, Gogotov IN (2005) Appl Biochem Microbiol 41:631
58. Hehre EJ, Suzuki H (1966) Arch Biochem Biophys 113:675
59. Genghof DS, Hehre EJ (1972) Proc Soc Exp Biol Med 140:1298
60. Binder TP, Robyt JF (1983) Carbohydr Res 124:287
61. Tsuchiya HM (1960) Bull Soc Chim Biol 42:1777
62. Binder TP, Cote GL, Robyt JF (1983) Carbohydr Res 124:275
63. Robyt JF, Kimble BK, Walseth TF (1974) Arch Biochem Biophys 165:634
64. Robyt JF, Eklund SH (1982) Bioorg Chem 11:115
65. Robyt JF, Eklund SH (1983) Carbohydr Res 121:279
66. Robyt JF, Taniguchi H (1976) Arch Biochem Biophys 174:129
67. DeBelder AN (1993) Dextran. In: Whistler RL, BeMiller JN (eds) Industrial gums, polysaccharides and their derivatives, 3rd edn. Academic, San Diego, p 399
68. Hehre EJ (1941) Science 93:237
69. Jeanes A (1965) Dextrans. In: Whistler RL, BeMiller JN, Wolfrom ML (eds) Methods in carbohydrate chemistry, vol 5, general polysaccharides. Academic, New-York, p 118
70. Koepsell HJ, Tsuchiya HM (1952) J Bacteriol 63:293
71. Tsuchiya HM, Koepsell HJ, Corman J, Bryant G, Bogard MO, Feger VH, Jackson RW (1952) J Bacteriol 64:521
72. Hamdy MK, Gardner E, Stahly GL, Weiser HH, Van Winkle Q (1954) Ohio J Sci 54:317
73. Sabatie J, Choplin L, Moan M, Doublier JL, Paul F, Monsan P (1988) Carbohydr Polym 9:87
74. Hellman NN, Tsuchiya HM, Rogovin SP, Lamberts BI, Tobin R, Glass CA, Stringer CS, Jackson RW, Senti FR (1955) Ind Eng Chem 47:1593
75. Braswell E, Goodman A, Stern KG (1962) J Polym Sci 61:143
76. Jeanes A, Wilham CA, Tsuchiya HM, Haynes WC (1957) Arch Biochem Biophys 71:293

77. Miller AW, Robyt JF (1984) Biochim Biophys Acta 785:89
78. Nigam M, Goyal A, Katiyar SS (2006) J Food Biochem 30:12
79. Wolff IA, Mehltretter CL, Mellies RL, Watson PR, Hofreiter BT, Patrick PL, Rist CE (1954) Ind Eng Chem 46:270
80. Zief M, Brunner G, Metzendorf J (1955) Ind Eng Chem 48:119
81. Eckelt J, Sugaya, Wolf BA (2006) Carbohydr Polym 63:205
82. Oriol E, Auriol D, Monsan P (1986) Carbohydr Res 149:433
83. Fagerson SI (1967) J Agric Food Chem 17:747
84. Ruckel ER, Schuerch C (1966) J Org Chem 31:2233
85. Kakuchi T, Kusuno A, Miura M, Kaga H (2000) Macromol Rapid Commun 21:1003
86. Ruckel ER, Schuerch C (1967) Biopolymers 5:515
87. Ruckel ER, Schuerch C (1966) J Am Chem Soc 88:2605
88. Lemarchand C, Gref R, Couvreur P (2004) Eur J Pharm Biopharm 58:327
89. Jordan A, Scholz R, Wust P, Hermann Schirra H, Schiestel T, Schmidt H, Roland Felix R (1999) J Magn Magn Mater 194:185
90. Berry CC, Wells S, Charles S, Curtisa ASG (2003) Biomaterials 24:4551
91. Sinha J, Dey PK, Panda T (2000) Appl Microbiol Biotechnol 54:476
92. Matsumoto U, Ban M, Shibusawa Y (1984) J Chromatogr A 285:69
93. Maeda Y, Ito H, Izumida R, Kitano H (1997) Polymer Bull 38:49
94. Flodin P (1998) Polym Eng Sci 38:1220
95. Haff LA, Easterday RL (1978) J Liq Chromatogr 1:811
96. Aehle W (ed) (2004) In: Enzymes in industry, production and applications, 2nd edn. Wiley, Weinheim, p 62
97. Denizli BK, Can HK, Rzaev ZMO, Guner A (2004) Polymer 45:6431
98. Hovgaard L, Bronsted H (1995) J Controlled Release 36:159
99. Kim SH, Won CY, Chu CC (1999) Carbohydr Polym 40:183
100. Grönwall A, Ingelman B, Mosimann H (1945) Uppsala Läkarförening Forh 51:397
101. Sato T, Nishimura-Uemura J, Shimosato T, Kawai Y, Kitazawa H, Saito T (2004) J Food Prot 67:1719
102. Suzuki F, Ishida N, Suzuki M, Sato T, Suzuki S (1975) Proc Soc Exp Biol Med 149:1069
103. Suzuki M, Mikami T, Matsumoto T, Suzuki S (1977) Carbohydr Res 53:223
104. Suzuki M, Mikami T, Kadowaki M, Matsumoto T, Suzuki S (1994) Biosci Biotechnol Biochem 58:178
105. Mumper RJ, Hoffman AS (2000) AAPS PHarmSciTech 1:article 3
106. Whistler RL, Spencer WW (1961) Arch Biochem Biophys 95:36
107. Mauzac JM, Jozefonvicz J (1984) Biomaterials 5:301
108. Chaubet F, Champion J, Maiga O, Mauray S, Jozefonvicz J (1995) Carbohydr Polym 28:145
109. Chaubet F, Huynh R, Champion J, Jozefonvicz J, Letourneur D (1999) Polym Int 48:313
110. Papy-Garcia D, Barbier-Chassefiere V, Rouet V, Kerros ME, Klochendler C, Tournaire MC, Barritault D, Caruelle JP, Petit E (2005) Macromolecules 38:4647
111. Ingelman B (1947) Ark Kemi Min Geol 248:4
112. Walton KW (1952) Br J Pharmacol 7:370
113. Ricketts CR (1952) Biochem J 51:129
114. Ricketts CR, Walton KW (1953) Br J Pharmacol 8:476
115. Ricketts CR, Walton KW (1954) Br J Pharmacol 9:1
116. Ricketts CR, Walton KW, VanLeuven BD, Birbeck A, Brown A, Kennedy AC, Burt CC (1953) Lancet 265:1004

117. Kindness G, Long WF, Williamson FB (1979) Thromb Res 15:49
118. Eto N, Kojima I, Uesugi N, Inagi R, Miyata T, Fujita T, Johnson RJ, Shankland SJ, Nangaku M (2005) J Am Soc Nephrol 16:2997
119. Gemsa D, Seitz M, Kramer W, Till G, Resch K (1978) J Immunol 120:1187
120. Minchin SA, Leitenberg D, Stunz LL, Feldbush TL (1990) J Immunol 145:2427
121. Kettman J, Söderberg A, Lefkovits I (1984) Cell Immunol 88:129
122. Burger R, Hadding U, Schorlemmer HU, Brade V, Bitter-Suermann D (1975) Immunology 29:549
123. Yoshida T, Nakashima H, Yamamoto N, Uryu T (1993) Polym J 25:1069
124. Ito M, Baba M, Sato A, Pauwels R, DeClercq E, Shigeta S (1987) Antiviral Res 7:361
125. Ueno R, Kuno S (1987) Lancet 1:1379
126. Mizumoto K, Sugawara I, Ito W, Kodama T, Hayami M, Mori S (1988) Jpn J Exp Med 58:145
127. Baba M, Schols D, De Clercq E, Pauwels R, Nagy M, Gyorgyi-Edelenyi J, Low M, Sandor G (1990) Antimicrob Agents Chemother 34:134
128. Batinic D, Robey FA (1992) J Biol Chem 267:6664
129. Mitsuya H, Looney DJ, Kuno S, Ueno R, Woong-Staal F, Broder S (1988) Science 226:172
130. Baba R, Snoeck R, Pauwels R, DeClercq E (1988) Antimicrob Agents Chemother 32:1724
131. Piret J, Lamontagne J, Bestman-Smith J, Roy S, Gourde P, Désormeaux A, Omar RF, Juhasz J, Bergeron MG (2000) J Clin Microbiol 38:110
132. Abrams DI, Kuno S, Wong R, Jeffords K, Nash M, Molaghan JB, Gorter R, Ueno R (1989) Ann Intern Med 110:183
133. Flexner C, Barditch-Crovo PA, Kornhauser DM, Farzadegan H, Nerhood LJ, Chaisson RE, Bell KM, Lorentsen KJ, Hendrix CW, Petty BG (1991) Antimicrob Agents Chemother 35:2544
134. Hartman NR, Johns DG, Mitsuya H (1990) AIDS Res Hum Retroviruses 6:805
135. Comper WD, Tay M, Dawes J (1994) Biochem J 297:31
136. Comper WD (2006) US Patent 2004-542743P 20040206
137. Usher TC, Patel N, Tele CG, Wolk IL (1995) WO Patent 9505199
138. Hagiwara A, Sawai K, Sakakura C, Shirasu M, Ohgaki M, Imanishi T, Yamasaki J, Togawa T, Takahashi T (1997) Anti-Cancer Drugs 8:894
139. Suemasu K, Watanabe K, Ishakawa S (1971) Gann 62:331
140. Zimmermann K, Preinl G, Ludwig H, Greulich K (1983) J Cancer Res Clin Oncol 105:189
141. Udabage L, Brownlee GR, Stern R, Brown TJ (2004) Glycoconjugate J 20:461
142. Beringue V, Adjou KT, Lamoury F, Maignien T, Deslys JP, Race R, Dormont D (2000) J Virol 74:5432
143. Hjort P, Stormorken H, Gilje O (1957) Scand J Clin Lab Invest 9:29
144. Hint HC, Richert AW (1958) Br J Pharmacol 13:109
145. Oda M, Tanahashi N, Niimi H, Nakamura M, Asada M, Matsui H, Hibi N, Tsuchimoto K, Inoue J, Oda M (2006) Clin Hemorheol Microcirc 34:193
146. Mennigen RB, Bruewer M, Kersting S, Krieglstein CF (2005) Chirurgisches Forum fuer Experimentelle und Klinische Forschung 167
147. Araki Y, Sugihara H, Hattori T (2006) Int J Mol Med 17:331
148. Decher G (1996) In: Sauvage JP, Hosseini MW (eds) Comprehensive supramolecular chemistry, vol 9. Pergamon, Oxford, p 507
149. Serizawa T, Yamaguchi M, Matsuyama T, Akashi M (2000) Biomacromolecules 1:306
150. Akashi M, Yamaguchi M, Tachabonyakiat W, Serizawa T (2002) Adv Chitin Sci 5:278

151. Serizawa T, Yamaguchi M, Akashi M (2002) Macromolecules 35:8656
152. Sakiyama T, Takata H, Kikuchi M, Nakanishi K (1999) J Appl Polym Sci 73:2227
153. Fuentes M, Pessela BCC, Maquiese JV, Ortiz C, Segura RL, Palomo JM, Abian O, Torres R, Mateo C, Fernandez-Lafuente R, Guisan JM (2004) Biotechnol Prog 20:1134
154. Motomura T, Miyashita Y, Ohwada T, Onishi M, Yamamoto N (1997) US Patent 5667684
155. Schatz C, Domard A, Viton C, Pichot C, Delair T (2004) Biomacromolecules 5:1882
156. Balabushevich NG, Sukhorukov GB, Larionova NI (2005) Macromol Rapid Commun 26:1168
157. Louie AY, Jarrett BR (2006) WO Patent 2006012201
158. Yuan GL, Kuramoto N (2002) Chem Lett 5:544
159. Heinze TT, Liebert T, Koschella A (2006) Springer laboratory: esterification of polysaccharides. Springer, Berlin Heidelberg New-York
160. Liebert T, Hornig S, Hesse S, Heinze T (2005) J Am Chem Soc 127:10484
161. Sanchez-Chaves M, Arranz F (1983) Angew Makromol Chem 118:53
162. Zhang J, Pelton R, Wagberg L (1998) Colloid Polym Sci 276:476
163. Sanchez-Chaves M, Arranz F (1985) Makromol Chem 186:17
164. Lu M, Albertson PA, Johansson G, Tjerneld F (1994) J Chromatography A 668:215
165. Lu M, Johansson G, Albertson PA, Tjerneld F (1995) Bioseparation 5:351
166. Lee K, Na K, Kim Y (1999) Polym Prepr 40:359
167. Usmanov TI, Karimova UG (1990) Vysokomol Soedin, Ser A 32:1871
168. Novak LJ, Tyree JT (1960) US Patent 2954372
169. Sanchez-Chavez M, Arranz F, Diaz C (1989) Makromol Chem 190:2391
170. Ramirez JC, Sanchez-Chavez M, Arranz F (1994) Polymer 35:2651
171. Arranz F, Sanchez-Chavez M, Ramirez JC (1992) Angew Makromol Chem 79:194
172. Arranz F, Sanchez-Chavez M, Ramirez JC (1993) Polymer 34:1908
173. Kim SH, Won CY, Chu CC (2000) WO Patent 2000012619
174. Kim SH, Won CY, Chu CC (1999) J Biomed Mater Res 46:160
175. Kim SH, Chu CC (2000) J Biomed Mater Res 49:517
176. Yamaoka T, Tanihara M, Mikami H, Kinoshita H (2003) JP Patent 2003252936
177. VanDijk-Wolthuis WNE, Tsang SKY, Kettenes-VanDenBosch JJ, Hennink WE (1997) Polymer 38:6235
178. Larsen C (1989) Adv Drug Delivery Rev 3:103
179. Mehvar R (2000) J Control Release 69:1
180. Sanchez-Chaves M, Arranz F (1997) Polymer 38:2501
181. Larsen C (1989) J Pharm Biomed Anal 7:1173
182. Harboe E, Johansen M, Larsen C (1988) Farmaci Sci Edn 16:73
183. Larsen C, Jensen BH, Olesen HP (1991) Acta Pharm Nord 3:71
184. Larsen C, Johansen M (1989) Acta Pharm Nord 1:57
185. Vasil'ev AE, Khachatur'yan AA, Rozenberg GY (1971) Khim Prir Soedin 7:698
186. Kol'tsova GN, Khachatur'yan AA, Doronina TN, Vasil'ev AE, Rozenberg GY (1974) Chem Nat Compd 265:8
187. Haslam E (1980) Tetrahedron 36:2409
188. Fenselau AH, Moffatt JG (1966) J Am Chem Soc 88:1762
189. Bamford CH, Middleton IP, Al-Lamee KG (1986) Polymer 27:1981
190. Kochetkov NK, Khachatur'yan AA, Vasil'ev AE, Rozenberg GY (1969) Khim Prir Soedin 5:427
191. Azhigirova MA, Vasil'ev AE, Gerasimovskaya LA, Khachatur'yan AA, Rozenberg GY (1977) J Gen Chem USSR 47:464
192. Nichifor M, Carpov A (1999) Eur Polym J 35:2125

193. Nichifor M, Stanciu MC, Zhu XX (2004) React Funct Polym 59:141
194. Vermeersch J, Vandoorne F, Permentier D, Schacht E (1985) Bull Soc Chim Belges 94:591
195. Larsen C (1986) Acta Pharm Suec 23:279
196. Larsen C, Kurtzhals P, Johansen M (1988) Acta Pharm Suec 25:1
197. Staab HA (1962) Angew Chem 74:407
198. Kol'tsova GN, Khachatur'yan AA, Doronina TN, Vasil'ev AE, Rozenberg GY (1972) Khim Prir Soedin 3:266
199. Williams AS, Taylor G (1992) Int J Pharm 83:233
200. Chung JT, Vlugt-Wensink KDF, Hennink WE, Zhang Z (2005) Int J Pharm 288:51
201. VanThienen TG, Lucas B, Flesch FM, VanNostrum CF, Demeester J, DeSmedt SC (2005) Macromolecules 38:8503
202. DeGeest BG, Urbanski JP, Thorsen T, Demeester J, DeSmedt SC (2005) Langmuir 21:10275
203. Huang X, Lowe TL (2005) Macromolecules 6:2131
204. DeGeest BG, Stubbe BG, Jonas AM, VanThienen T, Hinrichs WLJ, Demeester J, DeSmedt SC (2006) Biomacromolecules 7:373
205. Hornig S, Heinze T, Hesse S, Liebert T (2005) Macromol Rapid Commun 26:1908
206. Gref R, Rodrigues J, Couvreur P (2002) Macromolecules 35:9861
207. Ferreira L, Gil MH, Dordick JS (2002) Biomaterials 23:3957
208. Norman B (1968) Acta Chem Scand 22:1381
209. Norman B (1968) Acta Chem Scand 22:1623
210. Haines AH (1976) Adv Carbohydr Chem Biochem 33:59
211. Hall LD, Preston CM (1976) Carbohydr Res 49:3
212. Hollo J, Laszlo E, Hoschke A (1968) Periodica Polytech Chem Eng 12:277
213. Rees DA, Richardson NG, Wright NJ, Hirst E (1969) Carbohydr Res 9:451
214. Hakness BR, Gray DG (1990) Macromolecules 23:1452
215. Ydens I, Rutot D, Degee P, Six JL, Dellacherie E, Dubois P (2000) Macromolecules 33:6713
216. Nouvel C, Ydens I, Degee P, Dubois P, Dellacherie E, Six JL (2002) Polymer 43:1735
217. Nouvel C, Dubois P, Dellacherie E, Six JL (2003) Biomacromolecules 4:1443
218. Nagy J, Borebely-Kuszmann A, Becker-Palossy K, Zimonyi-Hegedus E (1973) Makromol Chem 165:335
219. Huynh R, Chaubet F, Jozefovicz J (1998) Angew Makromol Chem 254:61
220. Krentsel L, Chaubet F, Rebrov A, Champion J, Ermakov I, Bittoun P, Fermandjian S, Litmanovich A, Plate N, Jozefonvicz J (1997) Carbohydr Polym 33:63
221. Krentsel L, Ermakov I, Yashin V, Rebrov A, Litmanovich A, Plate N, Chaubet F, Champion J, Jozefonvicz J (1997) Vysokomol soedin 39:83
222. Maiga-Revel O, Chaubet F, Jozefonvicz J (1997) Carbohydr Polym 32:89
223. DeBelder AN, Norrman B (1969) Carbohydrate Research 10:391
224. Kesler CC, Hjermstad ET (1964) Methods Carbohydr Chem 4:304
225. Bonnet J, Choe TB, Lee CH, Masse P, Verdier A (1986) Studies Environ Sci 29:421
226. Georgieva M, Kabaivanov V (1969) Doklady Bolgarskoi Akademii Nauk 22:1393
227. Rogovin ZA, Virnik AD, Khomyakov KP, Laletina OP, Penenzhik MA (1972) J Macromol Sci Chem 6:569
228. Flodin P, Granath KA, Ingelman BG-A, Johansson C (1965) FR Patent M3515 9651004
229. Rotureau E, Leonard M, Dellacherie E, Durand A (2004) Phys Chem Chem Phys 6:1430
230. Rotureau E, Dellacherie E, Durand A (2005) Macromolecules 38:4940
231. Rotureau E, Chassenieux Ch, Dellacherie E, Durand A (2005) Macromol Chem Phys 206:2038

232. Durand A, Marie E, Rotureau E, Leonard M, Dellacherie E (2004) Langmuir 20:6956
233. Rouzes C, Durand A, Leonard M, Dellacherie E (2002) J Colloid Interface Sci 253:217
234. Sadtler V, Imbert P, Dellacherie E (2002) J Colloid Interf Sci 254:355
235. Imbert P, Sadtler V, Dellacherie E (2002) Colloids Surf A: Physicochem Eng Aspects 211:157
236. Laletina OP, Virnik AD, Rogovin ZA (1968) Vysokomol Soedin, Ser B 10:708
237. Rogovin ZA, Virnik AD, Khomiakov KP, Laletina OP, Penenzlik MA (1972) J Macromol Sci-Chem A 6:569
238. Mora M, Pato J (1990) Makromol Chem 191:1051
239. Kernan WMc, Ricketts CR (1993) J Appl Polym Sci 47:815
240. Kubota N, Kikuchi Y (1993) J Appl Polym Sci 47:815
241. Hanselmann R, Burchard W (1995) Makromol Chem Phys 196:2259
242. Kikuchi Y, Kubota N (1988) Makromol Chem Rapid Commun 9:731
243. Gubensek F, Lapange S (1968) J Macromol Sci Chem A 2:1045
244. Fox RM, Mynderse JF, Goulian M (1977) Biochem 16:4470
245. Uchmann G, Chrzanowska M (1980) Farmacja Polska 36:471
246. Nichifor M, Lopes S, Bastos M, Lopes A (2004) J Phys Chem B 108:16463
247. Bai G, Santos LMNBF, Nichifor M, Lopes A, Bastos M (2004) J Phys Chem B 08:405
248. Bai G, Nichifor M, Lopes A, Bastos M (2005) J Phys Chem B 109:518
249. Bai G, Nichifor M, Lopes A, Bastos M (2005) J Phys Chem B 109:21681
250. Ghimici L, Nichifor M (2005) J Polym Sci B Polym Phys 43:3584
251. Francis M, Piredda M, Cristea M, Winnik FM (2003) Polym Mater Sci Eng 89:55
252. Francis MF, Lavoie L, Winnik FM, Leroux JC (2003) Eur J Pharm Biopharm 56:337
253. Parkinson TM (1967) Nature 215:415
254. Rosemeyer H, Seela F (1984) Makromol Chem 185:687
255. Ceska M (1971) Experientia 27:1263
256. Ceska M (1972) Experientia 28:146
257. Hodge JE, Karjala SA, Hilbert GE (1951) J Am Chem Soc 73:3312
258. Klemm D, Philipp B, Heinze T, Heinze U, Wagenknecht W (1998) Comprehensive cellulose chemistry, functionalization of cellulose, vol 2. Wiley, Weinheim
259. Croon J (1959) Acta Chem Scand 13:1235
260. DeBelder AN, Lindberg B, Theander O (1962) Acta Chem Scand 16:2005
261. Landoll LM (1982) J Polym Sci, Part A: Polym Chem 20:443
262. Rouzes C, Gref R, Leonard M, De Sousa-Delagado A, Dellacherie E (2000) J Biomed Mat Res 50:557
263. DeSousa-Delgado A, Leonard M, Dellacherie E (2001) Langmuir 17:4386
264. Fournier C, Leonard M, LeCoq-Leonard I, Dellacherie E (1995) Langmuir 11:2344
265. Fournier C, Leonard M, Dellacherie E, Chikhi M, Hommel H, Legrand AP (1998) J Colloid Interface Sci 198:27
266. Rouzes C, Leonard M, Durand A, Dellacherie E (2003) Colloids Surf B 32:125
267. Lewis M (1990) In: Chasin M, Langer R (eds) Biodegradable polymers as drug delivery systems. Marcel Dekker, New York, p 1
268. Francis MF, Cristea M, Winnik FM (2004) Pure Appl Chem 76:1321
269. Francis MF, Cristea M, Yang Y, Winnik FM (2005) Pharm Res 22:209
270. Horter D, Dressman JB (2001) Adv Drug Deliv Rev 46:75
271. Wiedemann TS (2002) J Pharm Sci 91:1743
272. Lasic DD (1992) Nature 355:279
273. Kwon GS, Okano T (1996) Adv Drug Deliv Rev 21:107
274. Kataoka K, Harada A, Nagasaki Y (2001) Adv Drug Deliv Rev 47:113
275. Zhao CL, Winnik MA, Riess G, Croucher MD (1990) Langmuir 6:514

276. Winnik FM, Regismond STA (1998) Fluorescence methods in the study of polymer-surfactant systems. In: Kwak ICT (ed) Polymer-surfactant systems. Marcel Dekker, New York, p 267
277. Kalyanasundaram K, Thomas JK (1977) J Am Chem Soc 99:2039
278. Nagarajan R, Ganesh K (1998) Macromolecules 22:4312
279. Francis MF, Cristea M, Winnik FM (2005) Biomacromolecules 6:2462
280. Khalil MI, Hashem A, Hebeish A (1990) Starch/Staerke 42:60
281. Logeart-Avramoglou D, Jozefonvicz J (1999) J Biomed Mater Res 48:578
282. Jozefonvicz J, Jozefowicz M (1992) Pure Appl Chem 64:1783
283. DeRaucourt E, Mauray S, Chaubet F, Maiga-Revel O, Jozefowicz M, Fischer AM (1998) J Biomed Mater Res 41:49
284. Maaroufi RM, Jozefowicz M, Tapon-Bretaudiere J, Jozefonvicz J, Fischer AM (1997) Biomaterials 18:359
285. Crepon B, Maillet F, Kazatchkine M, Jozefonvicz J (1987) Biomaterials 8:248
286. Thomas H, Maillet F, Letourneur D, Jozefonvicz J, Fischer E, Kazatchkine MD (1996) Mol Immununol 33:643
287. Letourneur D, Champion J, Slaoui F, Jozefonvicz J (1993) In Vitro Cell Dev Biol 29A:67
288. Letourneur D, Logeart D, Avramoglou T, Jozefonvicz J (1993) J Biomater Sci Polym Ed 4:431
289. Logeart D, Avramoglou T, Jozefonvicz J (1994) Colloids Surf, B 2:315
290. Mauzac M, Maillet F, Jozefonvicz J, Kazatchkine MD (1985) Biomaterials 6:61
291. Jozefowicz M, Jozefonvicz J (1997) Biomaterials 18:1633
292. Huynh R, Chaubet F, Jozefonvicz J (2001) Carbohydr Res 332:75
293. Hennink WE, Klerx JPAM, van Dijk H, Feijen J (1984) Thromb Res 36:281
294. Sache E, Maillard M, Bertrand M, Maman M, Kunz M, Chosy J, Fareed J, Messmore M (1982) Thromb Res 25:443
295. Sederel LC, VanDerDoes L, VanDuijl JF, Beugeling T, Bantjes A (1981) J Biomed Mater Res 15:819
296. Morere JF, Letourneur D (1992) Anti-Cancer Drugs 3:629
297. Bagheri-Yarmand R, Morerere JF, Letourneur D, Jozefonvicz J, Israel L, Crepin M (1992) Anticancer Res 12:1641
298. Bagheri-Yarmand R, Bittoun P, Champion J (1994) In Vitro Cell Dev Biol 30A:822
299. Meddahi A, Lemdjabar H, Caruelle JP, Barritault D, Hornebeck W (1995) Biochimie 77:703
300. Meddahi A, Benoit J, Ayoub N, Sezeur A, Barritault D (1996) J Biomed Mater Res 31:293
301. Sibikina OV, Iozep AA, Passet (2004) Zh Prikl Khim 77:1161
302. Iozep AA, Il'ina TY, Passet BV (1994) Zh Prikl Khim 67:470
303. Iozep AA, Sibikina OV, Kuznetsova TE, Passet BV (1995) Zh Prikl Khim 68:307
304. Krasnikova AV, Iozep AA (2004) Zh Prikl Khim 77:801
305. Götze T, Gansau C, Buske N, Roeder M, Görnert M, Bahr M (2002) J Magn Magn Mater 252:399
306. Pouliquen D, LeJeune JJ, Perdrisot R, Ermias A, Jallet P (1991) Magn Resson Imaging 9:275
307. Schwalbe M, Jörke C, Buske N, Höffken K, Pachmann K, Clement JH (2005) J Magn Magn Mater 293:433
308. Wagner K, Kautz A, Röder M, Schwalbe M, Pachmann K, Clement JH, Schnabelrauch M (2004) Appl Organometal Chem 18:515
309. Brusentsov NA, Baiburtskii FS, Tarasov VV, Komissarova LK (2002) Pharm Chem J 36:197

310. Soma CE, Dubernet C, Barratt G, Nemati F, Appel M, Benita S, Couvreur P (1999) Pharm Res 16:1710
311. Ohya Y, Masunaga T, Baba T, Ouchi T (1996) Pure Appl Chem 33:1005
312. Hattori M, Nagasawa K, Ametani A, Kaminogawa S, Takahashi K (1994) J Agric Food Chem 42:2120
313. Hattori M, Nagasawa K, Ohgata K, Sone N, Fukuda A, Matsuda H, Takahashi K (2000) Bioconjugate Chem 11:84
314. Nagasawa K, Ohgata K, Takahashi K, Hattori M (1996) J Agric Food Chem 44:2538
315. Hattori M (2002) Food Sci Technol Res 8:291
316. Kobayashi K, Hirano A, Ohta A, Yoshida T, Takahashi K, Hattori M (2001) J Agric Food Chem 49:823
317. Kobayashi K, Yoshida T, Takahashi K, Hattori M (2003) Bioconjugate Chem 14:168
318. Harada M, Sakakibara H, Yano T, Suzuki T, Okuno S (2000) J Controlled Release 69:399
319. Chau Y, Tan FE, Langer R (2004) Bioconjugate Chem 15:931
320. Peppas NA, Bures P, Leobandung W, Ichikawa H (2000) Eur J Pharm Biopharm 50:27
321. Lee KY, Mooney DJ (2001) Chem Rev 101:1869
322. Qiu Y, Park K (2001) Adv Drug Deliv Rev 53:321
323. Chiu HC, Hsiue GH, Lee YP, Huang LW (1999) J Biomater Sci, Polym Ed 10:591
324. Chiu HC, Lin YF, Hsu YH (2002) Biomaterials 23:1103
325. Zhang R, Tang M, Bowyer A, Eisenthal R, Hubble J (2005) Biomaterials 26:4677
326. Tomihata K, Ikada Y (1997) J Biomed Mater Res 37:243
327. Nakajima N, Ikada Y (1995) Bioconjugate Chem 6:123
328. Markland P, Zhang YH, Amidon GL, Yang VC (1999) J Biomed Mater Res 47:595
329. Risbud MV, Hardikar AA, Bhat SV, Bhonde RR (2000) J Controlled Release 68:23
330. Soldani G, Maccheroni M, Martelli F, Mengozzi G, Cardini G (1987) Internat J Obesti 11:201
331. Montanari G, Gianfranceschi G, Franceschini G, Bertoli M, Sirtori CR (1985) Proc 7th Int Symp Atherosclerosis 141
332. DiLuigi L, DalLago A, Vita F, Isadori A (1986) Clin Ther 117:37
333. Bandini S, Comparini L, Mancini G, Salvadori M, Tosi PL, Sodi A (1990) Clin Trial J 27:30
334. Fiala M, Satzman B (1969) Appl Microbiol 17:190
335. Liptay S, Weidenbach H, Adler G, Schmid RM (1998) Digestion 59:142
336. Schenborn ET (2000) Methods Mol Biol 130:91
337. Schenborn ET, Goiffon V (2000) Methods Mol Biol 130:147
338. Mack KD, Wei R, Elbagarri A, Abbey N, McGrath MS (1998) J Immunol Methods 211:79
339. Gibson PD, Higgins J, Woodward JR (1992) Analyst 117:1293
340. Kikuchi Y, Koda T (1979) Bull Chem Soc Japan 52:880
341. Miyazaki Y, Yakou S, Nagai T, Takayama K (2003) Drug Develop Industr Pharmacy 29:795
342. Kikuchi Y, Sasayama S (1982) Makromol Chem 183:2153
343. Kikuchi Y, Kubota N (1985) J Appl Polym Sci 30:2565
344. Kikuchi Y, Kubota N, Maru K, Goto Y (1987) Makromol Chem 188:263
345. Soyez H, Schacht E, Vanderkerken S (1996) Adv Drug Delivery Rev 21:81
346. Fagnani R, Hagan MS, Bartholomew R (1990) Cancer Res 50:3638
347. Mikolajczyk SD, Meyer DL, Fagnani R, Hagan MS, Law KL, Starling JJ (1996) Bioconjugate Chem 7:150
348. Fuentes M, Mateo C, Fernandez-Lafuente R, Guisan JM (2005) Enzyme Microb Technol 36:510

349. Fuentes M, Segura RL, Abian O, Betancor L, Hidalgo A, Mateo C, Fernandez-Lafuente R, Guisan JM (2004) Proteomics 4:2602
350. Betancor L, Fuentes M, Dellamora-Ortiz G, Lopez-Gallego F, Hidalgo A, Alonso-Morales N, Mateo C, Guisan JM, Fernandez-Lafuente R (2005) J Mol Catal B: Enzym 32:97
351. Dellacherie E, Bonneaux F (1993) Polym Bull 31:145
352. Axen R, Porath J, Ernback S (1967) Nature 214:1302
353. Dimitriu S, Chornet E (1998) Polysaccharides as support for enzyme and cell immobilization. In: Dimitriu S (ed) Polysaccharides, structural diversity and functional versatility. Marcel Dekker, New York, p 637
354. Pozsgay V (2002) US Patent 2002051788
355. Baker SA, Disney HM, Somers PJ (1972) Carbohydr Res 25:237
356. Ramirez JC, Sanchez-Chavez M, Arranz F (1995) Angew Makromol Chem 225:123
357. Vandoorne F, Vercauteren R, Permentier D, Schacht E (1985) Makromol Chem 186:2455
358. Rutili G, Arfors KE (1976) Microvasc Res 12:221
359. DeBelder AN, Granath K (1973) Carbohydr Res 30:375
360. Haugland RP (1992) In: Larison KD (ed) Handbook of fluorescent probes and research chemicals, 5th edn. Molecular probes. p 185
361. Noguchi T, Mabuchi I (2002) Mol Biol Cell 13:1263
362. Shih LB, Primus FJ, Goldenberg MD (1987) US Patent 4699784
363. Arrio-Dupont M, Foucault G, Vacher M, Devaux PF, Cribier S (2000) Biophys J 78:901
364. Andrieu M, Loing E, Desoutter JF, Connan F, Choppin J, Gras-Masse H, Hanau D, Dautry-Varsat A, Guillet JG, Hosmalin A (2000) Eur J Immunol 30:3256
365. Kozlowski DJ, Murakami T, Ho RK, Weinberg ES (1997) Biochem Cell Biol 75:551
366. Kempers R, VanBel AJE (1997) Planta 201:195
367. Koschella A, Leermann T, Brackhagen M, Heinze T (2006) J Appl Polym Sci 100:2142
368. Kolova AF, Komar VP, Skornyakov IV, Virnik AD, Zhbanov RG, Rogovin ZA (1978) Cellul Chem Technol 12:553
369. Klotz IM, Stryker VH (1959) Biochem Biophys Res Commun 1:119
370. Frazier RA, Matthijs G, Davies MC, Roberts CJ, Schacht E, Tendler SJB (2000) Biomaterials 21:957
371. Frazier RA, Davies MC, Matthijs G, Roberts CJ, Schacht E, Tasker S, Tendler SJB (1996) The self-assembly and inhibition of protein adsorption by thiolated dextran monolayers at hydrophobic metal surfaces. In: Ratner BD, Castner DG (eds) Surface modification of polymeric biomaterials. Plenum, New York, p 117
372. Frazier RA, Davies MC, Matthijs G, Roberts CJ, Schacht E, Tendler SJB, Williams PM (1997) Langmuir 13:7115
373. Frazier RA, Davies MC, Matthijs G, Roberts CJ, Schacht E, Tendler SJB, Williams PM (1997) Langmuir 13:4795
374. Holmberg A, Meurling L (1993) Bioconjugate Chem 4:570
375. Pawlowski A, Kallenius G, Svenson SB (1999) Vaccine 17:1474
376. Schuyten HA, Weaver JW, Reid JD, Jürgens JF (1948) J Am Chem Soc 70:1919
377. Sweeley CC, Bentley R, Makita M, Wells WW (1963) J Am Chem Soc 85:2497
378. Shibata M, Asahina M, Teramoto N, Yosomiya R (2001) Polymer 42:59
379. DeVos R, Goethals E (1986) J Polym Bull 15:547
380. Li Y, Nothnagel J, Kissel T (1997) Polymer 38:6197
381. Teramoto Y, Nishio Y (2003) Polymer 44:2701
382. Ohya Y, Maruhashi S, Ouchi T (1998) Macromolecules 31:4662
383. Ohya Y, Maruhashi S, Ouchi T (1998) Macromol Chem Phys 199:2017

384. Youxin L, Volland C, Kissel T (1998) Polymer 39:3087
385. Nouvel C, Dubois P, Dellacherie E, Six JL (2004) J Polym Sci, Part A: Polym Chem 42:2577
386. Nouvel C, Frochot C, Sadtler V, Dubois P, Dellacherie E, Six JL (2004) Macromolecules 37:4981
387. Ydens I, Degee P, Nouvel C, Dellacherie E, Six JL, Dubois P (2005) e-Polymers 046
388. Lemarchand C, Couvreur P, Besnard M, Constantini D, Gref R (2003) Pharm Res 20:1284

Author Index Volumes 201–205

Subject Index

BC